Regulation of Nutrient Uptake by Plants

Gyanendra Nath Mitra

Regulation of Nutrient Uptake by Plants

A Biochemical and Molecular Approach

Springer

Gyanendra Nath Mitra
Department of Soil Science and Agricultural Chemistry
Orissa University of Agriculture and Technology
Bhubaneswar, Odisha, India

ISBN 978-81-322-2333-7 ISBN 978-81-322-2334-4 (eBook)
DOI 10.1007/978-81-322-2334-4

Library of Congress Control Number: 2015938782

Springer New Delhi Heidelberg New York Dordrecht London

Printed on acid-free paper

Springer (India) Pvt. Ltd. is part of Springer Science+Business Media (www.springer.com)

Preface

During the last 25–30 years, intensive research has been carried out globally to explain nutrient uptake by plants at the molecular level and a large volume of information has been generated. This is primarily due to the use of modern technology developed in biotechnological research, instrumentation, modern computation facilities, bioinformatics, large volume of information generated from the use of various 'omics' and, of course, the dedicated hard work of the researchers.

The Nobel Prize in Chemistry 2003 was awarded to Peter Agre and Roderick MacKinnon for their pioneering discoveries concerning water and ion channels of cells.

To quote from the Introduction of this book, 'Recent research indicates that nutrient uptake and its transport and redistribution in plants are under genetic control. There are groups of genes for every nutrient which encode transporter proteins whose functions are to acquire the specific nutrient from the soil and transport it across the plasma membrane of the root hair cells for use in plant metabolism. Deficiency or sufficiency of a plant nutrient induces different groups of genes to produce m-RNA transcripts for translation of transporter proteins. There are also evidences which suggest post-translational regulation of transporter proteins in response to changes in the nutrient status of the plants. A large number of metabolic enzymes are up- or down-regulated in response to deficiency of plant nutrients. Amino acids, plant growth regulators, intermediate metabolites and the nutrients themselves are involved in the induction or repression of transporter-encoding genes as well as post- transcriptional modification of transporter proteins'.

Excellent review articles on uptakes of individual nutrients are published in journals, annual reviews and as chapters of some books. A book containing comprehensive information on all the nutrients taken up by plants incorporating recent developments at biochemical and molecular levels is probably missing.

This book is intended

1. To be used as a reference manual by the researchers working in the area of nutrient uptake by plants and in related fields
2. To provide teachers offering courses which include nutrient uptake by plants, with latest information to update their courses
3. To update knowledge of students and create interest in them to undertake research in this area

Additional information on some of the specific topics has been given in boxes. Terminology used has been explained as far as possible in the text, mostly within parenthesis. The readers can always cross-check the information given in this book from the extensively cited original papers in the text. The sources of these papers are given under 'References' at the end of each chapter.

Bhubaneswar, India Gyanendra Nath Mitra

Acknowledgements

I gratefully acknowledge all the researchers whose publications I have consulted while preparing the draft of this book. I have duly cited their papers in the references. I acknowledge adapting and redrawing some of the diagrams on water and ion uptake from the 'Information for the public – The Nobel Prize in Chemistry (2003), The Royal Swedish Academy of Sciences, 8 October, 2003' and including them in Chap. 1. I have acknowledged the source under each of the diagrams. I acknowledge that the chemical structure of two compounds, Pterin and Molybdenum cofactor have been taken from Wikipedia as cited under them.

I sincerely thank Dr. M. L. Lodha, former Professor and Head, Division of Biochemistry, Indian Agricultural Research Institute, New Delhi, and former editor of the *Journal of Plant Biochemistry and Biotechnology*, and Dr. G. P. Srivastava, former Professor and Head, Department of Biochemistry, C. Z. Azad Agricultural University, Kanpur, and former editor of the *Indian Journal of Agricultural Biochemistry*, for going through parts of the manuscript and for their valuable suggestions.

I thank Springer for agreeing to publish this book. I am indebted to the editorial team for guiding me to suitably modify the manuscript in the prescribed format of the publisher.

Gyanendra Nath Mitra

Contents

About the Author

Gyanendra Nath Mitra Honorary Professor in the Orissa University of Agriculture and Technology, Bhubaneswar, India, was formerly Dean, Faculty of Agriculture, and Professor and Head, Department of Agricultural Chemistry, Soil Science and Biochemistry. His academic qualifications include M.Sc. (Chemistry), Associate I.A.R.I. (Soil Science and Agricultural Chemistry) and Ph.D. (Biochemistry). Dr. Mitra has offered courses and carried out research in diverse fields with about 100 peer-reviewed publications to his credit. Post superannuation he has authored three books. The book *Regulation of Nutrient Uptake by Plants: A Biochemical and Molecular Approach* is a result of his intense effort in a span of 10 years to study thousands of research publications on the subject as they got published and use them to write different sections of 20 chapters of the book. He has delivered a number of lectures at the faculty level and in various seminars and symposia and published articles in journals to attract attention of the research workers and students to get informed on the topic. The author hopes that wide circulation of the book would spur more research on the subject and help in developing crop varieties with superior nutrient use efficiency and beneficial quality parameters.

List of Boxes

List of Figures

List of Tables

1 Introduction and Uptake of Water and Nutrient Ions

Abstract

Mineral nutrients required for optimal plant growth and development generally exist at a relatively low concentration and show seasonal variation in arable soils. To cope with wide variations in mineral concentrations in soil, plants have evolved mechanisms so that net intake of a nutrient depends on the plant's need for this element rather than its concentration in the rooting medium.

The plasma membrane of cells contains a large number of pores or channels, which are specific for water, ions or other molecules and restrict any other type to pass through them. Such selectivity is caused by intrinsic transmembrane transporter proteins with fixed topology, lodged inside the channels. Cellular ion channel proteins are large molecules with multiple transmembrane α-helices. Channels alternate between open and closed conformations (gating) and allow water, ions and other molecules to pass through them.

1.1 Introduction

Mineral nutrients required for optimal plant growth and development generally exist at a relatively low concentration in soil. To cope with wide variations in mineral concentrations in soil, plants have evolved mechanisms so that net intake of a nutrient depends on the plant's need for this element rather than its concentration in the rooting medium (Imsande and Touraine 1994).

Recent research indicates that nutrient uptake and its transport and redistribution in plants are under genetic control (Orsel et al. 2002; Hammond et al. 2004; Rodriguez-Navarro and Rubio 2006). There are groups of genes for every nutrient, which encode transporter proteins whose functions are to acquire the specific nutrient from the soil and transport it across the plasma membrane of the root hair cells for use in plant metabolism. Deficiency or sufficiency of a plant nutrient induces different groups of genes to produce mRNA transcripts for translation of transporter proteins. There are also evidences, which suggest post-translational regulation of transporter proteins in response to changes in nutrient status of the plants. A large number of metabolic enzymes are up- or downregulated in response to deficiency of plant nutrients. Amino acids, plant

G.N. Mitra, *Regulation of Nutrient Uptake by Plants: A Biochemical and Molecular Approach*,
DOI 10.1007/978-81-322-2334-4_1,

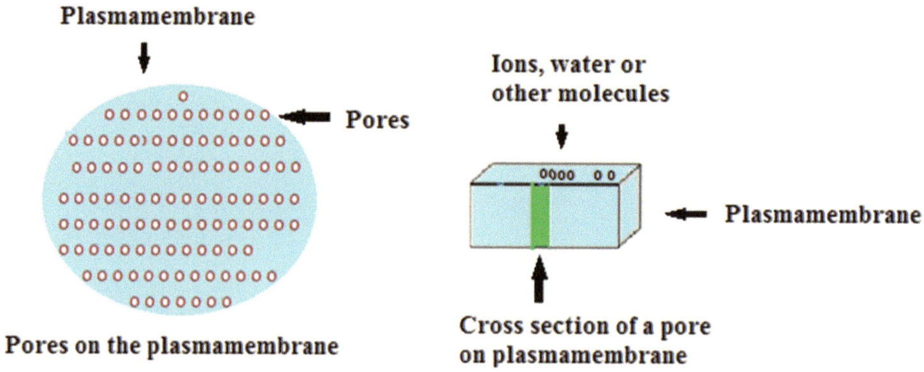

Fig. 1.1 Pores on plasma membrane and a cross section of a pore across plasma membrane (Adapted from 'Information for the public', The Nobel Prize in Chemistry (2003), The Royal Swedish Academy of Sciences, 8th October, 2003. (*Diagram redrawn and modified*))

growth regulators, intermediate metabolites and the nutrients themselves are involved in induction or repression of transporter encoding genes as well as post-transcriptional modification of transporter proteins (Miller et al. 2008).

1.2 Mechanism of Water and Ion Uptake

The mechanism of water and nutrient ion uptake by living cells (including plants) has been elegantly explained by the Royal Swedish Academy of Sciences in their 'Information for the Public, October 8, 2003', while awarding Nobel Prize in Chemistry, 2003, to Peter Agre and Roderick MacKinnon for their pioneering discoveries concerning water and ion channels of cells.

The plasma membrane of cells contains a large number of pores or channels, which are specific for water, ions or other molecules and restrict any other type to pass through them (Fig. 1.1).

1.2.1 Water Channels

Water movement across the channel is controlled by proteins named as aquaporins (AQPs), which consist of a large protein family found in both eukaryotes and prokaryotes. The peptide sequence of a number of aquaporins (AQPs), their three-dimensional structures and the corresponding DNA sequence have been determined.

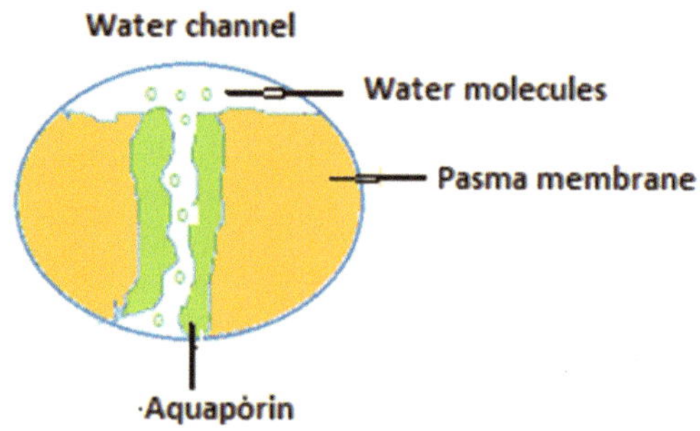

Fig. 1.2 Water channel, with aquaporins lining the walls of the pore across the plasma membrane (Adapted from 'Information for the public'. The Nobel Prize in Chemistry (2003), The Royal Swedish Academy of Sciences, 8th October, 2003. (*Diagram redrawn and modified*))

The narrow channels selectively allow water molecules in proper orientation to pass through them and prevent passage of protons (H^+) or oxonium ions (H_3O^+) due to their positive charges. Smaller uncharged molecules such as glycerol and urea are allowed passage through the water channels (Fig. 1.2).

1.3 Plant Aquaporins

Plants have to maintain water balance under extreme water regimes such as drought and flooding superimposed by weather conditions involving fluctuations in light, temperature and nutrient stress. Plant genomes therefore contain a large number of aquaporin genes to facilitate and regulate water transport across cell membranes.

Table 1.1 No. of homologues of subfamilies of MIPs found in some plants with their intra-group amino acid sequence identities (within parenthesis %)

Name of the plant	PIPs	TIPs	NIPs	SIPs	XIPs
Arabidopsis	13 (71.8–97.8)	11 (44.3–93.1)	11 (38.9–84.7)	3 (28.3–71.2)	–
Maize	14 (64–100)	13 (16–35)	5 (16–35)	3 (16–35)	
Soybean	22	23	13	6	–
Barley	10	10	3	2	–
Wheat	24	11	–	–	–

1.3.1 Genes of Aquaporins in Different Plants

Arabidopsis has 38 aquaporin genes of 2–3 kb size (Johanson et al. 2001; Quigley et al. 2002), maize 33 (Chaumont et al. 2001), barley 23 (Katsuhara et al. 2002), rice 34 (Nguyen et al. 2013), wheat 35 (Forrest and Bhave 2008) and soybean 66 (Zhang et al. 2014) AQP genes.

1.3.2 Major Intrinsic Proteins

The proteins coded by aquaporin genes in plants are diverse and belong to a major intrinsic protein (MIP) family. MIPs are classified into five subfamilies.

PIP: Plasma membrane intrinsic protein
TIP: Tonoplast intrinsic protein
NIP: NOD 26-like intrinsic protein
SIP: Small basic intrinsic protein
XIP: Recently discovered X intrinsic protein of unknown function

All the members of the subfamilies however are not always located as their names signify in specific locations, such as TIP and PIP in tonoplast and plasma membrane, respectively. They may be found elsewhere.

Arabidopsis: MIPs consist of 13 homologues of PIPs (with 71.8–97.8 % amino acid sequence identity), 11 TIPs (44.1–93.1 % intra-group identities), 11 NIPs (38.9–84.7 % identities) and 3 SIPs (28.1–71.2 % identities). The amino acid sequence identities among subfamilies are low (22.1–33.1 %), which indicates significant functional differences (Quigley et al. 2002) (Table 1.1).

Maize has 14 PIPs, 13 TIPs, 5 NIPs and 3 SIPs (Chaumont et al. 2001). The PIP proteins are more closely related to each other with 64–100 % identity. The members of the other three groups are poorly related with only 16–35 % conserved amino acid sequences (Chaumont et al. 2001).

Soybean: MIPs (GmMIPs) consist of 22 genes of GmPIPs, 23 genes of GmTIPs, 13 genes of GmNIPs, 6 genes of GmSIPs and 2 genes of GmXIPs (Zhang et al. 2014). There is high amino acid sequence similarity between GmPIPs and GmTIPs. The amino acid sequences of GmNIPs and GmSIPs are diverse.

Barley has 10 PIPs, 10 TIPs, 3 NIPs and 2 SIPs (Katsuhara et al. 2002).

In *wheat*, 24 PIPs and 11 TIPs have been identified. The PIP proteins show high degree of conservation of signature sequences, whereas TIPs are more diverse (Forrest and Bhave 2008).

PIPs have been subdivided into two groups, PIP1 and PIP2. In *Arabidopsis* PIP2;2 and PIP2;4 appear to be exclusively expressed in roots and siliques (Quigley et al. 2002). The abundance of transcripts of the genes TIP1;2 (230 ESTs), TIP1;1 (180 ESTs), TIP2;1 (79 ESTs) and TIP2;2 (70 ESTs) is highest in roots. In barley roots HvPIP1s have been detected in the vicinity of the xylem and cortex, HvPIP2;2 in the epidermis, also in the stele (Katsuhara et al. 2002). In rice at both early tillering (21 days after germination) and panicle formation (56 days) stages, six genes including OsPIP2;4 and OsPIP2;5 have been found to be predominantly expressed in roots and 14 genes including OsPIP2;7 and OsTIP1;2 expressed in leaf blades. The 8 genes including OsPIP1;1 and OsTIP4;1 are evenly expressed in leaf blades roots and anthers. High water channel activity is found, when OsPIP2;4 or OsPIP2;5 are expressed in yeast. This does not happen when OsPIP1;1 or OsPIP1;2 are similarly expressed in yeast (Sakurai et al. 2005).

In maize ZmPIP2;5 has been found to be a good water channel, but ZmPIP1;1 and ZmPIP1;2 and other ZmPIP1 homologues show a poor water transport activity in oocytes (Chaumont et al. 2000).

1.3.3 Structure of Aquaporins

Arabidopsis AQPs are predicted to be between 240 (SIP1;1) and 323 (NIP3;1) amino acid long and have six putative transmembrane (TM) helices. The pore-defining NPA motifs are conserved among the predicted amino acid sequences of PIPs and TIPs but vary in the NIPs and SIPs.

The AQP proteins of maize have been reported to contain 243–302 amino acids. All sequences have six putative transmembrane helices (TM1–TM6). Most of them have the double NPA (Asn-Pro-Ala) motif in two of the loops (B and E loops) connecting the domains (Chaumont et al. 2001). Many GmMIPs have high sequence similarity but functionally diverse roles (Zhang et al. 2014).

Plant AQPs appear to have the same general structure as mammalian AQP1 (Daniels et al. 1999). AQPs generally exist as tetramers. Each of the four monomers independently operate as water pores, but tetramerisation gives them a synergistic benefit along with forming a central pore, which allows passage of gas molecules as observed in AQP. The central pore may conduct ions through cGMP-mediated activation. This is probably caused by arginine-rich cytoplasmic D loop (The Nobel Prize in Chemistry 2003).

1.3.4 Functions of AQPs

Several reports indicate that the functions of AQPs are not solely transport of water. Other molecules such as glycerol, urea, ammonia, other uncharged molecules and possibly carbon dioxide may pass through membranes containing aquaporins (Quigley et al. 2002). Glycerol molecules, which are much larger than water molecules, appear to move in a single file through the narrow amphipathic channel where NPA motifs play a critical role (Chaumont et al. 2001).

Functionally GmMIPS consist of true aquaporins, glyceroporins, aqua-glyceroporins and mixed transport facilitators (Zhang et al. 2014).

1.3.5 Regulation of Water Uptake Under Adverse Water Regime

Passage of water through aquaporins is a passive process and occurs through osmosis across the membranes. Plant aquaporins have however developed special mechanisms to regulate water flow under adverse water regimes such as drought, flooding or salt stress. Such conditions trigger certain cellular signals (dephoshphorylation and change of pH) which close the channel and restrict water flow. A study on cellular mechanism of water flow through membranes under adverse water regime in spinach has shown that a cytoplasmic loop occludes and physically blocks the entry of water through the pore. Phosphorylation removes the loop from the entrance of the pore and allows re-entry of water. Hydrophobic amino acids are proposed to be involved in this process (Törnroth-Horsefield et al. 2006).

1.4 Ion Channels and Transporters

Many of the pores on the plasma membrane are adapted to allow passage of one specific ion or molecule and not others (Fig. 1.3). The ions are transported by proteins, which are too large to move across the membrane. They are intrinsic transmembrane proteins with fixed topology. The transporter proteins are divided into two classes:

1. Ion channels
2. Ion transporters (carriers)

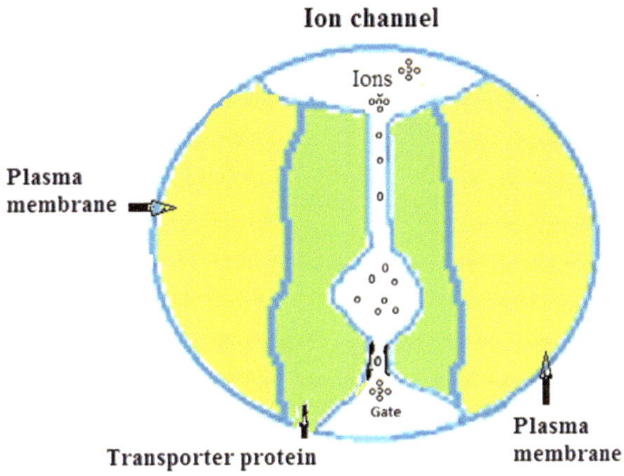

Fig. 1.3 Ion channel (Adapted from 'Information for the public'. The Nobel Prize in Chemistry (2003), The Royal Swedish Academy of Sciences, 8th October, 2003. (*Diagram redrawn and modified*))

1.4.1 Ion Channels

Cellular ion channel proteins are large molecules with multiple transmembrane α-helices. Channels alternate between open and closed conformations (gating). Control of channel gating (opening and closing) is a form of allosteric regulation. There is a conformational change of the channel protein caused by any one of the extrinsic factors such as (1) changes in membrane potential, (2) binding of a small regulatory molecule or (3) membrane stretch (e.g. via a link to the cytoskeleton) (Dubyak 2004; Diwan 2007). These factors determine if the channel is in a gated state (open for ion transport) or closed state (incapable of ion transport). The extrinsic factors control the accessibility of ions to the pore domain, which acts as a pathway for movement of ions from one side of the membrane to the other side. Since there is no energetic interaction involved between channel protein and transported ion, the rate of transport of ions through channels is many times faster than by carrier-type transporter proteins (Dubyak 2004). Acquisition of K^+ by plants through channels, which operates at relatively higher concentrations of 1 mM or above, is considered as a low-affinity system in contrast to high-affinity K^+ acquisition by transporters operating in micromolar range (See Sect. 4.1.1).

All channels mediate passive transport of ions down their chemical or electrochemical gradient across the membrane due to difference in concentrations of ions on each side of the membrane as well as any electrical potential across the membrane.

1.4.2 Ion Transporters (Carriers)

Ion transporters (carriers) according to Dubyak (2004) are 'vectoral' enzymes whose functioning involves:

1. A selective recognition/binding of the ion to be transported
2. Conformational changes in the carrier protein due to binding of the ion
3. Physical movement of the ion across the membrane caused by such conformational change

Ion transporter can catalyse movement of ions against their electrochemical gradient (not ion channels) deriving energy from ATP hydrolysis, for example, (PM)H^+ATPase (see Fig. 2.1), (Sect. 5.4.3.2) and P-type Ca^{2+}ATPases (see Sect. 5.4.3.2).

There are three types of ion transporters:

Uniporters They transport one type of ion across the membrane, such as P-type Ca^{2+} ATPases of *Arabidopsis*, ECAs, ACAs, etc. (Kudla et al 2010). (see Sect. 5.4.3.2)

Symporter (Cotransporter) The transporter binds more than one type of ion and transports them across the membrane, such as cation–chloride co-transporter (CCC family) represented in plants by AtCCC in *Arabidopsis*, which catalyses coordinated symport of (K^+), (Na^+) and (Cl^-) (Colmenero-Flores et al. 2007). (See Sect. 17.4.3.2)

Antiporters (Exchangers) There is exchange of one ion for the other, which moves in opposite direction such as calcium proton antiporters (see Sect. 5.4.3.1). Also known as calcium exchangers (CAX) are a group of proteins coded by six genes present in *Arabidopsis*. They regulate homeostasis of Ca^{2+} and other divalent cations such as Mn, Zn, Cd, Hg and Ni (Zhao et al. 2008; Kudla et al. 2010). A steep pH gradient exists across the vacuolar membrane, the tonoplast. While its cytosolic side maintains the physiologic pH, the vacuole has a significantly lower pH of 4–5. A pH gradient is established across tonoplast by proton pumps such as H^+-ATPase or H^+-pyrophosphatase. The CAX transporters take advantage of this pH gradient to move cations from cytoplasm to vacuole in exchange for H^+, which is present abundantly inside it (Kamiya and Maeshima 2004) (Fig. 1.4).

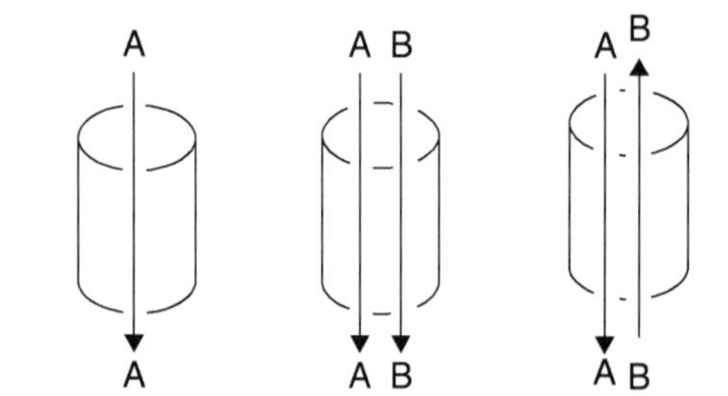

Fig. 1.4 Types of transporters (mechanisms of uptake of different nutrient ions by plants and regulation of their uptake by ion channels and transporters are discussed in Chaps. 2, 3, 4, 5, 6, 7, 8, 9, 10, 11, 12, 13, 14, 15, 16, 17, 18, 19 and 20)

References

Chaumont F, Barrieu F, Jung R, Chrispeels MJ (2000) Plasma membrane intrinsic proteins from maize cluster in two sequence subgroups with differential aquaporin activity. Plant Physiol 122:1025–1034

Chaumont F, Barrieu F, Wojcik E, Chrispeels MJ, Jung R (2001) Aquaporins constitute a large and highly divergent protein family in maize. Plant Physiol 125:1206–1215

Colmenero-Flores JM, Martinez G, Gamba G, Vazquez N, Iglesias DJ, Brumos J, Talon M (2007) Identification and functional characterization of cation-chloride cotransporters in plants. Plant J 50(2):272–298

Daniels MJ, Chrispeels MJ, Yeager M (1999) Projection structure of a plant vacuole membrane aquaporin by electron cryo-crystallography. J Mol Biol 294:1337–1349

Diwan JJ (2007) Membrane transport, Molecular Biochemistry-I, Copyright 1998–2007 by Joyce J. Diwan. All rights reserved

Dubyak GR (2004) Ion homeostasis, channels, and transporters: an update on cellular mechanisms. Adv Physiol Educ 28(1–4):143–154

Forrest KL, Bhave M (2008) The PIP and TIP aquaporins in wheat form a large and diverse family with unique gene structures and functionally important features. Funct Integr Genomics 8(2):115–133

Hammond JP, Broadle MR, White PJ (2004) Genetic responses to phosphorus deficiency. Ann Bot 94(3):323–332. doi:10.1093/aob/mch156

Imsande J, Touraine B (1994) N demand and the regulation of nitrate uptake. Plant Physiol 105:3–7

Johanson U, Karlsson M, Johansson I, Gustavsson S, Sjovall S, Fraysse L, Weig AR, Kjellbom P (2001) The complete set of genes encoding major intrinsic proteins in Arabidopsis provides a framework for a new nomenclature for major intrinsic proteins in plants. Plant Physiol 126:1358–1369

Kamiya T, Maeshima M (2004) Residues in internal repeats of the rice cation/H+ exchanger are involved in the transport and selection of cations. J Biol Chem 279:812–819

Katsuhara M, Akiyama Y, Koshio K, Shibasaka M, Kasamo K (2002) Functional analysis of water channel in barley roots. Plant Cell Physiol 43:885–893

Kudla J, Batistic O, Hashimoto K (2010) Calcium signals: the lead currency of plant information processing. Plant Cell 22:541–563

Miller AJ, Fan X, Shen Q, Smith SJ (2008) Amino acids and nitrate as signals for the regulation of nitrogen acquisition. J Exp Bot 59(1):111–119

Nguyen MX, Moon S, Jung KH (2013) Genome-wide expression analysis of rice aquaporin genes and development of a functional gene network mediated by aquaporin expression in roots. Planta 238(4):669–681. doi:10.1007/s00425-013-1918-9. Epub 2013 Jun 26

Orsel M, Filleur S, Fraisier V, Daniel-Vedele F (2002) Nitrate transport in plants: which gene and which control? J Exp Bot 53(370):825–833

Quigley F, Rosenberg JM, Shachar-Hill Y, Bohnert HJ (2002) From genome to function: the *Arabidopsis* aquaporins. Genome Biol 3(1):research0001.1–research0001.17

Rodriguez-Navarro A, Rubio F (2006) High-affinity potassium and sodium transport systems in plants. J Exp Bot 57:1149–1160

Sakurai J, Ishikawa F, Yamaguchi T, Uemura M, Maeshima M (2005) Identification of 33 rice aquaporin genes and analysis of their expression and function. Plant Cell Physiol 46:1568–1577

The Nobel Prize in Chemistry (2003) The Royal Swedish Academy of Sciences, Information for the Public, 8 Oct 2003

Törnroth-Horsefield S, Wang Y, Hedfalk K, Johanson U, Karlsson M, Tajkhorshid E, Neutze R, Kjellbom P (2006) Structural mechanism of plant aquaporin gating. Nature 439:688–694. doi:10.1038/nature0431

Zhang DY, Ali Z, Wang CB, Xu L, Yi JX et al (2014) Genome-wide sequence characterization and expression analysis of major intrinsic proteins in soybean (Glycine max L.). PLoS One 9(1):10.1371

Zhao J, Barkla BJ, Marshall J, Pittman JK, Hirschi KD (2008) The Arabidopsis cax3 mutants display altered salt tolerance, pH sensitivity and reduced plasma membrane H+-ATPase activity. Planta 227:659–669

2 Nitrogen (N) Uptake

Abstract

Globally indigenous nitrogen in soil cannot meet the crop requirement at contemporary production levels. Synthetic nitrogenous fertilisers along with other nutrients have to be applied to sustain existing production and, in many countries, further increase crop production commensurate with their population growth. Nitrogen use efficiency of crops is abysmally low (25–50 %) under uncontrolled field conditions. This not only is an economic loss, but the unutilised nitrogen also causes environmental pollution.

Nitrogen is taken up by plants as NO_3^- and NH_4^+. It has been recently found that uptake of both the forms is strictly under genetic control. There are high-affinity transporters, which carry the ions across the plasma membrane of root cells when their concentrations in the growth medium are low as well as low-affinity transporters when the concentrations are high. Many of these transporters have been characterised and mechanism of their action is known.

Biotechnological approach to improve nitrogen use efficiency includes overexpression of transporters, manipulation of genes involved in N-uptake, N-assimilation and N-translocation. Transgenic GDH-rice plants have been found to have larger number of tillers, spikelet numbers per panicle, higher biomass production, higher grain yield as well as higher NUE than the control plants. AlaAT transgenic rice shows improved NUE at medium and high N-supply.

2.1 Occurrence of Nitrogen (N) and Soil Reactions

Nitrogen (N) is a major plant nutrient required in relatively large quantities along with phosphorus and potassium for proper plant growth and metabolism. Increasing food demand of the world with increase in population necessitated developing high-yielding crop varieties, which need high nutrient application to produce optimum yield since globally indigenous plant nutrient status of most of the soils is inadequate. The

G.N. Mitra, *Regulation of Nutrient Uptake by Plants: A Biochemical and Molecular Approach*, DOI 10.1007/978-81-322-2334-4_2, © Springer India 2015

nitrogen demand of the world as estimated by FAO (2011) is 112.9 million tonnes for 2015 (ftp.fao.org) at an annual growth rate of 1.7 %. Availability of applied nitrogen from soil to crops is 35–50 % under best of conditions. About 60–75 % of applied nitrogen is lost through leaching, surface run off under flooded conditions due to heavy downpour or river floods and cause environmental pollution of ground water or open waterbodies. Farmers, who pay for the entire cost of fertiliser, get benefit of less than 30 % of it for crop production. Fertilisers around the world are heavily subsidised by governments to keep the food cost low. Since the subsidy given by the governments is sourced from the taxes paid by common man, the burden of inefficient fertiliser use is borne by every human being in the world. A proper understanding of mechanism of nutrient uptake and developing suitable methods for their efficient use would be of benefit to every human being.

Atmosphere, which contains 78 % of N_2, is the ultimate source of N used by plants. Higher plants cannot use atmospheric N_2 directly. It is converted to forms, which can be used by plants through:

1. Symbiotic nitrogen fixation by root legume bacteria
2. Non-symbiotic nitrogen fixation by free-living microorganisms
3. Atmospheric electrical discharge, which converts N_2 into oxides of N that is incorporated into the soil along with rain water
4. Synthetic N-fertilisers manufactured from atmospheric N_2

Biological nitrogen fixation (symbiotic and non symbiotic) accounts for addition of 130–180 × 10^6 metric tons of N to the soil globally. About 50 % of this is fixed by *Rhizobia,* which inhabit the roots of the legumes.

Total soil N is in the range of <0.02 % in subsoils and > 2.5 % in organic soil. Surface soils generally contain 0.03–0.4 % of total N. Soil N consists of inorganic N and organic N. About 95 % of N in surface soils is organic N.

2.2 Nitrogen Content of Plants

Plants contain about 1–6 % of N by weight. Plants generally take up nitrogen in two major forms, NO_3^- or NH_4^+. Under conditions of sufficient nitrate availability, nitrate concentration in roots and shoots may be about 100 mM, most of it stored in the vacuole (Crawford and Glass 1998). Nitrate-specific microelectrode studies indicate that cytoplasmic nitrate concentration, which is more constant, is limited to about 2–5 mM (3 mM in maize and 4–5 mM in barley) and 5–75 mM inside the vacuole (Miller and Smith 1996). NH_4^+ is toxic and is not allowed to accumulate within the plants. It is oxidised to NO_3^-, assimilated to form amino acids or converted to amides.

2.3 Functions of Nitrogen in Plants

Nitrogen is a component of amino acids, proteins, purine and pyrimidine rings of nucleic acids, chlorophyll and enzymes (which are also proteins). All of these compounds are involved in plant metabolism and growth. Under conditions of adequate N-supply, there is vigorous vegetative growth and high photosynthetic activity leading to a dark green colour of the foliage. Nitrogen deficiency causes impaired photosynthetic activity leading to degradation of chloroplasts, which appears first as yellowing of older leaves (chlorosis), while younger and growing leaves remain green.

2.4 Mechanism of Nitrogen Uptake by Plants

Acquisition, uptake, transport and redistribution of NO_3^- and NH_4^+, the two major forms in which plants take up N, are under strict genetic control (Siddiqi et al. 1990; King et al. 1993).

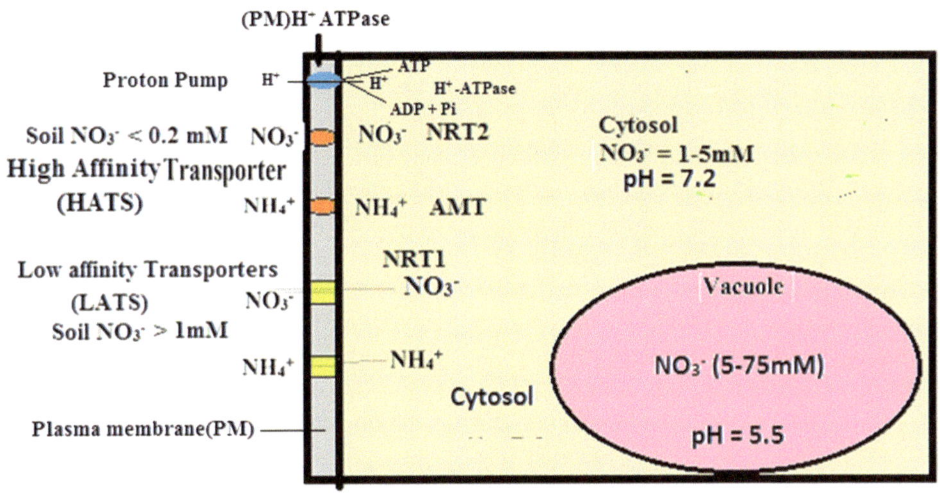

Fig. 2.1 High-affinity and low-affinity transport systems (HATS and LATS)

The primary event of NO_3^- uptake is its active transport through the plasma membrane of root epidermal and cortical cells (Fig. 2.1). This is carried out by a favourable H^+ electrochemical gradient created by the plasma membrane(PM) H^+-ATPases (proton pumps) (Miller and Smith 1996; Quaggiotti et al. 2003; Sperandio et al. 2014). (PM) H^+-ATPase activity maintains membrane potential ($\Delta\Psi$) and proton motive force (Δp) necessary for ion transport. NO_3^- uptake takes place by symport of $2H^+$/ NO_3^- both for high- and low-affinity transport systems (Crawford and Glass 1998).

2.4.1 Nitrate Transporters

There are at least three different nitrate transport systems in plants. An essentially unregulated, low-affinity transport system (LATS) constitutively expressed operates when the external NO_3^- concentration is high (1–50 mM) (Crawford and Glass 1998). At low external NO_3^- concentration (<0.2 mM), two high-affinity transport systems (HATS) operate, some constitutively expressed (cHATs) and the others induced by NO_3^- (iHATs) (Fig. 2.1).

The NRT1 genes encode low-affinity nitrate transporters (LATS), when NO_3^- concentration in the soil is high (>1 mM, Orsel et al. 2002). The NRT2 genes encode high-affinity nitrate transporters (HATS), when NO_3^- concentration is low (<0.2 mM). Some of the NRT2 genes are inductive (iHATS) and others constitutive (cHATs). CHL1 (At NRT1;1) is a dual affinity nitrate transporter, switched off and on by phosphorylation/dephosphorylation of threonine T101 (Liu et al. 1999). The CBL (calcineurin B-like, Sect. 5.4.3.3.3)-interacting protein kinase CIPK23 (SnRK3;23) phosphorylate T101 under low nitrate conditions, allowing NRT1;1 to function as a high-affinity nitrate transporter (Ho et al. 2009). Phosphorylated NRT1;1 is a high-affinity nitrate transporter and the dephosphorylated NRT1;1 is a low-affinity transporter. There are 53 NRT1 and 7 NRT2 genes in *Arabidopsis*.

The NRT1 and NRT2 proteins share the same structural topology, with 12 transmembrane domains, distributed in two sets of six helices connected by a cytosolic loop. Beside their common structure, no primary sequence homology is found between the NRT1 and NRT2 classes. In higher plants, the *NRT2* genes isolated so far are preferentially expressed in roots (Tsay et al. 2007). Lin et al. (2000) identified OsNRT1 from rice a homologue of the *Arabidopsis* CHL1 (AtNRT1) protein. *OsNRT1* is a constitutive component of a low-affinity nitrate uptake system in rice and is expressed in epidermal cells of roots.

The transporters involved in nitrate uptake in *Arabidopsis* are two from NRT1 family, AtNRT1;1 and AtNRT1;2, and two from NRT2 family, AtNRT2;1 and AtNRT2;2. The NRT2;1 proteins are localised on the plasma membrane (Chopin et al. 2007) and constitute major component of the HATS, when the external nitrate concentration is low (Filleur, et al. 2001; Orsel et al. 2004). The NRT2;1 protein requires a second protein, named NAR2 (AtNRT3;1), as found in barley (Tong et al. 2005) or in *Arabidopsis* (Okamoto et al. 2006; Orsel et al. 2006). NAR2 is possibly involved in NRT2;1 stability (Wirth et al. 2007). Yong et al. (2010) reported that NAR2 and NRT2;1 form a tetramer consisting of two subunits each of AtNRT2;1 and AtNAR2;1 at the plasma membrane to form the active nitrate transporter. According to Li et al. (2007) AtNRT2;1 is responsible for 72 % of HATS activity. AtNRT2;2 makes only a small contribution to the system, except when AtNRT2;1 is lost (Li et al. 2007; Dechorgnat et al. 2011). AtNRT2;4 is involved in petiole nitrate storage. AtNRT2;7 is involved in nitrate accumulation.

2.4.1.1 Rice

Four high-affinity nitrate transporters, OsNRT2;1, OsNRT2;2, OsNRT2;3 and OsNRT2;4, and two NAR proteins, OsNAR2;1 and OsNAR2;2, have been identified in rice. OsNRT2;1, OsNRT2;2, OsNRT2;3 and OsNAR2;1 are primarily expressed in roots. OsNAR2;1 is essential for nitrate uptake by OsNRT2;1 and OsNRT2;2 (Feng et al. 2011; Sperandio et al. 2014). OsNRT2;3a is also involved in nitrate uptake in association of OsNAR2;1, but has a tenfold lower affinity for nitrate than OsNRT2;1 and OsNRT2;2 (Yan et al. 2011).

2.4.1.2 Maize

In maize two high-affinity nitrate transporters have been identified. ZmNRT2;1 is involved in influx activity and ZmNRT2;2 in Xylem loading process (Trevisan et al. 2008).

Some members of NRT1 are peptide transporters called OPTs and PTRs. The oligopeptide transporters (OPTs) are involved in transport of tetra- and pentapeptides. Some of the OPTs transport glutathione, glutathione conjugates, phytochelatins and metals (Tsay et al. 2007). The PTRs are di- and tripeptide transporters called NRT1/PTRs. Several families of NRT1/PTRs have been identified. They transport nitrate, di- and tripeptides, auxins or carboxylates (Fan et al. 2014). SP1 (short panicle 1), a putative PTR gene abundantly expressed in the young panicle branches of rice, determines panicle size (Li et al. 2009). OsPTR6 transports gly-His and Gly-His-Gly and shows substrate selectivity for di-/tripeptides.

During seed germination of barley, HvPTR1, a transporter expressed in plasma membrane of scutellar epithelial cells, transports peptides produced by hydrolysis of endosperm storage proteins to the developing embryo.

2.4.2 Ammonium Transporters

The AMT1 family of high-affinity NH_4^+ transporters contains five members, of which AtAMT1;1, AtAMT1;2 and AtAMT1;3 from *Arabidopsis* have been studied in detail. All the three genes are expressed in roots. Only AtAMT1;1 is expressed in significant amounts in roots. In response to N deprivation, it was found that transcript abundance of AtAMT1;1 (in leaves) increased fivefold within 72 h, whereas in roots transcripts of AtAMT1;3 increased twofold, but there was no change in the abundance of AtAMT1.2 (Gazzarini et al. 1999). Ludewig et al. (2002) characterised LeAMT1;1, a root hair ammonium transporter from tomato (*Lycopersicon esculentum*), and suggested that NH_4^+, rather than NH_3, is the true transport substrate.

2.4.2.1 Rice

Suenaga et al. (2003) identified four ammonium transporter genes from rice roots, OsAMT1, OsAMT2, OsAMT3 and OsAMT4, based on their phylogenetic relationship. OsAMT1s (OsAMT1;1, OsAMT1;2, OsAMT1;3) have high sequence similarities with putative 11 transmembrane spanning domains and differ from the others of the OsAMT families (Sonoda et al. 2003).

OsAMT1;1 is more constitutive and expressed in shoots and roots. OsAMT1;2 is root specific and inducible by NH_4^+. OsAMT1;3 is root specific and derepressed by nitrogen application. Rice AMTs differ from those in tomato or *Arabidopsis* by their distinct nitrogen-dependent regulation (Sonoda et al. 2003).

2.4.2.2 Maize

Two rhizodermis localised transporters, ZmAMT1;1 and ZmAMT1;3, have been identified from maize, which probably constitute high-affinity ammonium transporters in maize roots. A specific regulatory feature of these transporters is their induction by ammonium rather than upregulation by nitrogen deficiency (Gu et al. 2013).

2.4.3 Regulation of Nitrate Transporters

Some of the genes encoding NO_3^- transporters are subjected to transcriptional regulation through inductive effects of NO_3^-, while both encoding NO_3^- and NH_4^+ transporters are subject to downregulating effects of glutamine (Anthony et al. 2002). Miller et al. (2008) suggested that the internal pools of amino acids within plants might indicate nitrogen status by providing a signal that can regulate nitrate uptake by the plant. In support of this idea, both nitrate and ammonium influx and transporter transcripts were shown to decrease in root tissue treated with exogenously applied amino acids especially glutamine. The feedback regulation occurs by changing the expression of transporters, but may also involve the post-translational modification of proteins. According to Ho et al. (2009) and Wang et al. (2009), CHL1 (At NRT1.1), the dual affinity nitrate transporter with a phosphorylation switch (to sense a wide range of nitrate concentration), functions as ion sensor in higher plants. Nitrate signalling mediated by the NRT1;1 nitrate transporter antagonises L-glutamate-induced changes in root architecture. Nitrate stimulates primary root growth, both directly and by antagonising the inhibitory effect of glutamate, which stimulates root branching (Walch-Liu and Forde 2008). Nitrate and glutamate concentration constitutes an intricate N regulatory network at the root tip that is responsible for orchestrating changes in root growth rate and root architecture to accommodate variations in the extrinsic and intrinsic supply of N (Forde and Walch-Liu 2009). Gojon et al. (2011) used the term 'transceptor' to describe the dual function of NRT1.1 as NO_3^- transporter and signalling molecule in analogy with the term used in animals and yeast. Cai et al. (2007) reported that while exogenous ABA and Gln induced HATS genes in the roots of wheat, they did not induce nitrate influx. Some of the cytoplasmic enzymes responsible for nitrate assimilation are regulated by phosphorylation and binding of a 14-3-3 protein. Both high-affinity and low-affinity nitrate transporters are induced by deficiencies of other nutrients such as phosphate, potassium and iron (Wang et al. 2001). [*14-3-3 proteins* are binding proteins that interact with a wide array of enzymes involved in primary biosynthetic and energy metabolism in plants. Most of the interactions are phosphorylation dependent with their binding partners found in the cytosol (Huber et al. 2002)].

2.4.4 Regulation of Ammonium Transporters

Ammonium transporters are oligomeric proteins, essential for ammonium uptake by plants. Conformational coupling among the monomers provides a mechanism for tight regulation, for increasing the dynamic range of sensing and memorising prior events and may be a general mechanism for regulation of ammonium transporters. Rapid shut-off mechanisms are required to prevent toxic accumulation of NH_4^+. The crystal structure of an *Archaeoglobus fulgidus* ammonium transporter (AMT) suggests that the C-terminus interacts physically with cytosolic loops of the neighbouring subunit. Phosphorylation of conserved sites in the C-terminus is proposed as the cognate control mechanism (Loque et al. 2007). Employing the co-expressed transporters AMT1;1 and AMT1;3 from *Arabidopsis thaliana* as a

Fig. 2.2 Methods of calculation of agronomic efficiency (AE), apparent nitrogen recovery (ANR) and production efficiency (PE)

$$\text{Agronomic efficiency (AE= kg of grain/kg of N applied)} = \frac{\text{Grain yield in the fertilised plot (kg/ha)} - \text{Grain yield in the contol plot (kg/ha)}}{\text{Fertiliser N applied (kg/ha)}}$$

$$\text{Apparent N recovery(ANR) (\%)} = \frac{\text{N uptake in fertilised plot (kg/ha)} - \text{N uptake in control plot (kg/ha)}}{\text{Fertiliser N applied (kg/ha)}} \text{x100}$$

$$\text{Production efficiency (PE) (kg of grain/ kg of N absorbed)} = \frac{\text{Grain yield in the fertilised plot (kg/ha)} - \text{Grain yield in the contol plot (kg/ha)}}{\text{N uptake in fertilised plot (kg/ha)} - \text{N uptake in control plot (kg/ha)}}$$

model, Yuan et al. (2013) have reported that these two isomers form functional homo- and heterotrimers in yeast and plant roots. AMT1;3 containing a phosphomimetic residue in its C-terminus negatively regulates both homo- and hetero-trimer formation in vivo.

2.4.5 Nitrate or Ammonium

When nitrogen is applied to nitrogen-starved plants, both nitrate and ammonium transporters are rapidly downregulated (Glass 2003). At higher levels of nitrogen supply, there is efflux of nitrate and ammonium. Application of higher levels of ammonium blocks nitrate uptake by roots. The expression levels of genes encoding ammonium and nitrate transporters and rates of nitrate and ammonium uptake are substantially reduced, when incident radiation is low either due to diurnal variation of light intensity or under heavy cloud cover. The expression of high-affinity nitrate transporters is reduced at low temperature, when growth and nitrogen demands by plants are low (Malagoli et al. 2004).

2.4.6 Biotechnological Approach to Improve Nitrogen Use Efficiency (NUE)

2.4.6.1 The Problem of Low NUE

NUE is essentially a measure of units of edible (or useful) parts of a crop produced per unit of applied nitrogen (agronomic efficiency). Apparent nitrogen recovery is nitrogen uptake by the plants expressed as percent of nitrogen applied. Production efficiency is a measure of units of edible parts produced per unit of nitrogen absorbed. This gives an estimate of nitrogen utilised by the plants from adsorbed nitrogen. The method of calculation is given in Fig. 2.2.

Most of the crops respond admirably to nitrogen application (except those with symbiotic N-fixation mechanisms) since arable soils around the world are generally deficient in nitrogen and cannot meet the N-requirement of crops. Commonly used nitrogenous fertilisers, such as urea, ammonium sulphate, diammonium phosphate (containing both N and P), etc., are highly water soluble. About 25–50 % of applied nitrogen is utilised by crops under field conditions. The unutilised nitrogen forms a considerable source of environmental pollution. A large part of N-fertiliser is simply washed away from the field if there is heavy rainfall after its application or due to over irrigation and finds its way to nearest waterbodies. Considerable quantities are leached beyond the rhizosphere of the crops in sandy soils and pollute sub-surface water. Soil microorganisms lock up a large part of applied N for their own metabolism. Microbial denitrification of applied N releases gaseous N_2 and oxides of nitrogen (NO_x) (N_2O has 300 times higher greenhouse gas effects than CO_2) to the atmosphere. The low NUE causes enormous financial loss to the farmers as their crops utilise only a small part of the applied N-fertiliser. The governments throughout the world suffer financial loss as they give large subsidies on fertiliser and food prices.

2.4.6.2 Agronomic Approach to Improve NUE

Soil scientists have developed various technologies to reduce N loss from applied N-fertilisers and improve NUE. Instead of applying the entire dose of N at the time of planting, N-fertilisers are applied in smaller doses at specific stages of crop growth according to their need (based on research results) to enable better utilisation and minimise loss. Slow-acting nitrogenous fertilisers have been developed, which release N at a lower concentration in a sustained manner to provide greater scope for roots for uptake. Placement of N-fertilisers closer to rhizosphere is another technology adopted to increase NUE. Most of these technologies are yet to be adopted by farmers, and the NUE of nitrogenous fertilisers hovers around 35 % or less for waterlogged rice. It is necessary to develop crop varieties with higher NUE to prevent environmental pollution and minimise economic loss to the farmers. The current biotechnological approaches are aimed at realising higher yields with application of lower doses of N-fertilisers.

2.4.6.3 Biotechnological Approach

The current biotechnological approach to improve NUE includes manipulation of genes involved in:

1. N-Uptake
2. N-Assimilation
3. N-Translocation to the edible/useful parts of the crops

2.4.6.4 N-Uptake

Overexpression of Nitrate Transporters

Malagoli et al. (2004) studied the quantitative effects of different factors such as day/night cycle, ontogenetic stages, root temperature, photosynthetically active radiation and soil nitrate availability on nitrogen uptake by different nitrate transporters in oilseed rape through simulation studies and correlating them with N-uptake under field conditions. They observed that the high-affinity transport system accounted for about 89 % of total NO_3^- uptake (18 and 71 % for cHATS and iHATS, respectively) when no fertiliser was applied. In the treatment without N-application, LATS accounted for a minor proportion of total N-uptake, and its activity was restricted to the early phase of the growth cycle. However, application of N-fertilisers increased LATS activity for an extended period of time, and its contribution to nitrate uptake was more. Activities of HATS also increased with N-application, but its contribution to total N-uptake was reduced. The diurnal pattern showed two peaks of nitrate uptake one at 9 AM and another at 6 PM. Nitrate influx by HATS and HATS + LATS decreased by 66 % from light to dark period. The transition from vegetative to reproductive phase was characterised by a significant decrease in activities of HATS and HATS + LATS (Malagoli et al. 2004).

Overexpression of OsPTR6 (di-/tripeptide transporter) in rice has been reported to show significant increase in growth, high total N accumulation, OsAMT1 gene expression and GS activities, when treated with high ammonium, but there was significant decrease in NUE (Fan et al. 2014).

OsNRT2;1, OsNRT2;2 and OsNAR2;1, which are induced by nitrate, are suggested to be potential candidates for use in breeding programmes to develop varieties with higher nitrate uptake efficiency in rice (Arak and Hasegawa 2006; Sperandio et al. 2014).

Overexpression of Ammonium Transporters

OsAMT1.1 in rice is the most active and most responsive gene for high-affinity ammonium uptake. Hoque et al. (2006) studied the $^{13}NH_4^+$-uptake and its subsequent assimilation under different nitrogen regimes of transgenic lines overexpressing *OsAMT1-1* produced by *Agrobacterium*-mediated transformation of two rice cultivars, Taipei 309 and Jarrah, with an *OsAMT1-1* cDNA gene construct driven by the maize ubiquitin promoter. Roots of the transgenic plants showed increased ammonium uptake and ammonium content. The biomass of the transgenic lines decreased at early stages of growth.

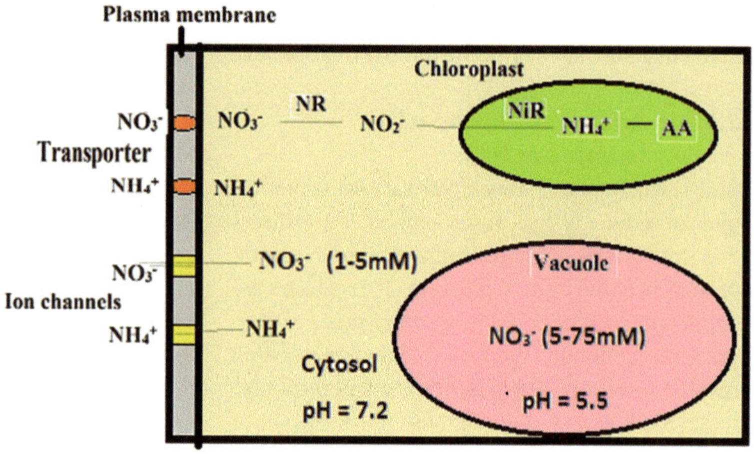

Fig. 2.3 Transport of NO_3^- and NH_4^+ across plasma membrane and reduction of nitrate to nitrite in cytosol and nitrite to ammonium in chloroplast. *NR* nitrate reductase, *NiR* nitrite reductase, *AA* amino acids. Concentration of NO_3^- in cytosol is 1–5 mM and vacuole, 5–75 mM. pH of cytosol is 7.2 and of vacuole 5.5

This was probably due to mismatch between accumulation of ammonium in the roots and its subsequent assimilation. Kumar et al. (2006) reported similar results.

Interaction of Ammonium and Nitrate in Rice Roots

Ammonium rather than NO_3^- is considered to be the major source of N in lowland rice soils. However, very little of free NH_4^+ translocate to the shoot in rice (Kronzucker et al. 1998). Results of recent research have shown that lowland rice is exceptionally efficient in absorbing NO_3^- formed by nitrification in the rhizosphere. Rice roots excrete oxygen and aerate the rhizosphere. About one third (34 %) of the total nitrogen absorbed by the rice plant is in the form of NO_3^- (Kirk and Kronzucker 2005; Duan et al. 2006). Duan et al. (2007) conducted solution culture experiments with two rice cultivars, Nanguang (high NUE) and Elio (low NUE). Nitrogen was applied as NH_4^+ and NO_3^- in the ratios of 75–25 and 100–0 respectively. Uptake experiments with ^{15}N showed that NO_3^-increased NH_4^+ uptake efficiency in 'Nanguang (High-NUE)' by increasing Vmax (14 %), but there was no effect on Km. This indicates that partial replacement of NH_4^+ by NO_3^- (PNN) can increase the number of the ammonium transporters but does not affect the affinity of the transporters for NH_4^+. Real-time PCR showed that expression of OsAMT1s in 'Nanguang' improved by PNN, while that in 'Elio (Low-NUE)' did not change. This is in accordance with the differing responses of these two cultivars to PNN. These results indicate a relationship between NUE and PNN. Similar results have been observed by Li et al. (2008), who compared an *indica* rice variety (Yangdao 6) with high NUE with *japonica* rice variety (Nongken 57) with low NUE. They observed that rhizosphere nitrification activity, NO_3^- concentration and abundance of ammonia-oxidising bacteria associated with Yangdao 6 (high NUE) were always higher than Nongken 57 (low NUE).

2.4.6.5 Nitrogen Assimilation

Nitrate taken up by the plants is first converted to nitrite by nitrate reductase (NR). Nitrite translocate from the cytoplasm to the chloroplast, where it is reduced by nitrite reductase (NiR) to form ammonium (Fig. 2.3). A number of studies have been conducted to improve NUE through expression of NR and NiR genes.

Overexpression of NR and NiR Genes

A number of endogenous and environmental factors in plants regulate expression of NR and NiR genes at the transcriptional, translational and post-translational levels (Meyer and Stitt 2001). Several studies have been performed on plants where the NR and NiR genes have been overexpressed using either constitutive or inducible promoters (Good et al. 2004). Expression of NR gene by 35S promoter allows a deregulated transcription of the gene, but a post-translational phosphorylation mechanism through binding of a

14-3-3 protein inhibits the enzymatic activity of NR protein (Provan et al. 2000). Introduction of deregulated NR gene under the control of the CaMV 35S promoter into transgenic potato increases nitrate levels in the tubers but does not show either increase in yield or number of tubers (Dejannane et al. 2002).

Overexpression of nitrite reductase (NiR) in *Arabidopsis* and tobacco does not show any phenotypic differences in transgenic plants, suggesting that post-transcriptional regulation is operating on NiR expression (Crété et al. 1997).

Overexpression of either the NR or the NiR gene in plants increases mRNA levels and often affects N-uptake but does not seem to increase the yield or growth of the plants regardless of the nitrogen source available.

Glutamine Synthetase (GS) and Glutamate Synthase (GOGAT)

There are two GS isoforms, located either in the cytosol (GS1) or in the plastids (GS2). They have specific functions in assimilating or recycling ammonium. GOGAT catalyses the reductant-dependent conversion of glutamine and 2-oxaloglutarate to two molecules of glutamate. There are two distinct isoforms, one ferredoxin dependent (Fd-GOGAT) and the other NADH dependent (NADH-GOGAT). Fd-GOGAT is the predominant form and plays an important role in leaf photo-respiratory ammonium assimilation. NADH-GOGAT is found primarily in non-photosynthetic tissue, where it is the major form of GOGAT and combines with GS1 to assimilate NH_4^+ produced by nitrogen-fixing bacteria.

Several studies conducted with overexpression of GS showed that GS activity was not always consistent with decreases in plant N and biomass accumulation (Ortega et al. 2001; Oliveira et al. 2002; Fei et al. 2003). Increased growth rates in transgenics overexpressing GS1 have been reported under low N-conditions (Good et al 2004). Chichkova et al. (2001) reported that overexpression of NADH-glutamate synthase in transgenic tobacco plants resulted in higher carbon and nitrogen content but did not show any phenotype associated with this trait. Overexpression of NADH-GOGAT gene in transgenic lines of *indica* rice 'Kasalath' under the control of NADH-GOGAT promoter of 'Sasanishiki', a *japonica* rice, showed an increase in grain weight (80 % as a maximum), indicating the importance of NADH-GOGAT in nitrogen utilisation and grain filling of rice (Yamaya et al. 2002).

Glutamate Dehydrogenase

Naohiro et al. (2009) introduced fungal NADP-dependent glutamate dehydrogenase (NADP-GDH) gene into potato and rice plants under the control of the CaMV-35S promoter. The transgenic GDH-potato plants showed a higher photosynthetic rate, a larger number of stolons and tubers, a higher tuber yield and a higher NUE for tuber dry weight. In transgenic GDH-rice plants, number of tillers and spikelet numbers per panicle increased as compared to the control plants, resulting in a higher biomass production and a higher grain yield. NUE was also higher in the GDH-rice plants than the control plants. The higher productivity and NUE in the GDH-transgenic plants was more remarkable in soil with lower nitrogen application. Further in ^{15}N labelling experiments, in combination with the use of an inhibitor for plant glutamine synthetase, they observed that ammonium ions were assimilated directly in the roots of GDH-rice seedlings by the introduced NADP-GDH. These results indicate that transgenic plants overexpressing NADP-GDH activities possess a high NUE, particularly under low nitrogen condition due to direct effect of introduced NADP-GDH on nitrogen assimilation.

Alanine Amino Transferase

Alanine is the major storage amino acid under anaerobic conditions (e.g. flooding) since pyruvate is predominantly available as the carbon skeleton under these conditions (Vanlerberghe and Turpin 1990). Good et al. (2007) introduced a barley AlaAT cDNA driven by a canola root-specific promoter (btg26) into canola plants to overexpress alanine amino transferase (AlaAT). As compared to wild canola, the transgenic canola plants produced higher biomass and seed yield both under laboratory and field conditions.

This was however evident only under low N-conditions. The transgenics had a higher influx of nitrate. Application of N-fertilisers could be decreased by 40 % without any loss of yield. Shrawat et al. (2008) introduced a barley AlaAT (alanine amino transferase) cDNA driven by a rice tissue-specific promoter (OsAnt1) into rice plants. There were significant increases in the biomass and grain yield as compared to control plants when plants were well supplied with nitrogen. Transgenic rice plants also demonstrated significant changes in key metabolites and total nitrogen content, indicating increased nitrogen uptake efficiency. Recently Beatty et al. (2013) have reported that AlaAT transgenic rice shows improved NUE at medium and high N-supply with upregulation of transcripts associated with photosynthesis, non-mevalonate pathway secondary metabolites, protein degradation and many unknown function transcripts.

2.4.6.6 Nitrogen Remobilisation for Grain Filling

The capacity to remobilise photosynthates stored before anthesis in the vegetative parts and to translocate them to sinks such as grains, fruits, tubers, roots, etc., constitutes the major yield determining factor and nutrient use efficiency of a crop. This has to be tightly integrated with nutrient uptake. The two processes differ among different plant species. In rice, wheat and barley about 90 % of N stored in vegetative parts before anthesis is mobilised for translocation to grains. In maize 35–55 % of grain N is derived from N-uptake from soils after anthesis.

According to Masclaux-Daubresse et al. (2008), factors, which govern N-remobilisation for grain filling, are (1) seed sink strength; (2) transfer processes located in the source leaves, stems and in reproductive structures; and (3) phloem pathway efficiency.

Seed Sink Strength

All emerging plant parts serve as sinks and trigger nitrogen remobilisation from older organs. Once fully developed, they turn into source and supply nutrients to the newer parts. After anthesis, seed filling takes precedence and remobilisation of nutrients is triggered from leaves to leaves and leaves to seeds. The high transport of nitrogen from leaves to seeds causes leaf senescence. Genes encoding catabolic enzymes are activated, which gradually dismantles cellular components that mainly reside in the chloroplast. The number of chloroplast decreases. At the ultrastructural level, there is reduction of thylakoid membrane system, loosening of grana stacks, swelling of intra-thylakoid space and increase in the number of plastoglobuli (containing degradation products of thylakoids) (Gregersen et al. 2008). However, the basic metabolic activities are kept intact until cell death, which ensures the process of degradation of high molecular weight components and the subsequent export of the products and minerals to the phloem. A generic senescence programme exists across monocot and dicot species (Gregersen et al 2008).

Transfer Processes Located in the Source Leaves, Stems and in Reproductive Structures

In C_3 plants about 50 % of proteins present in leaves consist of RuBisCo (ribulose-1,5-biphosphate carboxylase/oxygenase), which constitutes a major source of protein (Mae et al. 1983). Small glycoproteins called vegetative storage proteins (VSP) accumulate to high levels in the leaves, stems and pod walls of legumes and constitute an additional source of proteins that can be remobilised (Lansing and Franceschi 2000). The exact mechanism of RuBisCo degradation is not well understood and is still under intense investigation. CND41, an aspartate protease encoded by a nuclear genome and located in chloroplast, has RuBisCo proteolytic activity, but cannot act on an intact molecule unless its structure is denatured (Kato et al. 2004). This is probably caused by ROS (reactive oxygen species) formed during various degradative reactions, which occur during senescence.

Phloem Loading

Concentration of phloem sap-free amino acid is in the range of 50–200 mM for most plant species, except for *Brassica* sp. which contain up to 650 mM free amino acids in their phloem sap

(Tilsner et al. 2005). Asparagine is the major phloem-translocated amino acid in pea. Glutamate/glutamine is the preferred N-transporter in tomato, cereals, *Arabidopsis*, tobacco and a number of other plant species. The major amino acid transported in barley and wheat is glutamate, followed by varying levels of aspartate, glutamine, threonine and serine (Caputo et al. 2001; Kichey et al. 2006). The rate of amino acid loading for apoplasmic phloem loaders depends on amino acid transporters located in sieve tubes/companion cell complexes (Tilsner et al. 2005).

A large number of amino acid transporters, which are upregulated during leaf senescence, have been catalogued by Masclaux-Daubresse et al. (2008). Some of them as reported by various workers are amino acid permeases (neutral and basic and aromatic amino acid transporters, AAP3, AAP4; peptide transporters, PPT; lysine and histidine transporters, LHT1; proline transporter, ProT2; oligopeptide transporters, OPT4, etc.)

While phloem loading process is a key step in N-remobilisation, seed sink strength could also be another factor. Many amino acid and peptide transporters are located in the seeds and are probably involved in cotyledon loading of N-compounds (Rentsch et al. 2007; Tsay et al. 2007). These include PTRs (peptide transporters), CAT6 (cationic amino acid transporters), OPT5 (oligopeptide transporters), AAP3 (amino acid permeases), etc., as described by Masclaux-Daubresse et al. (2008).

Rolletschek et al. (2005) overexpressed *Vicia faba* amino acid permease gene VfAAP1 in pea (*Pisum sativum*) and *Vicia narbonensis* under the control of the legumin B4 promoter. This resulted in an increase in seed size by 20–30 % and in seed total N content by 10–25 %. However, seed yield per plant remained unaltered. They concluded from their studies on N-uptake of transgenic plants that VfAAP1 expression increased seed sink strength for nitrogen, improved plant nitrogen status, and lead to higher seed protein. Seed uptake activity for nitrogen was probably rate limiting for storage protein synthesis.

Amino Acid Biosynthesis for Phloem Loading During Leaf Senescence

Different proteolytic activities during leaf senescence break down proteins into amino acids, amides and ammonium. A major part of ammonium is used for synthesis of amino acids and a minor part is evaporated from the leaf canopy (Schjoerring et al. 1993). Glutamine synthetase (GS) activity appears to be of major importance for the re-assimilation of ammonium into exported amino acids during senescence, in particular the cytosolic GS forms (Habash et al. 2001; Miflin and Habash 2002). According to Masclaux-Daubresse et al. (2008), in young leaves, GS2 located in chloroplast and a ferredoxin-dependent glutamate synthase (Fd-GOGAT) play a major role in assimilation of ammonium generated from nitrate reductase activity and photo-respiration through GS/GOGAT pathway during periods of high photosynthetic activity. With the onset of senescence and degradation of chloroplasts, newly expressed cytosolic GS1 isoforms takes over. Mitochondrial GDH is also involved. Proteolysis of chloroplastic proteins releases a large number of amino acids, which through a series of transamination reactions produce glutamate. This serves as a substrate for GDH to produce a 2-oxoglutarate to support respiration and ammonium, which is utilised by GS1 to produce glutamine for export through phloem.

Hirel et al. (2001) and Gallais and Hirel (2004) conducted QTL studies on maize genome and reported that loci for yield traits and the gene encoding cytosolic GS (GS1) coincided. Obara et al. (2001) conducted similar studies with rice and observed that QTLs of a yield trait and structural gene of GS1 coincided. QTL studies in wheat indicate that the loci of GS1 and of grain nitrogen and grain weight coincide (Habash et al. 2007). A multigene family of GLN1 consisting of GLN1.1, GLN1.2, GLN1.3, GLN1.4 and GLN1.5 encodes GS1. Three GS1 encoding genes, OsGS1.1, OsGS1.2 and OsGS1.3, are found in rice (Ishiyama et al. 2004). The gene OsGS1.1, located in the companion cells and parenchyma cells of leaf tissues, is probably involved in generation of GS1 to synthesise glutamine for transport through phloem.

Manipulation of GS activity in leaves seems to be a useful instrument for achieving improved nitrogen remobilisation in cereals (Hirel et al. 2007). Preliminary result from wheat indicated that overexpression of a GS1 could increase grain yield (Habash et al. 2001).

Martin et al. (2006) overexpressed constitutively GLN1.3 an isoform of GS1 encoding gene in leaves of maize under suboptimal nitrogen feeding conditions and observed that kernel number increased by 30 %, which indicated its major role in kernel yield. They have assigned different roles to the five isoforms of Gln1 in maize, Gln.1.1, located in root cortex, synthesis of glutamine for vegetative growth; Gln.1.2, located in phloem, transport of gln.; Gln.1.3, located in mesophyll cells of leaves, increasing kernel numbers; Gln1.4, located in bundle sheath cells of leaves, increasing kernel size; and yet unknown function of Gln1.5.

2.4.6.7 Transcription Factors

GA-Insensitive TFs and Semidwarf Cereal Varieties

Transcription factors (TFs) are sequence-specific DNA-binding proteins, which interact with the promoter region of the target genes and regulate gene expression. TFs can be classified into different gene families based on DNA-binding domains and other conserved features. TFs from the same family often regulate similar physiological functions even among very different plant species. Regulation of most biological processes in the plant cell can be linked to one or more TF families (Century et al. 2008). *Arabidopsis* genome encodes more than 1,500 TFs, which constitute about 5 % of its genes (Riechmann 2002; Qu and Zhu 2006). The reduced height of semidwarf wheat varieties, which brought Green Revolution, is due to mutation in at least one of the two *Reduced height-1 loci (Rht-B1 and Rht-D1)*, which causes plants to respond abnormally to the hormone GA (Peng et al. 1999). *Rht-B1* and *Rht-D1* are orthologues of *Arabidopsis* GIBBERELLIN-INSENSITIVE (*GAI*) gene, a member of GRAS family of transcription factors, which functions as a transcriptional repressor of growth (Peng et al. 1997; Pysh et al. 1999). The mutations stabilised the protein causing a semi-dominant trait for GA insensitivity. Transfer of mutant *Arabidopsis gai* allele to Basmati rice (normally tall) resulted in the production of dwarf plants (Peng et al. 1999). According to Century et al. (2008), the results cited above confirm that GA insensitivity TFs were involved in the production of semidwarf wheat varieties and such technique could be used for other cereals to produce semidwarf varieties.

The development of first generation of transgenic crops was based in large part on simple monogenic traits, such as herbicide tolerance or insect resistance, which did not require manipulation of complex molecular pathways in the plant. The second generation of transgenic crops has to address more challenging traits relating to yield and yield stability, which are under complex polygenic control (Gutterson and Zhang 2004; Salmeron and Herrera-Estrella 2006). Transcription factors (TFs) naturally act as master regulators of cellular processes and are excellent candidates for modifying complex traits in crop plants. TF-based technologies are likely to be a prominent part of the development of next generation of transgenic crops (Century et al. 2008).

2.4.6.8 Dof TF and Nitrogen Metabolism

Nitrogen metabolism is inseparably connected to carbon metabolism and regulated tightly. In plants and other organisms, a metabolic pathway consists of many biochemical reactions catalysed by a number of enzymes. Transcription factors, which can regulate multiple gene expression, have been used to enhance nitrogen uptake and assimilation. Dof (DNA binding with one finger domain) proteins are plant-specific transcription factors, which contain a particular class of zinc finger DNA-binding domain (Lijavetzky et al. 2003). Dof proteins typically consist of multiple domains including a highly conserved N-terminal DNA-binding domain and a C-terminal domain for transcriptional regulation (Yanagisawa 2002). *Arabidopsis* has been reported to have 36 different Dof genes, rice 30 Dof genes (Lijavetzky et al. 2003) and wheat 31 TaDof genes (Shaw 2009). The first Dof TF identified in maize (ZmDof1) has been reported to be involved in transcriptional regulation of genes associated with carbon metabolism (Yanagisawa 2001). Yanagisawa et al. (2004) overexpressed the maize *Dof1* gene, which was known to be involved in organic acid metabolism, to create transgenic *Arabidopsis* plants that showed increases in free amino acid content and total nitrogen content (30 %), as well as improved growth, under low nitrogen conditions. Kurai et al. (2009) studied the effect of introduction of *ZmDof1* in *Oryza sativa* (cv. Nipponbare). *ZmDof1* cDNA, driven by the maize ubiquitin promoter, was introduced into rice by means of Agrobacterium. T-DNA insertion and *ZmDof1* expression was confirmed by PCR. A hydroponic experiment was conducted under two nitrogen regimes: 10 ppm (control) and 2.5 ppm (low-N). *ZmDof1*-expressing rice plants showed normal growth and phenotype. The dry weight of both shoots and roots was significantly increased. Total N contents also increased, yet the NC ratio remained the same. T/R ratio of Dof1 rice significantly decreased. Organic acid profile in leaves and amino acid profile in roots changed considerably. Soluble protein and chlorophyll concentration as well as PEP carboxylase activity in leaves increased. These results indicate Dof1 alters both nitrogen and carbon metabolism.

TaDof4 identified from wheat is probably associated with growth-related processes and TaDof5 with potential role in photoperiod responses associated with flowering (Shaw 2009).

References

Anthony DM, Glass D, Britto TD, Kaiser BN, James R, Kinghorn JR, Kronzucker HJ, Anshuman A, Okamoto M, Rawat S, Siddiqi MY, Unkles SF, Vidmar JJ (2002) The regulation of nitrate and ammonium transport systems in plants. J Exp Bot 53(370):855–864, Inorganic Nitrogen Assimilation Special Issue

Arak R, Hasegawa H (2006) Genes involved in high affinity nitrate transport during the period of nitrate induction. Breed Sci 56:295–302

Beatty PH, Carroll RT, Shrawat AK, Guevara D, Good AG (2013) Physiological analysis of nitrogen-efficient rice over-expressing alanine amino-transferase under different N regimes. Botany 91(12):866–883. doi:10.1139/cjb-2013-0171

Cai C, Zhao X-Q, Zhu Y-G, Li B, Tong Y-P, Li Z-S (2007) Regulation of the high-affinity nitrate transport system in wheat roots by exogenous abscisic acid and glutamine. J Integr Plant Biol 49(12):1719–1725

Caputo C, Fatta N, Barneix A (2001) The export of amino acids in the phloem is altered in wheat plants lacking the short arm of chromosome 7B. J Exp Bot 52:1761–1768

Century KT, Reuber K, Ratcliffe OJ (2008) Regulating the regulators: the future prospects for transcription-factor-based agricultural biotechnology products. Plant Physiol 147:20–29

Chichkova SJ, Arellano CP, Vance CP, Herna'ndez G (2001) Transgenic tobacco plants that over express alfalfa NADH-glutamate synthase have higher carbon and nitrogen content. J Exp Biol 52:2079–2087

Chopin F, Wirth J, Dorbe MF, Lejay L, Krapp A, Gojon A, Daniel-Vedele F (2007) The *Arabidopsis* nitrate transporter AtNRT2.1 is targeted to the root plasma membrane. Plant Physiol Biochem 45:630–635

Crawford NM, Glass ADM (1998) Molecular and physiological aspects of nitrate uptake in plants. Trends Plant Sci Rev 3(10):367–407

Crété P, Caboche M, Meyer C (1997) Nitrite reductase expression is regulated at the post-transcriptional level by the nitrogen source in *Nicotiana plumbaginifolia* and *Arabidopsis thaliana*. Plant J 11:625–634

Dechorgnat J, Nguyen CT, Amengaud P, Jossier M, Diatlof E, Filleur S, Daniel-Vedele F (2011) From the soil to the seeds: the long journey of nitrate in plants. J Exp Bot 62(4):1349–1359

Dejannane S, Chauvin JE, Quillere I, Meyer C, Chupeau Y (2002) Introduction and expression of a deregulated tobacco nitrate reductase gene in potato lead to highly reduced nitrate levels in transgenic tubers. Transgenic Res 11:175–184

Duan YH, Zhang YL, Shen QR, Wang SW (2006) Nitrate effect on rice growth and nitrogen absorption and assimilation at different growth stages. Pedosphere 16:707–717

Duan YH, Zhang YL, Ye LT, Fan XR, Xu GH, Shen QR (2007) Responses of rice cultivars with different nitrogen use efficiency to partial replaced nitrate nutrition. Ann Bot 99:1153–1160

Fan X, Xie D, Chan J, Lu H, Xu Y, Ma C, Xu G (2014) Over expression of OsPTR6 in rice increased plant growth at different nitrogen supplies but decreased nitrogen use efficiency at high ammonium supply. Plant Sci 227:1–11

Fei H, Chaillou S, Mahon JD, Vessey JK (2003) Over expression of a soybean cytosolic glutamine synthetase gene linked to organ-specific promoters in pea plants grown in different concentrations of nitrate. Planta 216:467–474

Feng H, Yan M, Li B, Shen Q, Miller AJ, Xu G (2011) Spatial expression and regulation of rice high affinity nitrate transporters by nitrogen and carbon status. J Exp Bot 62:2319–2332

Filleur S, Dorbe MF, Cerezo M, Orsel M, Granier F, Gojon A, Daniel-Vedele F (2001) An Arabidopsis T-DNA mutant affected in Nrt2 genes is impaired in nitrate uptake. FEBS Lett 489:220–224

Forde BG, Walch-Liu P (2009) Nitrate and glutamate as environmental cues for behavioural responses in plant roots. Plant Cell Environ 32(6):682–693

Gallais A, Hirel B (2004) An approach to the genetics of nitrogen use efficiency in maize. J Exp Bot 55(396):295–306.

Gazzarini S, Lejay L, Gojon A, Ninnemann O, Frommer WB (1999) Three functional transporters for constitutive, diurnally regulated, and starvation-induced uptake of ammonium into Arabidopsis roots. Plant Cell 11:937–947

Glass ADM (2003) Nitrogen use efficiency of crop plants: physiological constraints upon nitrogen absorption. Crit Rev Plant Sci 22(5):453–470

Gojon A, Krouk G, Perrine-Walker F, Laugier E (2011) Nitrate transceptor(s) in plants. J Exp Bot 62:2299–2308. doi:10.1093/jxb/erq419

Good AG, Shrawat AK, Muench DG (2004) Can less yield more? Is reducing nutrient input into the environment compatible with maintaining crop production? Trends Plant Sci 9(12):597–605

Good AG, Johnson SJ, Pauw MD, Carroll RT, Savidov N, Vidmar J, Lu Z, Taylor G, Stroeher V (2007) Engineering nitrogen use efficiency with alanine aminotransferase. Can J Bot 85(3):252–262. doi:10.1139/B07-019

Gregersen PL, Holm PB, Krupinska K (2008) Leaf senescence and nutrient remobilisation in barley and wheat. Plant Biol 10(Suppl 1):37–49

Gu R, Duan F, An X, Zhang F, von Wiren N, Yuan L (2013) Characterization of AMT-mediated high-affinity ammonium uptake in roots of maize (*Zea mays* L.). Plant Cell Physiol 54(9):1515–1524

Gutterson N, Zhang JZ (2004) Genomics applications to biotech traits: a revolution in progress? Curr Opin Plant Biol 7:226–230

Habash D, Massiah A, Rong H, Wallsgrove R, Leigh R (2001) The role of cytosolic glutamine synthetase in wheat. Ann Appl Biol 138:83–89

Habash D, Bernard S, Schondelmaier J, Weyen J, Quarrie S (2007) The genetics of nitrogen use in hexaploid wheat: N utilisation, development and yield. Theor Appl Genet 114:403–419

Hirel B, Bertin P, Quillere I et al (2001) Towards a better understanding of the genetic and physiological basis for nitrogen use efficiency in maize. Plant Physiol 125:1258–1270

Hirel B, Le Gouis J, Ney B, Gallais A (2007) The challenge of improving nitrogen use efficiency in crop plants: towards a more central role for genetic variability and quantitative genetics within integrated approaches. J Exp Bot 58:2369–2387

Ho C-H, Lin S-H, Hu H-C, Tsay Y-F (2009) CHL1 functions as a nitrate sensor in plants. Cell 138(6):1184–1194

Hoque MS, Masle J, Udvardi MK, Ryan PR, Upadhyaya NM (2006) Over-expression of the rice *OsAMT1-1* gene increases ammonium uptake and content, but impairs growth and development of plants under high ammonium nutrition. Funct Plant Biol 33(2):153–163. doi:10.1071/FP05165

Huber SC, MacKintosh C, Kaiser WM (2002) Metabolic enzymes as targets for 14-3-3 proteins. Plant Mol Biol 50(6):1053–1063

Ishiyama K, Inoue E, Watanabe-Takahashi A, Obara M, Yamaya T, Takahashi H (2004) Kinetic properties and ammonium-dependent regulation of cytosolic isoenzymes of glutamine synthetase in *Arabidopsis*. J Biol Chem 279:16598–16605

Kato Y, Murakami S, Yamamoto Y, Chatani H, Kondo Y, Nakano T (2004) The DNA-binding protease, CND41, and the degradation of ribulose-1,5-bisphosphate carboxylase / oxygenase in senescent leaves of tobacco. Planta 220:97–104

Kichey T, Heumez E, Pocholle D, Pageau K, Vanacker H, Dubois F, LeGouis J, Hirel B (2006) Combined agronomic and physiological aspects of nitrogen management in wheat highlight a central role for glutamine synthetase. New Phytol 169:265–278

King BJ, Siddiqi MY, Ruth TJ, Warner RH, Glass ADM (1993) Feedback regulation of nitrate influx in barley roots by nitrate, nitrite, and ammonium. Plant Physiol 102:1279–1286

Kirk GJD, Kronzucker HJ (2005) The potential for nitrification and nitrate uptake in the rhizosphere of wetland plants: a modeling study. Ann Bot 96:639–646

Kronzucker HJ, Schjoerring JK, Erner Y, Kirk GJD, Siddiqi MJ, Glass ADM (1998) Dynamic interactions between root NH_4^+ influx and long-distance N translocation in rice: insights into feedback processes. Plant Cell Physiol 39:1287–1293

Kumar A, Kaiser BN, Siddiqi MY, Glass A (2006) Functional characterization of OsAMT1.1 over-expression lines of rice (*Oryza sativa*). Funct Plant Biol 33(4):339–346

Kurai T, Wakayama M, Aoki N, Ohsugi R (2009) Over expression of *ZmDof1* in rice alters carbohydrate and nitrogen partitioning. Plant Biol, Abs # P51014

Lansing AJ, Franceschi VR (2000) The para-veinal mesophyll: a specialised path for intermediary transfer of assimilates in legume leaves. Aust J Plant Physiol 27:757–767

Li W, Wang Y, Okamoto M, Crawford NM, Siddiqi MY, Glass ADM (2007) Dissection of the AtNRT2.1,

AtNRT2.2 inducible high-affinity nitrate transporter gene cluster. Plant Physiol 143:425–433

Li Y, Rong X, Fan Q, Shen R (2008) The relationship between rhizosphere nitrification and nitrogen-use efficiency in rice plants. Plant Cell Environ 31(1):73–85

Li S, Qian Q, FU D, Meng DZ, Kyozuka J, Maekawa M, Zhu X, Zhang J, Wang LY (2009) Short Panicle 1encodes a putative PTR family transporter and determines rice panicle size. Plant J 58:592–605

Lijavetzky D, Carbonero P, Vicente-Carbajosa J (2003) Genome-wide comparative phylogenetic analysis of the rice and *Arabidopsis* Dof gene families. BMC Evol Biol 3:17

Lin CHM, Koh S, Stacey G, Yu SM, Lin TY, Tsay YF (2000) Cloning and functional characterization of a constitutively expressed nitrate transporter gene, OsNRT1, from rice. Plant Physiol 122:379–388

Liu KH, Huang CY, Tsay YF (1999) CHL1 is a dual-affinity nitrate transporter of *Arabidopsis* involved in multiple phases of nitrate uptake. Plant Cell 11:865–874

Loque D, Lalonde S, Looger LL, von Wiren N, Frommer WB (2007) A cytosolic trans-activation domain essential for ammonium uptake. Nature 446(7132):195–198

Ludewig U, von Wiren N, Frommer WB (2002) Uniport of NH4+ by the root hair plasma membrane ammonium transporter LeAMT1;1. J Biol Chem 277(16):13548–13555

Mae T, Makino A, Ohira K (1983) Changes in the amounts of ribulose bisphosphate carboxylase synthesized and degraded during the life span of a rice leaf (*Oryza sativa* L.). Plant Cell Physiol 24:1079–1086

Malagoli P, Laine P, Deunff EL, Rossato L, Ney B, Ourry A (2004) Modeling nitrogen uptake in oil seed rape Cv. Capitol during a growth cycle using influx kinetics of root nitrate transport system and field experimental data. Plant Physiol 134(1):388–400

Martin A, Lee J, Kichey T, Gerentes D, Zivy M, Tatout C, Dubois F, Balliau T, Valot B, Davanture M, Terce-Laforgue T, Quillere I, Coque M, Gallais A, Gonzalez-Moro M, Bethencourt L, Habash D, Lea P, Charcosset A, Perez P, Murigneux A, Sakakibara H, Edwards K, Hirel B (2006) Two cytosolic glutamine synthetase isoforms of maize are specifically involved in the control of grain production. Plant Cell 18:3252–3274

Masclaux-Daubresse C, Reisdorf-Cren M, Orsel M (2008) Leaf nitrogen remobilisation for plant development and grain filling. Plant Biol 10(Suppl 1):23–36

Meyer C, Stitt M (2001) Nitrate reduction and signaling. In: Morot-Gaudry JF, Lea PJ (eds) Plant nitrogen. Springer, Berlin, pp 37–59

Miflin B, Habash D (2002) The role of glutamine synthetase and glutamate dehydrogenase in nitrogen assimilation and possibilities for improvement in the nitrogen utilization of crops. J Exp Bot 53:979–987

Miller AJ, Smith SJ (1996) Nitrate transport and compartmentation in cereal root cells. J Exp Bot 47:843–854

Miller AJ, Fan X, Shen Q, Smith SJ (2008) Amino acids and nitrate as signals for the regulation of nitrogen acquisition. J Exp Bot 59(1):111–119

Naohiro A, Tomohito E, Akira K, Tomomi A, Masataka W, Haruto S, Ryu O (2009) Genetic modification of nitrogen use efficiency in potato and rice by introducing fungal glutamate dehydrogenase, American Society of Plant Biologists. Plant Biol, 2009, Honolulu, Hawaii, Abs # P39004

Obara M, Kajiura M, Fukuta Y et al (2001) Mapping of QTLs associated with cytosolic glutamine synthetase and NADH-glutamate synthase in rice (*Oryza sativa* L.). J Exp Bot 52:1209–1217

Okamoto M, Kumar A, Li W, Wang Y, Sidiqi MY, Crawford NM, Glass AD (2006) High-affinity nitrate transport in roots of *Arabidopsis* depends on expression of the NAR2-like gene AtNRT3.1. Plant Physiol 142:1304–1317

Oliveira IC, Brears T, Knight TJ, Clark A, Coruzzi GM (2002) Overexpression of cytosolic glutamate synthetase, relation to nitrogen, light, and photorespiration. Plant Physiol 129:1170–1180

Orsel M, Filleur S, Fraisier V, Daniel-Vedele F (2002) Nitrate transport in plants: which gene and which control? J Exp Bot 53(370):825–833

Orsel M, Eulenburg K, Krapp A, Daniel-Vedele F (2004) Disruption of the nitrate transporter genes AtNRT2.1 and AtNRT2.2 restricts growth at low external nitrate concentration. Planta 219:714–721

Orsel M, Chopin F, Leleu O, Smith SJ, Krapp A, Daniel-Vedele F, Miller AJ (2006) Characterization of a two-component high-affinity nitrate uptake system in *Arabidopsis*: physiology and protein–protein interaction. Plant Physiol 142:1304–1317

Ortega JL, Temple SJ, Sengupta-Gopalan C (2001) Constitutive overexpression of cytosolic glutamine synthetase (GS1) gene in transgenic alfalfa demonstrates that GS1 be regulated at the level of RNA stability and protein turnover. Plant Physiol 126:109–121

Peng J, Carol P, Richards DE, King KE, Cowling RJ, Murphy GP, Harberd NP (1997) The *Arabidopsis GAI* gene defines a signaling pathway that negatively regulates gibberellin responses. Genes Dev 11:3194–3205

Peng J, Richards DE, Hartley NM, Murphy GP, Devos KM, Flintham JE, Beales J, Fish LJ, Worland AJ, Pelica F et al (1999) 'Green revolution' genes encode mutant gibberellin response modulators. Nature 400:256–261

Provan F, Aksland LM, Meyer C, Lillo C (2000) Deletion of the nitrate reductase N-terminal domain still allows binding of 14-3-3 proteins but affects their inhibitory properties. Plant Physiol 123:757–764

Pysh LD, Wysocka-Diller JW, Camilleri C, Bouchez D, Benfey PN (1999) The GRAS gene family in *Arabidopsis*: sequence characterization and basic expression analysis of the *SCARECROW-LIKE* genes. Plant J 18:111–119

Qu LJ, Zhu YX (2006) Transcription factor families in *Arabidopsis*: major progress and outstanding issues for future research. Curr Opin Plant Biol 9:544–549

Quaggiotti S, Ruperti B, Borsa P, Destro T, Malagoli M (2003) Expression of a putative high-affinity NO3– transporter and of an H+–ATPase in relation to whole plant nitrate transport physiology in two maize genotypes differently responsive to low nitrogen availability. J Exp Bot 54(384):1023–1031

Rentsch D, Schmidt S, Tegeder M (2007) Transporters for uptake and allocation of organic nitrogen compounds in plants. FEBS Lett 581:2281–2289

Riechmann JL (2002) Transcriptional regulation: a genomic overview. In: Somerville CR, Meyerowitz EM (eds) The *arabidopsis* book. American Society of Plant Biologists, Rockville. doi:10.1199/tab.0085, http://www.aspb.org/publications/arabidopsis/

Riechmann JL, Ratcliffe OJ (2000) A genomic perspective on plant transcription factors. Curr Opin Plant Biol 3:423–434

Rolletschek H, Hosein F, Miranda M, Heim U, Götz KP, Schlereth A, Borisjuk L, Saalbach I, Wobus U, Weber H (2005) Ectopic expression of an amino acid transporter (VfAAP1) in seeds of *Vicia narbonensis* and pea increases storage proteins. Plant Physiol 137:1236–1249

Salmeron J, Herrera-Estrella LR (2006) Fast-forward genomics for improved crop production. Curr Opin Plant Biol 9:177–179

Schjoerring J, Kyllingsbaek A, Mortensen J, Byskov-Nielsen S (1993) Field investigations of ammonia exchange between barley plants and the atmosphere, I: concentration profiles and flux densities of ammonia. Plant Cell Environ 16:161–167

Shaw L (2009) The Dof transcription factor family in *Triticum aestivum*. PhD thesis, School of Biological Sciences, The University of Queensland

Shrawat AK, Carroll RT, DePauw M, Taylor GJ, Good AG (2008) Genetic engineering of improved nitrogen use efficiency in rice by the tissue-specific expression of alanine aminotransferase. J Plant Biotechnol 6(7):722–732

Siddiqi MY, Glass ADM, Ruth TJ, Rufty TW Jr (1990) Studies of the uptake of nitrate in barley, I. Kinetics of $^{13}NO_3$- influx. Plant Physiol 93:1426–1432

Sonoda Y, Ikeda A, Saiki S, von Wiren N, Yamaya T, Yamaguchi J (2003) Distinct expression and function of three ammonium transporter genes (OsAMT1;1–1;3) in rice. Plant Cell Physiol 44(7):726–734

Sperandio MVL, Santos LA, de Araujo OJL, Braga RP, Coelho CP, Nogueira EM, Fernandes MS, de Souza SR (2014) Response of nitrate transporters and PM H+–ATPase expression to nitrogen flush on two upland rice varieties contrasting in nitrate uptake kinetics. Aust J Crop Sci 8(4):568–576

Suenaga A, Moriya K, Sonoda Y, Ikeda A, von Wiren N, Hayakawa T et al (2003) Constitutive expression of a novel-type ammonium transporter OsAMT2 in rice plants. Plant Cell Physiol 44:206–211

Tilsner J, Kassner N, Struck C, Lohaus G (2005) Amino acid contents and transport in oilseed rape (*Brassica napus* L.) under different nitrogen conditions. Planta 221:328–338

Tong Y, Zhou JJ, Li Z, Miller AJ (2005) A two-component high-affinity nitrate uptake system in barley. Plant J 41:442–450

Trevisan S, Borsa P, Botton A, Varotto S, Malagoli M, Ruperti B, Quaggiotti S (2008) Expression of two maize putative nitrate transporters in response to nitrate and sugar availability. Plant Biol (Stuttg) 10:462–475

Tsay YF, Chiu CC, Tsai CB, Ho CH, Hsu PK (2007) Nitrate transporters and peptide transporters. FEBS Lett 581:2290–2300

Vanlerberghe GC, Turpin DH (1990) Anaerobic metabolism in the N-limited green algae *Selenastrum minutum*, II. Assimilation of ammonium by anaerobic cells. Plant Physiol 94:1124–1130

Walch-Liu P, Forde BG (2008) Nitrate signaling mediated by the NRT1.1 nitrate transporter antagonizes L-glutamate-induced changes in root architecture. Plant J 54(5):820–828. Epub 2008 Feb 7

Wang Y-H, Garvin DF, Kochian LV (2001) Nitrate-induced genes in tomato roots Array analysis reveals novel genes that may play a role in nitrogen nutrition. Plant Physiol 127:345–359

Wang R, Xing X, Wang Y, Tran A, Crawford NM (2009) A genetic screen for nitrate regulatory mutants captures the nitrate transporter gene *NRT1.1*. Plant Physiol 151(1):472–478

Wirth J, Chopin F, Santoni V, Viennois G, Tillard P, Krapp A, Lejay L, Daniel-Vedele F, Gojon A (2007) Regulation of root nitrate uptake at the NRT2.1 protein level in *Arabidopsis thaliana*. J Biol Chem 282:23541–23552

Yamaya T, Obara M, Nakajima H, Sasaki S, Hayakawa T, Sato T (2002) Genetic manipulation and quantitative-trait loci mapping for nitrogen recycling in rice. J Exp Bot 53:917–925

Yan M, Fan X, Feng H, Miller AJ, Shen Q, Xu G (2011) Rice OsNAR2.1 interacts with OsNRT2.1, OsNRT2.2 and OsNRT2.3a nitrate transporters to provide uptake over high and low concentration ranges. Plant Cell Environ 34:1360–1372

Yanagisawa S (2001) Dof1 and Dof2 transcription factors are associated with expression of multiple genes involved in carbon metabolism in maize. Plant J 21:281–288. doi:10.1046/j.1365-313x.2000.00685

Yanagisawa S (2002) The Dof family of plant transcription factors. Trends Plant Sci 7(12):555–560

Yanagisawa S, Akiyama A, Kisaka H, Uchimiya H, Miwa T (2004) Metabolic engineering with Dof1 transcription factor in plants: improved nitrogen assimilation and growth under low-nitrogen conditions. Proc Natl Acad Sci U S A 101:7833–7838

Yong Z, Kotur Z, Glass ADM (2010) Characterization of an intact two-component high-affinity nitrate transporter from *Arabidopsis* roots. Plant J 63:739–748

Yuan L, Gu R, Xuan Y, Smith-Valle E, Loque D, Frommer WB, von Wiren N (2013) Allosteric regulation of transport activity by heterodimerisation of *Arabidopsis* ammonium transporter complexes *in vivo*. Plant Cell 25(3):974–984

3 Phosphate (Pi) Uptake

Abstract

It is estimated that globally there is more than 30 % yield loss of crops due to P deficiency. Under conditions of P deficiency, plants adapt themselves suitably through modification of their roots and shoots so as to acquire more Pi from soil and use them frugally to support plant growth. Phosphate deficiency results in coordinated induction of hundreds of genes encoding enzymes, which maximise capacity of plants to acquire phosphate more efficiently from external sources and reprioritise internal use of phosphorus.

Phosphate (Pi) use efficiency (PUE) of crops is generally low (<15–20 %) due to various soil- and plant-related factors. Different approaches made to improve PUE including transgenic technologies have been discussed.

3.1 Occurrence of Phosphate (Pi) and Soil Reactions

Primary source of phosphate (Pi) is the phosphate rock, a non-renewable resource. There were speculations that production of phosphate rock will peak by 2033–2034 and decrease subsequently due to depletion of global reserve (Abelson 1999). Recent studies by IFDC (Van Kauwenbergh 2010) indicate that there are sufficient reserves of global phosphate rock concentrate to produce P fertiliser for the next 300–400 years.

The insoluble phosphate rocks, when added to soils, undergo slow dissolution due to weathering and form sparingly soluble secondary minerals such as iron, aluminium and calcium phosphates, the relative abundance of which depends on soil type and its composition. The red and laterite soils abound in iron and aluminium phosphates and the calcareous soils in calcium phosphates. Solubility of these minerals depends on various soil- and plant-related factors. Due to its strong interaction with soil components, Pi moves to root surface primarily by diffusion rather than mass flow (Hinsinger 2001). Rapid uptake of Pi by roots creates a depletion zone of 0.2–1 mm around the root surface (Barber et al. 1963; Holford 1997). While soil availability of Pi rarely exceeds 2 μM, the concentration of Pi in root cells is 2–20 mM (more than 10,000-fold higher than Pi in the soil solution; Bieleski 1973; Schachtman et al. 1998).

Phosphate (Pi) is primarily taken up as HPO_4^{2-} or $H_2PO_4^-$ ions, which are present in soil solution at very low concentrations (0.1–10 μm;

G.N. Mitra, *Regulation of Nutrient Uptake by Plants: A Biochemical and Molecular Approach*, DOI 10.1007/978-81-322-2334-4_3, © Springer India 2015

Hinsinger 2001). The forms of phosphate ions available to plants are $H_2PO_4^-$, HPO_4^{2-} and PO_4^{3-} based on pH around the rhizosphere. At pH 7.2, $H_2PO_4^- \approx HPO_4^{2-}$, above pH 7.2, $HPO_4^{2-} > H_2PO_4^-$, but below pH 7.2, $H_2PO_4^-$ dominates and is more than HPO_4^{2-}. Plant uptake of HPO_4^{2-} is much slower than $H_2PO_4^-$.

Organic P constitutes about 50 % of total soil P.

3.2 Pi Content of Plants

Along with N and K, P is considered as a primary nutrient required essentially for growth and metabolism of plants. Pi content of plants is considerably lower than N and K and is in the range of 0.05–0.5 % of dry weight (Vance et al. 2003). Cellular Pi concentration is in the range of 2–20 mM (Bieleski 1973; Schachtman et al. 1998).

3.3 Functions of P in Plants

P is a constituent of high-energy compounds, such as the nucleoside triphosphates such as ATP, GTP, CTP, UTP, phosphoenolpyruvate, creatine phosphate, etc., which supply energy to drive endergonic metabolic reactions. P is also a constituent of nucleic acids such as DNA and RNA, which are involved in protein synthesis. It is involved in many crucial metabolic processes such as cytoplasmic streaming, photosynthesis, membrane transport, respiration, lipid biosynthesis and protein and nucleic acid biosynthesis. Energy transfer and energy storage are the major biochemical functions of high-energy phosphorylated compounds.

3.4 Mechanism of Phosphate Uptake by Plants

Phosphate deficiency induces plants to adopt various morphological, physiological and metabolic mechanisms to acquire Pi from the soil and maintain Pi homeostasis within the plant through internal redistribution. The use of microarray technology reveals thousands of genes up- or downregulated due to Pi deficiency. Genetic intervention in some key steps to augment phosphate acquisition and metabolism may increase phosphate use efficiency (PUE).

3.4.1 Morphological Adaptation of Plants Due to Pi Deficiency

Under conditions of P deficiency, plants adapt themselves suitably through modification of their roots and shoots so as to acquire more Pi from soil and use them frugally to support plant growth.

3.4.1.1 Roots

Rhizosphere is a critical region around roots, where intense interactions among plant roots, soil and microorganisms take place. Plant roots exude organic acids, enzymes and mucilage, which determine various chemical and biological reactions around the root surface. These reactions control availability of plant nutrients, their uptake and use, resulting in higher productivity of crops.

Deficiency of Pi has profound effects on root growth and its architecture. These are modified suitably to explore a larger volume of soil so as to absorb more Pi to meet the P demand of plants.

Root Architecture

Plants tend to modify their root architecture under conditions of Pi deficiency. These include adventitious rooting, aerenchyma formation, basal root elongation, basal root growth angle, lateral rooting, root hair density and root hair length (Bates and Lynch 1996; Borch et al. 1999; Ma et al. 2001; Miller et al. 2003; Lynch 2011). Specific root length, defined as length of root per unit root weight, varies among species and cultivars and is associated with genetic difference in P-use efficiency (Sattelmacher et al. 1994; Lynch and Ho 2004). Under P-deficient conditions, there are restricted growth and elongation of primary roots and increase in density and elongation of laterals roots. There is proliferation of root hairs, which constitute about 77 % of root surface area in field crops (Parker et al. 2000). It has been reported that nutrients especially NO_3^- and PO_4^{3-} control root hair formation (Gilroy and Jones 2000; Jungk 2001). Root hair length and number is inversely proportional to P concentration in the plants of rape, spinach and tomato (Jungk 2001). P deficiency results in both increase in root hair length and density in legumes (Jungk et al. 1990). The varieties of common bean (*Phaseolus vulgaris*), which are more efficient in P uptake, have more roots in topsoil where the concentration of Pi is more (Lynch and Brown 2008). *Medicago truncatula* responds quickly to P deficiency with increased numbers and length of root hairs (Vance et al. 2003). There is a significant correlation between radius, length and number of root hairs and Pi uptake by plants of many species, which indicates a close relationship between root hairs and Pi uptake from soil (Fohse et al. 1991; Yan et al. 2004; Wang et al. 2010). In Pi-deficient soils, barley genotypes with variation in root hairs show significant changes in Pi uptake from soil and in grain yields (Gahoonia et al. 1997; Gahoonia and Nielson 2004a, b). Recent studies indicate that lateral root development decreased and there was no change in root architecture in *Medicago truncatula* until 28 days after planting under Pi-deficient conditions (Bucciarelli et al. 2006). Generally, N- and P-deficient plants allocate higher proportion of biomass to roots as compared to plants with optimum nutrient supply (Hill et al. 2006).

Root Cortical Aerenchyma

Root cortical aerenchyma induced under sub-optimal conditions of oxygen, water, Pi, N and S reduces carbon and nutrient cost of soil exploration (Lynch 2011). Simulated root modelling (Sim Root) studies (Lynch 2011) indicate that aerenchyma formation may increase growth by 70 % in maize under Pi-deficient conditions and by 14 % in bean due to lower Pi content of roots and reduced root respiration (Postma and Lynch 2010).

Root Development

Under conditions of Pi deficiency, a number of genes regulated by plant growth regulators are differentially expressed. Expression of 'auxin-responsive genes' including '*AIR1*, *AIR3*, *AIR9*, *AIR12*, *HRGP* and *LRP1*' (Hammond et al. 2004) which control lateral root development, root hair elongation and density increases due to Pi deficiency (Al-Ghazi et al. 2003; Casimiro et al. 2003; Uhde-Stone et al. 2003). P deficiency stimulates ethylene production in plant roots and is probably responsible in increase of root hair formation in P-deficient plants (Michael 2001). Expression of several genes involved in biosynthesis of ethylene (ACC oxidase, methionine synthase and *S*-adenosyl methionine synthetase) increases in P-deficient *Arabidopsis* roots and in the cluster roots of white lupine. Cytokinins act as negative regulators of root growth. Exogenous cytokinins suppress lateral root initiation in P-deficient *Arabidopsis* (López-Bucio et al. 2002). There is an increase in expression of genes encoding cytokinin oxidases, involved in breakdown of cytokinins to reduce its concentration under conditions of P deficiency, observed in the roots of *Arabidopsis* and cluster roots of white lupine (Hammond et al. 2004). There appears to be a coordinated role of auxin, ethylene and cytokinins on root development in P-deficient plants.

Root Architecture of Mono- and Dicotyledonous Plants

While there is significant difference between monocotyledonous and dicotyledonous plants in their root architecture, under Pi-deficient conditions, they undergo similar developmental modifications to forage for scarce Pi acquisition from the surface soils (Hochholdinger and Zimmermann 2008). Pi deficiency causes shortening of primary roots and changes in angle of growth and diameter of lateral root in both *Arabidopsis* and bean (Lynch and Brown 2001; Williamson et al. 2001). There is adaptive modification of the roots of wheat and barley due to Pi deficiency (Gahoonia et al. 1997). Pi deficiency causes change in postembryonic root system in maize and alters the angle, length and number of shoot-borne and lateral roots. Unlike *Arabidopsis*, the primary root system in maize is not affected in early stages of growth due to Pi deficiency since cereal grains are rich in P (O'Dell et al. 1972). Adaptation to Pi deficiency in different genotypes of common bean shows that root architecture, such as lateral root branching, length and growth angle of basal roots and root growth plasticity are correlated with efficiency of Pi uptake (Lynch 1998; Liao et al. 2001). A close relationship between root architecture and efficiency of Pi uptake from Pi-deficient soil has been observed in soybean genotypes. Genotypes with shallower root architecture with optimal three-dimensional root configuration and longer root lengths have been found to have higher Pi-uptake efficiency and give higher yield in P-deficient soils (Lin et al. 2008).

Cluster Roots

Cluster roots comprising of very densely packed determinate lateral rootlets formed on a parent axis, a highly coordinated modification of root development increasing the surface area by about 100 % as compared to normal roots, are formed under conditions of Pi deficiency in white lupin (*Lupinus albus*) and in many tree species (Vance et al. 2003). Plant species that form cluster roots usually do not form mycorrhizal associations (Skene 1998).

3.4.1.2 Shoot

Pi deficiency causes delayed leaf development, reduction in number of leaves and leaf expansion, decreased photosynthetic capacity, stunted growth (reduced auxiliary shoot emergence and elongation), impaired flower development and an increased root/shoot ratio of the plants (Vance 2010). Pi from lower and older leaves translocates to newer leaves. However, if there is enhanced uptake of Pi by roots and translocation to shoots, the excess Pi accumulates in older leaves and results in chlorosis and necrosis of leaf tips due to Pi toxicity (Dong et al. 1998; Aung et al. 2006).

3.4.2 Metabolic Mechanisms for Acquisition Pi from Soil

Plant roots exude a variety of organic compounds under normal conditions of growth. These include sugars, organic acids, amino acids, growth hormones, phenolics, proteins, etc., which affect rhizosphere chemistry and alter plant–microbe interaction, allelopathy and nutrient acquisition by plants.

3.4.2.1 Exudation of Organic Acids by Roots

An important mechanism through which plant roots respond to Pi deficiency is excretion of organic acids into rhizosphere. The resultant pH change in soil around the rhizosphere may be 2–3 units lower than the bulk of the soil. This may substantially affect dissolution of sparingly soluble soil P (Marschner 1995). The amount of carbon excreted as organic acids may range from 10 to more than 25 % of the total plant dry weight (Vance et al. 2003). The organic acids are derived from TCA cycle.

3.4.2.2 Activities of TCA Cycle Enzymes Due to P Deficiency

P deficiency has been correlated with increased synthesis of phosphoenolpyruvate carboxylase (PEPC), malate dehydrogenase and citrate synthase (Vance et al. 2003; Gregory et al. 2009). Overexpression of a PEPC maize gene in rice has been found to increase organic acid synthesis and excretion by rice roots (Fang et al. 2009). Overexpression of mitochondrial citrate synthase in either carrot or *Arabidopsis* results in increased excretion of citrate by roots of plants and results in an improved growth of plants when grown in medium containing aluminium phosphate or in P-deficient acidic soils (Fang et al. 2009). Organic acid excretion may also mobilise P bound to humic–metal complexes, increase the solubility of organic P and its dephosphorylation by purple acid phosphatases and promote the growth of symbiotic rhizosphere microbes involved in root Pi acquisition.

Mechanisms of Organic Acid Excretion

At cytoplasmic pH, organic acids are present in anionic form. The mechanism of organic acid excretion consists of two transport processes. One involves active H^+-ion efflux driven by a plasma membrane ATPase and the second passive efflux of organic acid anions by channel-like transporters (Yan et al. 2002; Diatloff et al. 2004). *Arabidopsis* genome contains 53 ATP-binding cassette (ABC) transporter genes (Martinoia et al. 2002). Walker et al. (2003) suggested involvement of ABC transporters (See Sect. 9.4.2.2) in exudation of organic acid by roots. Recently, a maize gene ZmALMT1 has been identified, which codes a 451-amino-acid transporter protein containing six transmembrane helices and is involved in organic acid exudation by roots (Piñeros et al. 2008).

Effects of Organic Acids on Solubilisation of Mineral Pi

While the protons excreted through organic acids lower the pH, the carboxylate anions react with Fe^{3+}, Al^{3+} and Ca^{2+} present in insoluble compounds of Pi-containing minerals. They form chelates with the cations and release Pi for uptake by the plants. This results in an increase of soil solution Pi concentration by about 1,000-fold (Plaxton and Tran 2011). Under conditions of edaphic stress of Pi deficiency and Al^{3+} toxicity, plant roots exude primarily malate and citrate. They may also cause organic P more susceptible to hydrolysis by acid phosphatases. Radish roots grown in nutrient solution under P-sufficient and P-deficient conditions primarily exude tartaric, malic and succinic acid, which increase by 15 times (succinic acid) to 60 times (malic acid), when grown under P-deficient conditions. In a sand culture, experiment radish has been found to utilise P from $AlPO_4$ more efficiently than $Ca_3(PO_4)_2$. Reverse is true for rape. This indicates preference for growth of radish in acid soils and rape in calcareous soils of China (Zhang et al. 1997). Soybean exudes malate and oxalate due to P deficiency with no effect on exudation of citrate. Toxicity of Al causes significant exudation of citrate, but there is virtually no effect on exudation of malate and oxalate. When there are both P deficiency and Al toxicity, there is a decrease in exudation of organic acids as compared to single stress indicating a possible antagonistic effect of P deficiency and Al toxicity (Dong et al. 2004). Metabolic profiling of rice root exudates under P deficiency has indicated that there is an increase in concentration of 39 compounds including organic acids such as malate, malonate, pyruvate and succinate (Tawaraya et al. 2009).

Exudation of Organic Acids by Cluster roots

Cluster roots, under P-deficient conditions, exude 20–40-fold more citrate and malate than P-sufficient roots. Carbon exuded by plants through these two organic acids may constitute 10 % to more than 25 % of plant dry weight while there is no apparent decrease in dry matter accumulation and nitrogen fixation until reproductive stage (Vance et al. 2003).

Interaction of exuded organic acids by plant roots with different soil types and their consequent effect on Pi availability needs further study. Concentration of organic acid in soil solution is generally low, about 1–50 μM (Strobel 2001). Even at a concentration of 10 mM, a solution of citrate at a pH of 7.5 could extract 1.9 μM of P and oxalate 0.8 μM P from soil, which are far below the growth demands of plants (Ström et al. 2005).

3.4.3 Acquisition of Pi from Soil with Pi Deficiency

Organic P constitutes more than 50 % of soil P and is a dominant Pi constituent of soil solution. Under conditions of Pi deficiency, plants can utilise different organic sources of P. Phosphate deficiency induces increased synthesis of ribonucleases (RNAses), nucleases, phosphodiesterases and APases (acid phosphatases), which are secreted into soil solution. A gene encoding secreted acid phosphatase, 'Ats,Apase', is strongly induced by P deficiency in *Arabidopsis* (Li et al. 2002). Secreted APases can hydrolyse Pi from a broad range of organic substrates such as RNA, DNA, ATP, 3-phospho-glycerate and hexose phosphates such as Glc-6-phosphate (Ticconi and Abel 2004; Richardson 2009; Liang et al. 2010; Plaxton and Tran 2011). *Arabidopsis* plants grown with RNA as their sole source of Pi grow as well as Pi-fertilised control plants. This indicates that nucleic acids present in decaying soil organic matter constitute an important source of P taken up by high-affinity phosphate transporters by P-deficient plants (Plaxton and Tran 2011). The largest group of non-specific plant APases induced due to Pi deficiency is purple acid phosphatase (PAPS; they show a distinct red/purple colour in solution due to charge transfer transition at about 560 nm from the metal coordinating tyrosine to the metal ligand Fe^{3+}; Tran et al. 2010). Due to Pi deficiency, 29 PAPS are encoded by *Arabidopsis* genome, out of which AtPAP26 is the predominant intracellular APase and plays an important role in metabolism of Pi-deficient *Arabidopsis* (Hurley et al. 2010).

Arabidopsis cannot acquire P from exogenous phytic acid (inositol hexaphosphate) due to the absence of an extracellular phytase. Transgenic plants overexpressing secreted phytases do not show any improvement in growth or Pi uptake, when grown on different agricultural soils (Richardson 2009).

3.4.4 Internal Redistribution of Pi in Plants Due to Pi Deficiency

Under conditions of Pi deficiency, plants recycle P from older tissues to new tissues. Plants also remobilise from non-essential uses to essential uses. P-deficient plants remobilise Pi from cellular metabolites through increased activities of phosphatases and nucleases (Hammond et al. 2004). Intracellular (vacuolar) acid phosphatases (with acidic pH optima) are upregulated by Pi deficiency, which remobilise Pi from internal phospho-monoesters and anhydrides. This causes cytoplasmic reduction of P metabolites under extended periods of P deprivation (Vance et al. 2003). Some of the P-rich organic constituents of cells are replaced and utilised to conserve Pi. Membrane phospholipids in Pi-starved plants are replaced by amphipathic sulfolipids and galactolipids (Plaxton and Tran 2011).

Both plant and animal cells secrete ATP into extracellular matrix, essential for maintaining plant cell viability. An Apase, PvPAP3, recently identified from Pi-deficient bean plant (Liang et al. 2010), can actively use ATP as its substrate.

3.4.5 Alternate Metabolic Pathways Caused by Pi Deficiency

Prolonged P deficiency results in increase in activity of a number of enzymes involved in alternate metabolic pathways to conserve P. Phosphate can cause allosteric activation or inhibition of many key enzymes involved in intermediary plant metabolism (Plaxton and Podesta 2006).

3.4.5.1 Starch Accumulation

Phosphate-deficient cells have been observed to accumulate starch. Phosphate causes allosteric inhibition of the enzyme ADP–Glc pyrophosphorylase, involved in starch biosynthesis in cells. Phosphate deficiency (up to 50-fold lower) in cellular Pi pool removes such allosteric inhibition. This results in starch accumulation in the cell (Vance et al. 2003).

3.4.5.2 Synthesis of Anthocyanins

A common symptom of Pi deficiency in plants is dark green or purple shoots. This is due to anthocyanin accumulation. Phosphate starvation causes induction of enzymes involved in synthesis of anthocyanins (Vance et al. 2003; Fang et al. 2009), which protect nucleic acid from UV damage and chloroplasts from photo-inhibitory damage (Zeng et al. 2010).

3.4.5.3 ATP Synthesis

Under severe Pi-deficient conditions, a large decline (up to 80 %) of ATP, ADP and other nucleoside phosphates occurs. Plants respond by adopting alternative metabolic pathways for cytoplasmic glycolysis, mitochondrial electron transport, tonoplast H^+ pumping to facilitate respiration and vacuolar pH maintenance. Critical roles are played by pyrophosphate-dependent glycolytic bypass enzymes and metabolic Pi-recycling systems (Plaxton and Tran 2011).

3.4.5.4 Glycolysis

Pi deficiency has been reported to significantly upregulate some of the glycolytic bypass enzymes such as pyrophosphate (PPi)-dependent phosphofructokinase, PPi-phosphoenol pyruvic kinase, pyruvate phosphate dikinase and the tonoplast H^+pyrophosphatase (Plaxton and Podesta 2006). In maize roots, Pi deficiency causes overexpression of genes encoding phosphoenolpyruvate carboxylase and phosphoenolpyruvate carboxykinase to sustain carbon supply through tricarboxylic acid cycle and provide carbon for C metabolism. Pi deficiency causes unequal distribution of C intermediates between root and shoot of maize. Glycolysis appears to be bypassed by avoiding those reactions requiring P (Calderon-Vazquez et al. 2008).

3.4.5.5 Electron Transport

Under Pi-deficient condition, there is severe depletion of intracellular Pi and ADP levels. This may impede respiratory electron flow through the cytochrome pathway especially in the steps involving coupled ATP synthesis. Plants adapt to such situations by increased use of nonenergy conserving pathways (rotenone or cyanide insensitive) of mitochondrial electron transport chain (Plaxton and Podesta 2006).

3.4.5.6 Nitrogen Metabolism

In maize roots, N-assimilation and amino acid metabolism are also affected by Pi deficiency. Moderate Pi deficiency causes significant reduction in glutamine synthetase and nitrate reductase enzymes (Calderon-Vazquez et al. 2008). Under Pi-deficient conditions, barley has been found to utilise amino acids as a source of C, especially in its roots (Huang et al. 2008).

3.4.5.7 Lipid Metabolism

Pi deficiency also affects lipid metabolism. Expression of 59 transcripts of genes involved in lipid metabolism is affected by Pi deficiency in maize (Calderon-Vazquez et al. 2011). There is an increase in β-oxidation along with fatty acid synthesis possibly to provide C intermediates for energy production (Calderon-Vazquez et al. 2008). Alteration in lipid metabolism in rice roots due to Pi deficiency has been reported by Wasaki et al. (2006).

3.4.6 Genetic Response to Phosphate Deficiency

Phosphate deficiency results in coordinated induction of hundreds of genes encoding enzymes, which maximise capacity of plants to acquire phosphate more efficiently from external sources and reprioritise internal use of phosphorus (Plaxton and Tran 2011). Micro- and macroarray analyses of P-stressed plants have shown transcript abundance of a number of genes with homology to Pi transporters, organic acid synthesis, purple acid phosphatase, multidrug and toxin efflux (MATE), transcription factors, signalling and defence (Tesfaye et al. 2007).

3.4.6.1 Phosphate Transporters

Plants have both low- and high-affinity Pi-transport systems encoded by corresponding genes. The high-affinity system operates at low Pi concentration and low-affinity system at higher Pi concentration. High-affinity phosphate transporters are located primarily in plasma membrane of root hairs. The high-affinity transporters are induced when Pi is deficient. Low-affinity transporters are constitutive in nature.

The transporters, which are $H_2PO_4^-/H^+$ symporters, move Pi against the steep concentration gradient of about 10,000-fold or higher through active transport with energy derived from ATP. The movement from root surface to xylem is symplastic and is at a rate of about 2 mM h^{-1} (Bieleski 1973). Transport of Pi to aboveground parts is through xylem flow and to cells in tissues through symplastic transport. Movements of Pi through plasma membrane into cells and into vacuole within cells are carried out by $H_2PO_4^-/H^+$ symporters with energy derived from ATP (Ullrich and Novacky 1990).

3.4.6.2 Structure of Phosphate Transporters

Molecular characterisation of plant Pi transporters indicates that they have a MW of ≈ 58 kD and have 525–550 amino acid residues. They have 12 transmembrane domains that occur in two groups (6+6) connected by a hydrophobic domain of 60 amino acids (Schachtman et al. 1998; Smith et al. 2000). Studies through computer modelling indicate that both N and C terminals in case of high-affinity Pi transporters are located on the inner surface of plasma membrane. For low-affinity Pi transporters, these are located on the exterior surface of the plasma membrane (Vance et al. 2003).

3.4.6.3 Genes Involved in Pi Acquisition and Transportation

Genes of four transporter families, PHT1, PHT2, PHT3 and PHT4, are found in *Arabidopsis*. Members of PHT1 gene family are expressed in root epidermal cell, and the encoded transporters are located on the plasma membrane (Lin et al. 2009). They are high-affinity $H_2PO_4^-/H^+$ symporter and function to acquire Pi from the rhizosphere. Members of PHT2 gene family are found in chloroplasts (Versaw and Harrison 2002). Members of PHT3 family are located in mitochondria (Poirier and Bucher 2002) and of PHT4 family located in non-photosynthetic plastids or the Golgi apparatus (Guo et al. 2008). According to Mudge et al. (2002) and Smith et al. (2003), the genes of PHT1 family (9 members in the *Arabidopsis* genome and at least 13 in rice) are involved in the uptake Pi from the soil solution and redistribution of Pi within the plant. The members of PHT1 family in *Arabidopsis* with 12 transmembrane domains have different functions: the Pht1;4 involved in Pi acquisition and Pht1;1 in Pi accumulation in shoots during Pi sufficiency due to its role in xylem loading process (Fang et al. 2009; Lin et al. 2009). Orthologous genes of Pht 1;1 have been found in barley, rice, maize, potato and *Medicago truncatula,* sharing the same expression pattern and their basic role in Pi uptake (Liu et al. 2008). Pht2;1 is the only member of PHT2 family in *Arabidopsis*, which is highly expressed in leaves but scarcely found in roots. It is located in chloroplast and encodes low-affinity Pi transporters.

3.4.6.4 Early and Late Genes

Genes that respond to P deficiency can be grouped into 'early genes' that respond rapidly and often non-specifically to Pi deficiency or 'late genes' that impact on the morphology, physiology or metabolism of plants upon prolonged Pi deficiency (Vance et al. 2003; Hammond et al 2004). The 'late' genes generally improve the acquisition of P or promote the efficient use of P within the plant. It is thought that the expression of these genes is coordinated by both general stress-related and low-P-specific signalling cascades (Hammond et al. 2003; Vance et al 2003; Franco-Zorrilla et al. 2004).

3.4.6.5 Early Genetic Response to Pi Deficiency

Under Pi-deficient conditions, the gene family TPSI1/Mt4/At4 is upregulated indicating its importance in the early stages of adaptation to low P availability (Hammond et al. 2004). The members of *TPSI*1/Mt4/At4 family have PHR1-binding sequences. Their transcripts are classified as non-coding RNAs and have no sequence homology among them except for a 22–24nt conserved sequence (Burleigh and Harrison 1999; Martín et al. 2000). TPSI is found in tomato, Mt4 in *Medicago truncatula* and At4 in *Arabidopsis*. The most significant induced gene, designated as OsPI1 (*Oryza sativa* phosphate limitation-inducible gene 1) with many properties similar to TPSI1/Mt4, has been isolated from rice roots and leaves, cloned and characterised. Resupply of Pi causes the most significant downregulation of OsPI1 gene. The number of genes downregulated in leaves is less than those in roots. The genes of the TPSI1/Mt4 family in roots are regulated by shoot P status (Wasaki et al. 2006). These genes respond rapidly to Pi deficiency and encode short non-conserved open reading frames. They probably act as ribo-regulators controlling the functions of RNA, DNA or proteins (Martín et al. 2000; Wasaki et al. 2003).

PHO Regulon Genes

There is a Pi-starvation-inducible rescue system in plants with their promoter region, the PHO regulon genes, under a common regulatory system (Goldstein et al. 1988). The promoter regions of PHO regulon genes, PHO2 and PHO4 in yeast, have two cis-regulatory elements, 'GCACGTGGG' and 'GCACGTTT' for PHO2 and PHO4, respectively, for binding and gene expression under conditions of Pi deficiency. The Pi-responsive genes, TPSI1 from tomato and Mt4 from *Medicago truncatula*, have cis-regulatory elements 'GCACG (G/T)' in their binding sites. The AtPHR1 gene from *Arabidopsis* encodes proteins that contain MYB DNA-binding domains and coiled-coil (CC) domains for protein–protein interactions during Pi-starvation stress. It has a motif, a *cis*-element 'GNATATNC', which is shared by several Pi-responsive genes. A genome-wide expression profiling of Pi-responsive genes revealed over-representation of the P1BS (PHR1-specific binding sequence, *cis*-element 'GNATATNC') in the promoter regions of Pi-starvation-induced genes encoding Pi transporters, phosphatases or translation-related proteins (Hammond et al. 2003; Misson et al. 2005). This motif 'P1BS' is recognised by the transcription factor PHR1 (phosphate starvation response 1), which binds as a dimer to the motif (Rubio et al. 2001). Overexpression of PHR1 results in increased concentration of Pi in the shoots along with an elevated expression of a large number of Pi deficiency genes encoding Pi transporters, phosphatases and RNase (Nilsson et al. 2007). PHR1 appears to be a key transcriptional activator, which controls Pi uptake and distribution within the plant, anthocyanin accumulation and carbon metabolism.

Two homologues of PHR1, identified in rice, OsPHR1 (*Oryza sativa* phosphate limitation-inducible gene 1) and OsPHR2, control expression of several Pi-starvation-induced genes (Zhou et al. 2008b). Overexpression of OsPHR2 results in increased Pi accumulation in shoot, root elongation and root hair proliferation in transgenic rice (Zhou et al. 2008b). Overexpression of PHR1 in *Arabidopsis* does not show such effects (Nilsson et al. 2007).

Data base search has resulted in identification of 26 potential phosphate transporter gene families in rice (Liu et al. 2011). Eight conserved motifs and 5–12 transmembrane segments, most of them conserved, have been found in them. At 2 kb upstream region of these genes, 237 putative *cis* elements have been found, most of which are phosphate responsive or other stress-related regulatory *cis* elements, such as PHO-like, TATA-box-like, PHR1 or helix–loop–helix elements and WRKY1 and ABRE elements (Liu et al. 2011).

Early genetic response to Pi deficiency appears to be less specific. Many of the early response genes for deficiency of Fe-, K- and Pi are induced due to deficiency of any one of them as reported by Wang et al. (2002), from their studies on changes in transcript levels of related genes on deficiency of Fe-, K- and Pi- in tomato roots. There appears to be considerable crosstalk among them. The expression of a large number of genes is up- or downregulated by Pi deficiency even before its onset.

There appears to be redundancy in signal cascades that regulate response to Pi deficiency and other stress-response pathways by plants. For example, P and cold stress might share common regulatory signal cascade (Hammond et al. 2004).

3.4.6.6 Late Genetic Response to Pi Deficiency

PHR1 is involved in coordinated regulation of many 'late' Pi-starvation genes, such as RNases, phosphatases, TPSI/Mt4 family (Franco-Zorrilla et al. 2004; Hammond et al. 2004) and OPSI1 (Wasaki et al. 2006), which have PHR1 binding sites. PHR1 binds as a dimer to the promoter of 'late' Pi-starvation genes. Most of the Pi taken up by roots is subsequently transported through xylem to shoots. Phosphate transporters, OsPht1;2 and OsPht1;6, in rice are involved in Pi translocation from roots to shoots (Ai et al. 2009).

Response of gene expression to Pi-deficient conditions appears to be more dynamic in roots than in leaves of rice plants. The number of genes downregulated in leaves due to Pi deficiency is lower in rice leaves than in roots. The response to short- and long-term Pi deficiency on gene expression is different for leaves but similar in roots. The expression of the gene OPSI1 shows most significant increase in both roots and leaves under long-term Pi-deficient conditions (Wasaki et al. 2006).

3.4.7 Transcription Factors Involved in Expression of Pi Stress-Response Genes

Transcription factors (TFs) are sequence-specific DNA-binding proteins, which interact with the promoter region of the target genes and regulate gene expression. TFs are known to be involved in plant responses to biotic and abiotic stress. Several families of TFs, such as MYB, SCARE CROW, APETALA2 domain, homeobox, zinc fingers and WRKY, are involved in expression of Pi stress-response genes. Bioinformatic analysis of Pi-stressed tissues of legumes (*Medicago, Lupinus, Phaseolus and glycine*) indicates the presence of transcription factors, WRKY, MYB and zinc finger families of genes (Graham et al. 2006). Data base search has resulted in identification of 26 potential phosphate transporter gene families in rice (Liu et al. 2011). At 2 kb upstream region of these genes, 237 putative *cis* elements have been found, most of which are phosphate responsive or other stress-related regulatory *cis* elements, such as PHO-like, TATA-box-like, PHR1 or helix–loop–helix elements and WRKY1 and ABRE elements (Liu et al. 2011).

Under Pi-deficient conditions, maize root shows altered expression of transcription factors, 'SHORTROOT'- and 'SCARECROW-LIKE' TFs, which are involved in determining meristem identity and root morphology.

3.4.7.1 WRKY TF

WRKY proteins constitute the largest family of transcription factors, and their genes have so far been found only in plants (Rushton et al. 2010). These genes have strictly conserved DNA-binding domains, which code for WRKY (trypt-arg-lys-tyr). WRKY gene family consists of about 100 members in rice (Zhang and Wang 2005). Apart from Pi stress, these genes upon binding to their common W-box motif '(C/T) TGAC(C/T)' have been found to be upregulated by different stresses such as infection by pathogens, wounding and senescence (Eulgem et al. 2000).

3.4.7.2 bHLH (Basic Helix–Loop–Helix) TF

bHLH (basic helix–loop–helix) transcription factors consist of two α-helices, one small and the other larger connected by a loop. The larger helix contains the DNA-binding region, the E-box with a consensus sequence 'CACGTG' (palindromic) or non-palindromic sequences. There are 133 bHLH genes in *Arabidopsis* and 113 of them are expressed (Heim et al. 2003). A bHLH transcription factor involved in Pi stress in rice, OsPTF1 (*Oryza sativa phosphate starvation-induced transport factor1*), has been cloned and characterised (Yi et al. 2005). Normally, OsPTF1 is constitutively expressed in shoots of rice plant. Under Pi stress conditions, transcript accumulation of OsPTF1 is induced in roots.

3.4.7.3 The HD-Zip (Homeodomain Leucine Zipper) TF

The HD-Zip (homeodomain leucine zipper) TF proteins consist of a homeodomain formed by a number of proteins with a conserved DNA-binding domain which recognises pseudo-palindromic DNA sequence 'CAAT(A/T)ATTG' or 'CAAT(C/G)TTG' etc. for different classes of this TF. The leucine zipper (Zip) part of this TF is involved in protein homo- and hetero-dimerisation (Elhiti and Stasolla 2009). HD-Zip TF is involved in signalling expression of P-responsive genes in soybean (Tang et al. 2001).

3.4.7.4 MYB (Myeloblast) TF

MYB (myeloblast, first identified as an avian oncogene) transcription factors are found in eukaryotes. The C1 (Coloured1) gene transcription factor of maize, the first plant MYB TF to be identified, has significant structural homology with proto-oncogenic c-MYB TF of vertebrates (Martin and Paz-Ares 1997). Yanhui et al. (2006) identified 198 MYB TFs from the *Arabidopsis* genome. The *Arabidopsis* MYB TF has sequence homology with PHR1 (phosphate starvation response gene) of *Chlamydomonas reinhardtii* and binds to an imperfect palindromic consensus sequence '5-GNATATNC-3' (Rubio et al. 2001). Many Pi deficiency-induced genes such as LaPT1 and LaSAP1 of white lupin have 'GNATATNC' in their 5′ upstream region (Tesfaye et al. 2007).

3.4.7.5 Zinc Finger TF

Zinc finger transcription factors consist of proteins with an α-helix and an antiparallel β-sheet held as a coordination complex with a Zn ion linking two histidine and two cysteine residues located on the protein molecules. Using semi-quantitative reverse transcription PCR analysis of 13 ESTs (partially sequenced c-DNA inserts) encoding Zinc-finger transcription factors, it was observed that there was increased transcript abundance of two of the ESTs in Pi-starved roots of common bean (Tesfaye et al. 2007).

3.4.8 Sugar Signalling

Sucrose, the primary product of photosynthesis, is transported through phloem to different tissues of the plant for carbohydrate metabolism. Pi deficiency results in accumulation of starch and sugar in the leaves and higher phloem loading with sucrose, which is transported to the roots as a carbon source for root proliferation, and increase in their size relative to shoots. Sucrose also initiates the signalling cascade that alters the expression of genes involved in Pi acquisition from soil. There is increased expression of genes involved in transcription of inorganic phosphate transporters, secretion of acid phosphatases and organic acids from roots to release Pi from soil and optimisation of internal P use (Hammond and White 2008). Sucrose derived from photosynthesis appears to be a part of systemic signalling leading to Pi deficiency-induced lateral root formation and increase in root hair density (Zhou et al. 2008a; Vance 2010).

Inductions of a number of Pi-starvation-induced genes are also affected by sucrose. Müller et al. (2007) identified 149 transcripts of Pi-induced genes, which were regulated by the interaction between Pi deficiency and sucrose availability. A group of 47 genes with increased expression due to Pi deficiency had enhanced expression due to sucrose. Many of these genes encode proteins involved in carbohydrate metabolism and P remobilisation.

3.4.9 MicroRNA (miRNA)

Recent studies have indicated involvement of miRNAs in signalling Pi deficiency, Pi acquisition, allocation and remobilisation and thus regulating Pi homeostasis in plants (Kuo and Chiou 2011). (General information about miRNA and siRNA is given in Box 3.1.)

Plant microRNAs play critical roles in most of the biological processes such as development, differentiation and plant responses to biotic and abiotic stress (Lelandais-Brière et al. 2010). Majority of miRNA responsive to Pi deficiency target transcription factors involved in transcriptional regulation of genes, while others target genes coding proteins for biotic and abiotic stress.

The microRNA, miR399, has been identified in *Arabidopsis* and rice, first computationally and later verified experimentally (Jones-Rhoades and Bartel 2004). The *Arabidopsis* genome encodes six MIR399 genes, all of which are upregulated to different extent due to Pi deficiency. Phosphate deficiency causes upregulation of miR399, which decreases rapidly on Pi addition (Fujii et al. 2005; Bari et al. 2006). Overexpression of *Arabidopsis* miR399 in tomato results in increased accumulation of Pi. There is also augmented excretion of acid phosphatases and protons by roots, which facilitates Pi acquisition from soil (Gao et al. 2010).

Box 3.1: MicroRNA and SiRNA

MicroRNAs (miRNAs) containing 19–25 nucleotides are found in all animals and plants but not in fungi. They are posttranscriptional regulators encoded by specific genes, several at a time or by some portions of the introns of genes, whose mRNA they regulate. They either completely destroy the mRNA if their sequences exactly match (usually in plants) or repress the translation of mRNA if there is a partial match. In the later case, several of them simultaneously bind to the UTR (untranslated) region of mRNA. In plants they may target the coding region itself (He and Hannon 2004). About 40 % of miRNA genes may lie in the introns of protein-coding and non-coding genes and rarely in exons of non-protein-coding transcripts (Rodriguez et al. 2004). They usually but not exclusively are found in a sense orientation and share their expression and regulation with the promoter region of the host gene. About 42–48 % of miRNA genes with a common promoter originate from polycistronic units containing 2–7 discreet loops. Polymerase-II (Pol-II) transcribe miRNA from their corresponding *MIR* genes within the nucleus as a hairpin loop primary microRNA (pri-miRNA) containing hundreds of nucleotides with one to six miRNA precursors. A microprocessor complex then processes the pri-miRNA and the hairpin loop is spliced to form precursor microRNA (pre-miRNA). The enzyme Rnase III (dicer) cuts the loop joining 3′ and 5′ ends of pri-miRNA to form a double-stranded miRNA–miRNA* duplex with 21 nucleotides. Plants carry out this step inside the nucleus by a dicer-like homologue (DCL1). The duplex miRNA.miRNA* is methylated at the 3′ end and exported into the cytoplasm by Exportin5 homologue 'Hasty (HST)', where they separate out into single strands and mature into miRNA. A single strand of the pre-miRNA is used to form miRISC complex (miRNA-induced silencing complex). This complex contains Argonaute protein (AGO1), which orients miRNA and binds it with mRNA. However, miRNA defines the specificity of RISC complex to bind with the target mRNA, which it cleaves.

Another group of small RNAs found in plants are endogenous small interfering RNAs (siRNA), which regulate functions and processes in maintenance of genomic integrity, pattern of development and response to biotic and abiotic stress (Lelandais-Brière et al. 2010).

Homologues of miR399 have been found in rice, tomato, common bean (*Phaseolus vulgaris*) and *Medicago truncatula* (Kuo and Chiou 2011). Studies have clearly indicated that miR399 moves up the phloem sap and function as a long distance signal for Pi homeostasis (Lin et al. 2008; Pant et al. 2008). Apart from miR399, a number of microRNAs have been identified from plants of different species, which are involved in Pi deficiency syndrome, such as miR156, miR159, miR166, miR319, miR395, miR398, miR399, miR447 and miR827 (Kuo and Chiou 2011). Similar to miR399, most of these miRNAs are involved in signalling pathway for Pi deficiency. Some of the miRNAs involved in Pi deficiency have also been found to be affected by other plant nutrients. For example, microRNAs miR169, miR395 and miR398, which are down-regulated by Pi deficiency, are also similarly affected due to deficiency of N, K, Cu, Fe or S. This indicates that miRNAs involved in stress signal transduction pathways have considerable crosstalk with different nutrient homeostasis (Kuo and Chiou 2011).

3.4.10 Improving Phosphate Use Efficiency (PUE)

Phosphate (Pi) use efficiency (PUE) of crops is generally low (15–20 %) due to various soil- and plant-related factors. A large volume of research results is available, which documents the response of different plants to Pi deficiency. There is considerable difference to such responses among different plants and among different varieties of the same plant species.

3.4.10.1 Growing Suitable Plant Associations with High and Low Pi-Uptake Capacities

Growing suitable plant associations with high and low Pi-uptake capacities under Pi-deficient conditions as in mixed cropping can benefit both the crops. Selecting crop varieties capable of acquiring Pi under deficient conditions without compromising yield can be of help in better utilisation of Pi and improving PUE. Through traditional breeding, a wheat variety 'Xiaoyan 54' has been developed in China, which secretes more organic acid (malate and citrate) through its roots into the rhizosphere and is more efficient in P uptake from Pi-deficient soils (Li et al. 1995). Similarly, soybean cv 'BX10' with superior root traits is better adapted to Pi-deficient soils (Yan et al. 2006).

3.4.10.2 Facilitation of Pi Availability by One Crop to the Other Through Rhizosphere Acidification

Facilitation of Pi availability by one crop to the other through rhizosphere acidification has been observed especially in intercropping systems. In an intercropping experiment conducted for 4 years on Pi-deficient but N-rich soil with maize and faba bean (*Vicia faba* L.), Li et al. (2007) observed that maize outyielded by 43 % and faba bean by 26 % as compared to their sole crops. Exudation of organic acids and protons by faba bean facilitated Pi availability to the maize crop resulting in its enhanced yield. The increase in yield of faba bean is due to its different growing season and rooting depth.

3.4.10.3 Manipulating Expression of Genes Enabling Growth in Low-P Environments

It is suggested that manipulating the expression of genes enabling growth in low-P environments could improve the PUE of plants and reduce the P-fertiliser requirement of crops (Vance et al. 2003).

3.4.10.4 Developing Varieties with Expression of Modified Low-Affinity Pi Transporters

It has been observed that four barley genotypes, which differ in their phosphate acquisition efficiency (PAE), do not show any difference in the expression pattern of four paralogues of high-affinity Pi-transporter HvPHT1;1 among them, but the expression pattern of low-affinity Pi-transporter HvPHT1;6 and their close homologue HvPHT1;3 is correlated with their PAE and phosphate use efficiency (PUE). Further, PUE shows correlation with high root-to-shoot ratio under Pi-deficient conditions indicating a higher carbohydrate partitioning to the roots. However, high carbon partitioning to roots might result in low PAE of the crop. It is therefore necessary to develop varieties with expression of modified low-affinity Pi transporters, which can ensure high PUE with reasonable PAE suited to low-input agriculture (Huang et al. 2011).

3.4.10.5 Overexpression of the Gene OsPHR2 in Rice

Overexpression of the gene OsPHR2 in rice results in increased Pi accumulation in shoot, root elongation and root hair proliferation in transgenic rice (Zhou et al. 2008a, b).

3.4.10.6 Development of Transgenic Technologies Involving AVPI-Modified Crops

'Prototypical plant proton translocating pyrophosphatases' (H^+PPases) have 85 % or more amino acid sequence identities and are highly conserved (Drozdowicz and Rea 2001). There are two phylogenetically distinct H^+PPases in plants: type I and type II. Type I H^+PPases depend on cytosolic K^+ concentration for their activity and are moderately sensitive to inhibition by Ca^{2+} ions. Type II H^+PPases are insensitive K^+ ions but extremely sensitive to Ca^{2+} ions. Overexpression of type I H^+PPase 'AVPI' imparts drought and salt tolerance to *Arabidopsis* (AtAVPIOX) and other plants and results in increased root and shoot proliferation (Gaxiola et al. 2001, 2002; Yang et al. 2007). The AtAVPIOX overexpressed *Arabidopsis* plant excretes more organic acids in Pi-deficient medium and has higher scavenging capacity of Pi (Yang et al. 2007). Overexpression of the E229D gain-of-function mutant (AVPID) of the *Arabidopsis* H^+-PPase in transgenic tomato (LeAVPIDOX) in low Pi medium has been found to increase fruit dry weight by 82 % and Pi content by 30 % as compared to control (Yang et al. 2007). Transgenic rice with increased H^+PPase activity (OsAVPIDOX) shows robust root and shoot growth in Pi-deficient medium. Further development of transgenic technologies involving AVPI-modified crops is likely to increase PUE of crops (Gaxiola et al. 2011).

3.4.10.7 Transgenic Plants Overexpressing Secreted Phytases and PAPases

Out of 29 PAPs (purple acid phosphatases) encoded by *Arabidopsis* genome, AtPAP26 is the predominant intracellular PAP upregulated by Pi deficiency (Hurley et al. 2010). Wang et al. (2010) overexpressed *Arabidopsis* PAP gene AtPAP15 containing a carrot (*Daucus carota*) extracellular targeting peptide in soybean hairy roots and observed 1.5-fold increase in PAPase activity in transgenic hairy roots. Three homozygous transgenic lines of soybean overexpressing AtPAP15 and grown in sand culture with phytate as sole source of Pi were found to increase their plant dry weight by 117.8 %, 56.5 % and 57.8 % and plant P content by 90.1 %, 18.2 % and 62.6 %, respectively, as compared to wild-type plants grown on similar conditions. The transgenic soybean lines significantly increased PAPase and phytase activity in their root and leaf exudates. When grown on acid soils, the transgenic plants showed improvement of yields; the pod number per plant increased by 359 %, 41.0 % and 59.0 %; and seeds per plant increased by 46.0 %, 48.3 % and 66.7 %, respectively. However, according to Richardson (2009), transgenic plants overexpressing secreted phytases, when grown on different agricultural soils, do not show any improvement in growth or Pi uptake.

Overexpression of the transcription factor OsPTF1 in transgenic rice using the constitutive cauliflower mosaic virus 35S promoter enhances its tolerance to Pi deficiency (Yi et al. 2005). The transgenic rice so formed has a higher Pi content, 40 % increase in tiller number and 30 % increase in biomass under Pi-deficient conditions as compared to untransformed plants. According to Yi et al. (2005), this is probably due to increase in total root length, root surface area and Pi-uptake activity. OsPFT1 has been reported to upregulate several genes involved in Pi uptake and Pi homeostasis.

Current studies indicate that miRNAs play pivotal roles in controlling the adaptive responses to P deficiency. Thus, miRNA-based engineering or miRNA-implemented molecular breeding holds potential to improve crop yields with less application of fertiliser in the future. Overexpression of *Arabidopsis* miR399 in tomato (*Solanum lycopersicum*) not only results in increased accumulation of Pi but also enhances the secretion of acid phosphatase and proton in the roots, which facilitates the hydrolysis of soil organic P and dissolution of Pi (Gao et al. 2010).

Specific root length, defined as length of root per unit root weight, varies among species and cultivars and is associated with genetic difference in P-use efficiency (Sattelmacher et al. 1994; Lynch and Ho 2004).

References

Abelson PH (1999) A potential phosphate crisis. Science 283:2015

Ai P, Sun S, Zhao J, Fan X, Xin W, Guo Q, Yu L, Shen Q, Wu P, Miller AJ et al (2009) Two rice phosphate transporters, OsPht1;2 and OsPht1;6, have different functions and kinetic properties in uptake and translocation. Plant J 57:798–809

Al-Ghazi Y, Muller B, Pinloche S, Tranbarger TG, Nacry P, Rossignol M, Tardieu F, Doumas P (2003) Temporal responses of *Arabidopsis* root architecture to phosphate starvation: evidence for the involvement of auxin signaling. Plant Cell Environ 26:1053–1066

Aung K, Lin S-I, Wu C-C, Huang Y-T, Su C-I, Chiou TJ (2006) pho2 a phosphate over accumulator, is caused by a nonsense mutation in microRNA399 target gene. Plant Physiol 141:1000–1011

Barber SA, Walker JM, Vasey EH (1963) Mechanisms for the movement of plant nutrients from the soil and fertilizer to the plant root. J Agric Food Chem 11:204–207

Bari R, Pant BD, Stitt M, Scheible W-R (2006) PHO2, MicroRNA399, and PHR1 define a phosphate-signaling pathway in plants. Plant Physiol 141(3):988–999

Bates TR, Lynch JP (1996) Stimulation of root hair elongation in *Arabidopsis thaliana* by low phosphorus availability. Plant Cell Environ 19:529–538

Bieleski RL (1973) Phosphate pools, phosphate transport, and phosphate availability. Annu Rev Plant Physiol 24:225–252

Borch K, Bouma TJ, Lynch JP, Brown KM (1999) Ethylene: a regulator of root architectural responses to soil phosphorus availability. Plant Cell Environ 22:425–431

Bucciarelli B, Hanan J, Palmquist D, Vance CP (2006) A standardized method for analysis of *Medicago truncatula* phenotype development. Plant Physiol 142:207–219

Burleigh SH, Harrison MJ (1999) The down-regulation of Mt4-like genes by phosphate fertilization occurs systemically and involves phosphate translocation to the shoots. Plant Physiol 119:241–248

Calderon-Vazquez C, Ibarra-Laclette E, Caballero-Perez J, Herrera-Estrella I (2008) Transcript profiling of *Zea mays* roots reveals gene responses to phosphate deficiency at the plant- and species-specific levels. J Exp Bot 59:2479–2497

Calderon-Vazquez C, Sawers RJH, Herrera-Estrella I (2011) Phosphate deprivation in maize, genetics and genomics. Plant Physiol 156(3):1067–1077

Casimiro I, Beeckman T, Graham N, Bhalerao R, Zhang H, Casero P, Sandberg G, Bennett MJ (2003) Dissecting *Arabidopsis* lateral root development. Trends Plant Sci 8:165–171

Diatloff E, Roberts M, Sanders D, Roberts SK (2004) Characterization of anion channels in the plasma membrane of *Arabidopsis* epidermal root cells and the identification of a citrate-permeable channel induced by phosphate starvation. Plant Physiol 136:4136–4149

Dong B, Rengel Z, Delhaize E (1998) Uptake and translocation of phosphate by pho2 mutant and wild-type seedlings of *Arabidopsis thaliana*. Planta 205:251–256

Dong D, Peng X, Yan X (2004) Organic acid exudation induced by phosphorus deficiency and/or aluminium toxicity in two contrasting soybean genotypes. Physiol Plant 122(2):190–199

Drozdowicz YM, Rea PA (2001) Vacuolar H^{+}pyrophosphatases: from the evolution art backwaters into the mainstream. Trends Plant Sci 6:206–211

Elhiti M, Stasolla C (2009) Structure and function of homodomain-leucine zipper (HD-Zip) proteins. Plant Signal Behav 4(2):86–88

Eulgem T, Rushton PJ, Robatzek S, Somssich IE (2000) The WRKY superfamily of plant transcription factors. Trends Plant Sci 5:199–206

Fang ZY, Shao C, Meng YJ, Wu P, Chen M (2009) Phosphate signaling in *Arabidopsis* and *Oryza sativa*. Plant Sci 176:170–180

Fohse D, Claassen N, Jungk A (1991) Phosphorus efficiency of plants II. Significance of root radius, root hairs and cation anion balance for phosphorus influx in seven plant species. Plant Soil 132:261–272

Franco-Zorrilla JM, González E, Bustos R, Linhares F, Leyva A, Paz-Ares J (2004) The transcriptional control of plant responses to phosphate limitation. J Exp Bot 55:285–293

Fujii H, Chiou TJ, Lin SI, Aung K, Zhu JK (2005) A miRNA involved in phosphate-starvation response in *Arabidopsis*. Curr Biol 15:2038–2043

Gahoonia TS, Nielsen NE (2004a) Root traits as tools for creating phosphorus efficient crop varieties. Plant Soil 260:47–57

Gahoonia TS, Nielsen NE (2004b) Barley genotypes with long root hairs sustain high grain yields in low-P field. Plant Soil 262:55–62

Gahoonia TS, Care D, Nielson NE (1997) Root hairs and phosphorus acquisition of wheat and barley cultivars. Plant Soil 191:181–191

Gao N, Su Y, Min J, Shen W, Shi W (2010) Transgenic tomato over expressing ath-miR399d has enhanced phosphorus accumulation through increased acid phosphatase and proton secretion as well as phosphate transporters. Plant Soil 334:123–136

Gaxiola R, Li J, Undurraga S, Dang L, Allen G, Alper S, Fink G (2001) Drought and salt-tolerant plants result from over expression of the AVP1 H+–pump. Proc Natl Acad Sci U S A 98:11444–11449

Gaxiola RA, Fink GR, Hirschi KD (2002) Genetic manipulation of vacuolar proton pumps and transporters. Plant Physiol 129:967–973

Gaxiola RA, Edwards M, Elsre JJ (2011) A transgenic approach to enhance phosphorus use efficiency in crops as part of a comprehensive strategy for sustainable agriculture. Chemosphere 84:840–845

Gilroy S, Jones DL (2000) From form to function: development and nutrient uptake in root hairs. Trends Plant Sci 5:56–60

Goldstein AH, Baertlein DA, McDaniel RG (1988) Phosphate starvation inducible metabolism in *Lycopersicon esculentum* I. Excretion of acid phosphatase by tomato plants and suspension-cultured cells. Plant Physiol 87:711–715

Graham MA, Ramirez M, Valdes-López O, Lara M, Tesfaye M, Vance CP, Hernandez G (2006) Identification of candidate phosphorus stress induced genes in *Phaseolus vulgaris* through clustering analysis across several plant species. Funct Plant Biol 33:789–797

Gregory AL, Hurley BA, Tran HT, Valentine AJ, She YM, Knowles VL, Plaxton WC (2009) In *vivo* regulatory phosphorylation of the phosphoenolpyruvate carboxylase AtPPC1 in phosphate-starved *Arabidopsis thaliana*. Biochem J 420:57–65

Guo B, Jin Y, Wussler C, Blancaflor EB, Motes CM, Versaw WK (2008) Functional analysis of the *Arabidopsis* PHT4 family of intracellular phosphate transporters. New Phytol 177:889–898

Hammond JP, White PJ (2008) Sucrose transport in the phloem: integrating root responses to phosphorus starvation. J Exp Bot 59(1):93–109. doi:10.1093/jxb/erm221

Hammond JP, Bennett MJ, Bowen HC, Broadly MR et al (2003) Changes in gene expression in Arabidopsis shoots during phosphate starvation and the potential for developing smart plants. Plant Physiol 132:578–596

Hammond JP, Broadle MR, White PJ (2004) Genetic responses to phosphorus deficiency. Ann Bot 94(3):323–332. doi:10.1093/aob/mch156

He L, Hannon GJ (2004) Micro RNAs: small RNAs with a big role in gene regulation. Nat Rev Genet 5:522–531

Heim MA, Jacoby M, Werber M, Martin C, Weisshaar B, Bailey PC (2003) The basic helix–loop–helix transcription factor family in plants: a genome-wide study of protein structure and functional diversity. Mol Biol Evol 20(5):735–747

Hill JO, Simpson RJ, Moore AD, Chapman DF (2006) Morphology and response of roots of pasture species to phosphorus and nitrogen nutrition. Plant Soil 286:7–19

Hinsinger P (2001) Bioavailability of soil inorganic P in the rhizosphere as affected by root-induced chemical changes: a review. Plant Soil 237:173–195

Hochholdinger F, Zimmermann R (2008) Conserved and diverse mechanisms in root development. Curr Opin Plant Biol 11:70–74

Holford ICR (1997) Soil phosphorus: its measurement and its uptake by plants. Austr J Soil Res 35:227–239

Huang CY, Roessner U, Eickmeirer I, Gene Y, Callahan DL, Shirley N, Langridge P, Bacic A (2008) Metabolite profiling reveals distinct changes in carbon and nitrogen metabolism in phosphate-deficient barley plants (*Hordeum vulgare* L.). Plant Cell Physiol 49:691–703

Huang CY, Shirley N, Genc Y, Shi B, Langridge P (2011) Phosphate utilization efficiency correlates with expression of low-affinity phosphate transporters and non-coding RNA, *IPS1*, in barley. Plant Physiol 156(3):1217–1229

Hurley BA, Tran HT, Marty NJ, Park J, Snedden WA, Mullen RT, Plaxton WC (2010) The dual-targeted purple acid phosphatase isozyme AtPAP26 is essential for efficient acclimation of Arabidopsis to nutritional phosphate deprivation. Plant Physiol 153(3):1112–1122

Jones-Rhoades MW, Bartel DP (2004) Computational identification of plant MicroRNAs and their targets, including a stress-induced miRNA. Mol Cell 14(6):787–799

Jungk A (2001) Root hairs and the acquisition of plant nutrients from soil. J Plant Nutr Soil Sci 164:121–129

Jungk A, Asher CJ, Edwards DG, Meyer D (1990) Influence of phosphate status on phosphate uptake kinetics of maize (*Zea mays*) and soybean (*Glycine max*). Plant Soil 124:175–182

Kuo H-F, Chiou T-J (2011) The role of micro RNAs in phosphorus deficiency signaling. Plant Physiol 156(3):1016–1024

Lelandais-Brière C, Sorin C, Declerck M, Benslimane A, Crespi M, Hartmann C (2010) Small RNA diversity in plants and its impact in development. Curr Genomics 11(1):14–23

Li J, Liu X, Zhou W, Sun J, Tong Y, Liu W, Li ZS, Wang P, Yao S (1995) Technique of wheat breeding for efficiently utilizing soil nutrient elements. Sci China Ser B 38:1313–1320

Li D, Zhu H, Liu K, Liu X, Leggewie G, Udvardi M, Wang D (2002) Purple acid phosphatases of *Arabidopsis thaliana*: comparative analysis and differential regulation by phosphate deprivation. J Biol Chem 277:27772–27781

Li L, Li SM, Sun JH, Zhou LL, Bao XG, Zhang HG, Zhang FS (2007) Diversity enhances agricultural productivity via rhizosphere phosphorus facilitation on phosphorus deficient soils. Proc Natl Acad Sci U S A 104:11192–11196

Liang C, Tian J, Lam HM, Lim BL, Yan X, Liao H (2010) Biochemical and molecular characterization of PvPAP3, a novel purple acid phosphatase isolated from common bean enhancing extracellular ATP utilization. Plant Physiol 152:854–865

Liao H, Rubio G, Yan X, Cao A, Brown K, Lynch JP (2001) Effect of phosphorus availability on basal root shallowness in common bean. Plant Soil 232:69–79

Lin S-I, Chiang SF, Lin WY, Chen JW, Tseng CY, Wu PC, Chiou TJ (2008) Regulatory network of microRNA399 and PHO2 by systemic signaling. Plant Physiol 147:732–746

Lin W-Y, Lin S-I, Chiou T-J (2009) Molecular regulators of phosphate homeostasis in plants. J Exp Bot 60(5):1427–1438. doi:10.1093/jxb/ern303

Liu L, Liao H, Wang X, Yan X (2008) Adaptive changes of soybean genotypes with different root architectures to low phosphorus availability as related to phosphorus efficiency. Sci Agric Sin 41:1089–1099

Liu F, Chang X-J, Yea Y, Xie W-B, Wu P, Lian X-M (2011) Comprehensive sequence and whole-life-cycle expression profile analysis of the phosphate transporter gene family in rice. Mol Plant 4(6):1105–1122. doi:10.1093/mp/ssr058

López-Bucio J, Hernández-Abreu E, Sánchez-Calderón L, Nieto-Jacobo MF, Simpson J, Herrera-Estrella L (2002) Phosphate availability alters architecture and causes changes in hormone sensitivity in the *Arabidopsis* root system. Plant Physiol 129:244–256

Lynch JP (1998) Root architecture and phosphorus acquisition efficiency in common bean. In: Lynch JP, Deikman J (eds) Phosphorus in plant biology: regulatory roles in molecular, cellular, organismic, and ecosystem processes. American Society of Plant Physiologists, Rockville, pp 81–91

Lynch JP (2011) Root phenes for enhanced soil exploration and phosphorus acquisition: tools for future crops. Plant Physiol 156(3):1041–1046

Lynch JP, Brown KM (2001) Topsoil foraging: an architectural adaptation to low phosphorus availability. Plant Soil 237:225–237

Lynch JP, Brown KM (2008) Root strategies for phosphorus acquisition. In: White P, Hammond J (eds) The ecophysiology of plant-phosphorus interactions. Springer, Dordrecht, pp 83–116

Lynch JP, Ho M (2004) Rhizo-economics: carbon costs of phosphorus acquisition. Plant Soil 269:45–56

Ma Z, Bielenberg DG, Brown KM, Lynch JP (2001) Regulation of root hair density by phosphorus availability in *Arabidopsis thaliana*. Plant Cell Environ 24:459–467

Marschner H (1995) Mineral nutrition in plants, 2nd edn. Academic, San Diego

Martin C, Paz-Ares J (1997) MYB transcription factors in plants. Trends Genet 13:67–73

Martín AC, del Pozo JC, Iglesias J, Rubio V, Solano R, de la Peña A, Leyva A, Paz-Ares J (2000) Influence of cytokinins on the expression of phosphate starvation responsive genes in *Arabidopsis*. Plant J 24:559–567

Martinoia E, Klein M, Geisler M, Bovet L, Forestier C, Kolukisaoglu U, Muller-Rober B, Schulz B (2002) Multi functionality of plant ABC transporters: more than just detoxifiers. Planta 214:345–355

Michael G (2001) The control of root hair formation: suggested mechanisms. J Plant Nutr Soil Sci 164:111–119

Miller CR, Ochoa I, Nielson KL, Beck D, Lynch JP (2003) Genetic variation for adventitious rooting in response to low phosphorus availability: potential utility for phosphorus acquisition from stratified soils. Funct Plant Biol 30:973–985

Misson J, Raghothama KG, Jain A et al (2005) A genome-wide transcriptional analysis using *Arabidopsis thaliana* Affymetrix gene chips determined plant responses to phosphate deprivation. Proc Natl Acad Sci U S A 102:11934–11939

Mudge SR, Rae AL, Diatloff E, Smith FW (2002) Expression analysis suggests novel roles for members of the Pht1 family of phosphate transporters in *Arabidopsis*. Plant J 31:341–353

Müller R, Morant M, Jarmer H, Nilsson L, Nielsen TH (2007) Genome-wide analysis of the *Arabidopsis* leaf transcriptome reveals interaction of phosphate and sugar metabolism. Plant Physiol 143:156–171

Nilsson L, Muller R, Nielsen TH (2007) Increased expression of the MYB-related transcription factor, PHR1, leads to enhanced phosphate uptake in *Arabidopsis thaliana*. Plant Cell Environ 30:1499–1512

O'Dell B, de Boland A, Koirtyohann S (1972) Distribution of phytate and nutritionally important elements among the morphological components of cereal grains. J Agric Food Chem 20:718–721

Pant BD, Buhtz A, Kehr J, Scheible WR (2008) MicroRNA399 is a long-distance signal for the regulation of plant phosphate homeostasis. Plant J 53:731–738

Parker JS, Cavell AC, Dolan L, Roberts K, Grierson CS (2000) Genetic interactions during root hair morphogenesis in *Arabidopsis*. Plant Cell 12:1961–1974

Piñeros MA, Cancado GM, Maron LG, Lvi SM, Menossi M, Kochian LV (2008) Not all ALMT1-type transporters mediate aluminum-activated organic acid responses: the case of ZmALMT1 – an anion-selective transporter. Plant J 53(2):352–367. Epub 2007 Dec 6

Plaxton WC, Podesta FE (2006) The functional organization and control of plant respiration. Crit Rev Plant Sci 25:159–198

Plaxton WC, Tran HT (2011) Metabolic adaptations of phosphate-starved plants. Plant Physiol 156(3):1006–1015

Poirier Y, Bucher M (2002) Phosphate transport and homeostasis in Arabidopsis. In: Somerville CR, Meyerowitz EM (eds) The *Arabidopsis* book. The American Society of Plant Biologists, Rockville, pp 1–35

Postma J, Lynch JP (2010) Theoretical evidence for the functional benefit of root cortical aerenchyma in soils with low phosphorus availability. Ann Bot 107:829–841

Richardson AE (2009) Regulating the phosphorus nutrition of plants: molecular biology meeting agronomic needs. Plant Soil 322:17–24

Rodriguez A, Griffiths-Jones S, Ashurst JL, Bradley A (2004) Identification of mammalian micro RNA host genes and transcription units. Genome Res 14(10A):1902–1910

Rubio V, Linhares F, Solano R, Martín AC, Iglesias J, Leyva A, Paz-Ares J (2001) A conserved MYB transcription factor involved in phosphate starvation signaling both in vascular plants and in unicellular algae. Genes Dev 15:2122–2133

Rushton PJ, Somssich IE, Ringler P, Shen QJ (2010) WRKY transcription factors. Trends Plant Sci 15:247–258

Sattelmacher B, Horst WJ, Becker HC (1994) Factors that contribute to genetic variation for nutrient efficiency of crop plants. Z Pflanzenernaehr Bodenkd 157:215–224

Schachtman DP, Reid RJ, Ayling SM (1998) Phosphorus uptake by plants: from soil to cell. Plant Physiol 116:447–453

Skene KR (1998) Cluster roots: some ecological considerations. J Ecol 86:1060–1064

Smith FA, Jakobsen I, Smith SE (2000) Spatial differences in acquisition of soil phosphate between two arbuscular mycorrhizal fungi in symbiosis with *Medicago truncatula*. New Phytol 147:357–366

Smith FW, Mudge SR, Rae AL, Glassop D (2003) Phosphate transport in plants. Plant Soil 248:71–83

Strobel BW (2001) Influence of vegetation on low-molecular-weight carboxylic acids in soil solution – a review. Geoderma 99:169–198

Ström L, Owen AG, Godbold DL, Jones DL (2005) Organic acid behaviour in a calcareous soil implications for rhizosphere nutrient cycling. Soil Biol Biochem 37:2046–2054

Tang Z, Sadka A, Morishige DT, Mullet JE (2001) Homeodomain leucine zipper proteins bind to the phosphate response domain of the soybean *VspB* tripartite promoter. Plant Physiol 125:797–809

Tawaraya K, Horie R, Shinano T, Wagatsuma T, Saito K, Akira O (2009) Metabolite profiling of rice root exudate under phosphorus deficiency. In: The proceedings of the international plant nutrition colloquium XVI, Department of Plant Sciences, UC, Davis

Tesfaye M, Liu J, Allan DL, Vance CP (2007) Genomic and genetic control of phosphate stress in legumes. Plant Physiol 144:594–603

Ticconi CA, Abel S (2004) Short on phosphate: plant surveillance and countermeasures. Trends Plant Sci 9:548–555

Tran HT, Hurley BA, Plaxton WC (2010) Feeding hungry plants: the role of purple acid phosphatases in phosphate nutrition. Plant Sci 179:14–27

Uhde-Stone C, Zinn KE, Ramirez-Yáñez M, Li A, Vance CP, Allan DL (2003) Nylon filter arrays reveal differential gene expression in proteoid roots of white lupin in response to phosphorus deficiency. Plant Physiol 131:1064–1079

Ullrich C, Novacky A (1990) Extra- and intracellular pH and membrane potential changes induced by K+, Cl^-, $H2PO4^-$ and $NO3^-$ uptake and fusicoccin* in root hairs of Limnobium stoloniferum. Plant Physiol 94:1561–1567. (*The wilt-inducing toxin of* Fusicoccum amygdali, **fusicoccin (FC)** has long been used as a tool in plant physiology (Marré, 1979). FC instantaneously hyperpolarizes the membrane potential and produces a strong acidification of the apoplastic space. As a result of toxin application, the transport of many metabolites and ions across the plasmalemma may change and, in accord with the acid-growth hypothesis, cell elongation may also be stimulated)

Van Kauwenbergh SJ (2010) World phosphate rock reserves and resources. IFDC Tech Bull 75

Vance CP (2010) Sugars, and MicroRNAs: quaternaries in phosphate acquisition and use. Plant Physiol 154(2):582–588

Vance CP, Uhde-Stone C, Allan DL (2003) Phosphorus acquisition and use: critical adaptations by plants for securing a nonrenewable resource. New Phytol 157:423–447

Versaw WK, Harrison MJ (2002) A chloroplast phosphate transporter, PHT2;1, influences allocation of phosphate within the plant and phosphate-starvation responses. Plant Cell 14:1751–1766

Walker TS, Bais HP, Grotewold E, Vivanco JM (2003) Root exudation and rhizosphere biology. Plant Physiol 132(1):44–51

Wang Y-H, Garvin DF, Kochian LV (2002) Rapid induction of regulatory and transporter genes in response to phosphorus, potassium, and iron deficiencies in tomato roots, evidence for cross talk and root/rhizosphere-mediated signals. Plant Physiol 130:1361–1370

Wang X, Yan X, Liao H (2010) Genetic improvement for phosphorus efficiency in soybean: a radical approach. Ann Bot 106(1):215–222. doi:10.1093/aob/mcq029

Wasaki J, Yonetani R, Shinano T, Kai M, Osaki M (2003) Expression of the OsPI1 gene, cloned from rice roots using cDNA microarray, rapidly responds to phosphorus status. New Phytol 158:239–248

Wasaki J, Shinano T, Onishi K et al (2006) Transcriptomic analysis indicates putative metabolic changes caused by manipulation of phosphorus availability in rice leaves. J Exp Bot 57(9):2049–2059. doi:10.1093/jxb/erj158

Williamson LC, Ribrioux SP, Fitter AH, Leyser HM (2001) Phosphate availability regulates root system architecture in *Arabidopsis*. Plant Physiol 126:875–882

Yan F, Zhu YY, Müller C, Zörb C, Schubert S (2002) Adaptation of H^+-pumping and plasma membrane H^+ ATPase activity in proteoid roots of white lupin under phosphate deficiency. Plant Physiol 129:50–63

Yan X, Liao H, Beebe SE, Blair MW, Lynch JP (2004) QTL mapping of root hair and acid exudation traits and their relationship to phosphorus uptake in common bean. Plant Soil 265:17–29

Yan X, Wu P, Ling H, Xu G, Xu F, Zhang Q (2006) Plant nutriomics in China: an overview. Ann Bot 98:473–482

Yang H, Knapp J, Koirala P, Rajagopal D, Ann Peer W, Silbart LK, Murphy A, Gaxiola R (2007) Enhanced phosphorus nutrition in monocots and dicots over-expressing a phosphorus-responsive type I H+–pyrophosphatase. Plant Biotechnol J 5:735–745

Yanhui C, Xiaoyuan Y, Kun H, Meihua l, Jigang L, Zhaofeng G, Zhiqiang L, Yunfei Z, Xiaoxiao W, Xiaoming Q, Yunping S, Li Z, Xiaohui D, Jingchu L, Xing-Wang D, Zhangliang C, Hongva G, Li-Jia Q (2006) The MYB transcription factor superfamily of *Arabidopsis*: expression analysis and phylogenetic comparison with the rice MYB family. Plant Mol Biol 60(1):107–124

Yi K, Wu Z, Zhou J, Du L, Guo L, Wu Y, Wu P (2005) OsPTF1, a novel transcription factor involved in tolerance to phosphate starvation in rice. Plant Physiol 138:2087–2096

Zeng XQ, Chow WS, Su LJ, Peng XS, Peng XX (2010) Protective effect of supplemental anthocyanins on *Arabidopsis* leaves under high light. Physiol Plant 138:215–225

Zhang Y, Wang L (2005) The WRKY transcription factor superfamily: its origin in eukaryotes and expansion in plants. BMC Evol Biol 5:1

Zhang FS, Ma J, Cao YP (1997) Phosphorus deficiency enhances root exudation of low-molecular weight organic acids and utilization of sparingly soluble inorganic phosphates by radish (*Raghanus satiuvs* L.) and rape (*Brassica napus* L.) plants. Plant Soil 196:261–264

Zhou K, Yamagishi M, Osaki M, Masuda K (2008a) Sugar signaling mediates cluster root formation and phosphorus starvation-induced gene expression in white lupin. J Exp Bot 59:2749–2756

Zhou J, Jiao F, Wu Z, Li Y, Wang X, He X, Zhong W, Wu P (2008b) OsPHR2 is involved in phosphate-starvation signaling and excessive phosphate accumulation in shoots of plants. Plant Physiol 146:1673–1686

4 Potassium (K) Uptake

Abstract

Potassium is the most abundant plant nutrient present in the plants. Plants may accumulate K between 2 and 10 % of plant dry weight. Potassium content below 10 g kg^{-1} of dry weight may lead to deficiency symptoms in most of the plant species.

A large number of proteins encoded by their corresponding genes are involved in K^+ transport in plants. These transporters fall into several categories and have distinct functions. Location of the transporters in different parts of the plant, their subcellular localisation, structure and functions have been discussed.

4.1 Occurrence of Potassium and Soil Reactions

Potassium (K) is the fourth most abundant mineral and constitutes about 2.5 % of the lithosphere. Potassium-bearing minerals such as feldspars (orthoclase and microcline) and micas (muscovite, biotite and phlogopite) constitute 90–98 % of the total soil K. Concentration of K in soils varies widely and is in the range of 0.5–2.5 %. Potassium exists in four forms in soil, mineral K (0.5–2.5 %), non-exchangeable K (50–750 ppm), exchangeable K (40–600 ppm) and solution K (1–10 ppm). There is equilibrium among the different forms as follows.

$$\text{Mineral K}^+ \rightarrow \text{Non Exchangeable K}^+ \rightarrow \text{Exchangeable K}^+ \rightleftarrows \text{Solution K}^+$$

Release of K^+ from mineral K is extremely slow and becomes available after weathering. Non-exchangeable K^+ becomes available slowly, but exchangeable and solution K^+ are readily available for uptake by plants.

Due to fast uptake of K^+ by plants, its concentration in the rhizosphere is considerably lower than bulk of the soil solution and may be in the μM range (Grabov 2007). Plants have developed sophisticated mechanism of K^+ uptake from very low concentration of solution K^+ to acquire it against a concentration gradient of >1,000-fold.

4.2 Potassium Content of Plants

Potassium is the most abundant plant nutrient present in the plants. While K^+ concentration in soil solution is in the range of 0.1–6 mM, plants

G.N. Mitra, *Regulation of Nutrient Uptake by Plants: A Biochemical and Molecular Approach*, DOI 10.1007/978-81-322-2334-4_4, © Springer India 2015

accumulate large quantities of this element, which may constitute between 2 and 10 % of plant dry weight. Potassium content below 10 g kg^{-1} of dry weight may lead to deficiency symptoms in most of the plant species. Cytoplasmic concentration of K^+ is however maintained at approximately 100 mM although vacuole stocks up higher concentration of K^+ (20–200 mM) to be used under limiting supply (Gierth and Maser 2007). Apoplastic concentration of K^+ may vary between 10 and 200 mM but may increase up to 500 mM (White and Karley 2010; Wang et al. 2013).

4.3 Functions of Potassium in Plants

Potassium activates about 60 enzymes involved in various metabolic processes, such as photosynthesis, protein synthesis, oxidative metabolism, etc., and improves quality and stress tolerance of crops in addition to its role in osmoregulation, turgor-driven movements and maintenance of the plasma membrane potential. Within the cytosol, K^+ neutralises the soluble and insoluble macromolecular anions and stabilises pH at ≈ 7.2, the level optimal for most enzymatic reactions (Marschner 1995).

4.4 Mechanism of Potassium Uptake by Plants

In view of its large requirement, the plant roots have to take up adequate amount of K^+ from the soil solution and transport them to the aerial parts. Since all cells and organelles of the plant need K^+, it has to move across different types of membranes through elaborate mechanisms of transportation. A large number of proteins encoded by their corresponding genes are involved in K^+ transport in plants.

4.4.1 Classification of Potassium Transporters

According to recent Transporter Classification (TC) System, there are two major pathways for K^+ acquisitions by plants:

Class 1 Channels and other low-affinity systems effective at concentrations near 1 mM and above.

Class 2 High-affinity systems operative at K^+ concentrations in the micromolar range. Recent findings suggest, however, that channels also contribute to high-affinity K^+ uptake.

These pathways fall into several distinct categories (Maser et al. 2001, 2002; Very and Sentenac 2002; Shabala 2003; Gierth and Maser 2007; Grabov 2007) such as:

(a) K^+ channels consisting of three families: Shaker-type channels, KCO channels (a total of 15 genes in *Arabidopsis*) and cyclic nucleotide-gated channels (CNGC, 20 genes in *Arabidopsis*) (Very and Sentenac 2002)
(b) Trk/HKT transporters [Na^+/K^+ symporter] (Schachtman 2000), one gene in *Arabidopsis*
(c) KUP/HAK/KT transporters [H^+/K^+ symporter] (Kim et al. 1998), 13 genes in *Arabidopsis*
(d) K^+/H^+ antiporter homologue, six genes in *Arabidopsis*
(e) Glutamate receptors (GLRs), 20 genes in *Arabidopsis* (Very and Sentenac 2002)

4.4.2 Shaker Channels

Plant voltage-gated channels belonging to the Shaker family participate in sustained K^+ transport processes at the cell and whole plant levels, such as K^+ uptake from the soil solution, long-distance K^+ transport in the xylem and phloem and K^+ fluxes in guard cells during stomatal movements (Cherel 2004). The voltage-gated Shaker channels are either hyperpolarisation activated and thus inwardly rectifying (i.e. mediating an inward K^+ current) or depolarisation activated and thus outwardly rectifying (allowing K^+ efflux from the cell). Based on their voltage dependence, selectivity and sensitivity to pharmacological agents, they mediate most of the major K^+ currents described in the plasma membrane of plant cells. The Shaker family comprises nine members in *Arabidopsis*. Members of this family have also been identified in a number of other plant species (Very and Sentenac 2003).

4.4.2.1 SKOR and GORK Channels

The Shaker-type channels are further subdivided into SKOR and GORK channels (both depolarisation activated), KAT channels and AKT channels. AKT channels contain an ankyrin-binding motif, which is lacking in KAT type channels (Maser et al. 2001). An important feature of plant Shaker-like K^+ channels is that they can form hetero-tetrameric structures (Pilot et al. 2003), allowing plants to tune the K^+ transport activity in various cells, independently in each organ/tissue, in relation to environmental conditions. Subunit assembly is a prerequisite for channel function.

4.4.2.2 Genes Encoding Shaker Family Proteins

The genes encoding Shaker family proteins of *Arabidopsis* have been identified and their functions known. *AKT1* encodes an inwardly rectifying channel and plays a role in K^+ uptake from the soil solution. Once K^+ has been taken up, its secretion into the root xylem for delivery to the shoot involves the outwardly rectifying SKOR channel, which could mediate the delivery of up to 50 % of the K^+ in the xylem sap (Gaymard et al. 1998). At least two inwardly rectifying Shakers, KAT1 and KAT2, and one outwardly rectifying Shaker, GORK, are expressed in guard cells (Nakamura et al. 1995; Ache et al. 2000; Pilot et al. 2001; Szyroki et al. 2001). AKT2 is suggested to have a role in the control of the phloem cell membrane potential and in the regulation of sucrose loading/unloading into/from the phloem sap (Deeken et al. 2002). The AKT2 channel also accounts for about 50 % of K^+ permeability of mesophyll cells, AKT1 being responsible for the remaining 50 % (Dennison et al. 2001). The inwardly rectifying Shaker SPIK is specifically expressed in pollen where it mediates K^+ uptake. Pollen tube growth and, therefore, pollen competitive ability are dependent on the activity of this channel (Mouline et al. 2002).

4.4.2.3 Structure of Plant Shaker Polypeptides

Plant Shaker polypeptides typically display a rather short (*c.* 60-amino-acid) intra-cytoplasmic N-terminal domain, followed by a hydrophobic core composed of six transmembrane segments (S1 to S6, the pore domain being inserted between S5 and S6), and a long intra-cytoplasmic region representing more than half of the sequence. The fourth transmembrane segment harbours positively charged amino acids (R and K) and is expected to act as a voltage sensor. A well-conserved pore domain, carrying the hallmark GYGD/E motif of highly K^+ selective channels, is present between S5 and S6. The long C-terminal region harbours a putative cyclic nucleotide-binding domain and, in most Shaker channels, an ankyrin domain potentially involved in protein–protein interactions (Cherel 2004). (Ankyrin is a 33-residue repeating motif, which mediates specific macromolecular interactions with cytoskeletal, membrane and regulatory proteins.)

Shaker K^+ channel activity is regulated both at the transcriptional and post-translational levels. Different environmental and hormonal factors such as light, abscisic acid (ABA), auxin and salt stress may result in fluctuation of transcript levels (Very and Sentenac 2003). Channel activity at the post-translational level is controlled by membrane polarisation and intracellular factors such as H^+ (Hoshi 1995; Marten et al. 1999; Lacombe et al. 2000), calcium (Marten et al. 1999) and cyclic nucleotides (Hoshi 1995; Gaymard et al. 1996).

4.4.2.4 KCO Channels

The K^+ channel AtKCO1 from *Arabidopsis thaliana* is the prototype of a new family of plant K^+ channels. These are components of slow vacuolar (SV) double-pore K^+ channels located in the tonoplast. AtKCO1 promoter is active in various tissues and cell types, and the highest GUS (beta-glucuronidase) activity could be detected in mitotically active tissues of the plant. Promoter activity is strongly dependent on the presence of a 5′ leader intron. The same overall structure is identified in two genes encoding AtKCO1-like K^+ channels from *Solanum tuberosum*, StKCO1alpha and StKCO1beta (Schonknecht et al. 2002; Czempinski et al. 2002).

4.4.3 Cyclic Nucleotide-Gated Channels (CNGC)

The structure of CNGC is similar to Shaker channels. The plant CNGC in contrast to animals has overlapping binding domains of cyclic nucleotide (CN) and calmodulin (CaM) (Köhler et al. 1999; Arazi et al. 2000; Köhler and Neuhaus 2000), enabling crosstalk between CaM and CN signalling (Arazi et al. 2000). Their gating requires binding of ligand in the form of cGMP or cAMP. In *Arabidopsis*, AtCNGC1 and AtCNGC4 (HLM1) display equal permeability for Na^+ and K^+ (Hua et al. 2003; Balague et al. 2003; Bridges et al. 2005). Remarkably, AtCNGC2, characterised by a unique Ala-Asn-Asp selectivity filter, is highly selective for K^+ over Na^+ (Leng et al. 2002; Hua et al. 2003). Due to its K^+ permeability and appreciable expression in roots (Talke et al. 2003), AtCNGC2 may be directly involved in K^+ uptake.

4.4.4 Trk/HKT Transporters (TC: 2·A·38)

HKT (high-affinity K^+ transporter) proteins of plants are part of the Trk superfamily of cation transporters and are topologically related to K^+ channels. All known plant *HKT* genes contain two introns near the 3′ end. Plant HKT amino acid sequences are grouped into two subfamilies, and the genes of subfamily one have longer introns than those of subfamily two (Platten et al. 2006). Subfamily two exclusively contains monocot genes; subfamily one includes monocot genes and all known *HKTs* from dicots. The first HKT/Trk gene identified from plants, wheat *TaHKT1*, was found to function as a high-affinity K^+/Na^+ cotransporter that switched to low-affinity Na^+ uniport at high $[Na^+]/[K^+]$ (Rubio et al. 1995: Gassmann et al. 1996). In rice, OsHKT1 showed properties of a Na^+-selective uniporter similar to AtHKT1. A salt-tolerant cultivar of rice *Pokkali* contained OsHKT2, a K^+/Na^+ symporter similar to TaHKT1 (Horie et al. 2001; Garciadeblas et al. 2003). *OsHKT4* appeared to encode a low-affinity Na^+ transporter. Substrate specificity for OsHKT3, OsHKT6 and OsHKT9 are yet to be known.

4.4.5 KUP/HAK/KT Transporters (TC: 2·A·72)

All plant genomes contain genes encoding KT (potassium transporters)/KUP (potassium uptake permeases)/HAK (high-affinity potassium transporters) (given different acronyms by different research groups) transporters (not found in *Protista* and *Animalia*). These transporter genes are found primarily in organisms that acquire nutrients through absorption and not through ingestion forage on potassium-rich organic matter (Grabov 2007). KT/KUP/HAK transporters constitute a major high-affinity potassium acquisition system.

All KT/KUP/HAK transporters can be grouped into four distinct clusters (Rubio et al. 2000; Banuelos et al. 2002). All plants have Cluster I or Cluster II transporters. Cluster III genes are found only in *Arabidopsis* and rice. The smallest cluster is number IV, which comprises only four rice genes. There are a total of 13 genes in *Arabidopsis* and 27 in rice.

4.4.5.1 Cluster I Transporters

Cluster I transporters characterised so far have high affinity for the substrate and play a key role in potassium acquisition, when K^+ availability is low (Banuelos et al. 2002; Rodriguez-Navaro and Rubio 2006). HvHAK1, the main high-affinity potassium uptake transporter in barley roots, is induced by potassium starvation (Santa-Maria et al. 1997). Expression of *LeHAK5* in tomato and *AtHAK5* in *Arabidopsis* is activated by low external K^+ concentration (Wang et al. 2002; Ahn et al. 2004; Hampton et al. 2004; Gierth et al. 2005). Consistent with its function of K^+ acquisition, *AtHAK5* has been found to be expressed in the epidermis of main and lateral roots of *Arabidopsis* (Gierth et al. 2005). It has recently been reported that four transcription factors, *DDF2* (Dwarf and Delayed Flowering2), *JLO* (Jagged Lateral Organs), *TFII_A* (Transcription Initiation Factor II_A gamma chain) and bHLH 121 (basic Helix-Loop-Helix 121), can bind to the *AtHAK5* promoter in response to K^+ limitation and activate *AtHAK5* expression, allowing plants to adapt to nutrient stress (Hong et al. 2013).

4.4.5.2 Cluster II Transporters

Cluster II transporters probably facilitate the low-affinity K^+ transport complementing potassium channels (Senn et al. 2001; Garciadeblas et al. 2002). Some of these transporters are localised in the tonoplast and facilitate K^+ efflux from the vacuole. Under conditions of K^+ deprivation, export of K^+ from the vacuole can be mediated by a K^+/H^+ symporter with a 1:1 stoichiometry (Walker et al. 1996) for the maintenance of K^+ homeostasis in K^+-deprived plants.

KT/KUP/HAK transporters have been found to play important roles in some of the plant development processes. In turgor-dependent growth of rapidly expanding cotton fibres (*Gossypium hirsutum*), expression of the *GhKT1* correlates positively with build-up of turgor pressure (Ruan et al. 2001). In growing grapevine fruits (*Vitis vinifera*), expressions of *VvKUP1* and *VvKUP2* genes are required for the potassium-driven cell expansion in young grape berries (Davies et al. 2006).

4.4.6 K^+/H^+ Antiporter Homologues

The least-studied gene family in plants, KEA transporters, consists of six members in *Arabidopsis* and was identified through homology to bacterial K^+/H^+ antiporters (Maser et al. 2001). A member of the family Monovalent Cation: Proton Antiporter-2 (*CPA2*), which is also known as *CHX* (Cation/H^+ eXchanger), AtCHX17 is involved in K^+ acquisition and homeostasis rather than Na^+ transport (Cellier et al. 2004). Consistent with its function, AtCHX17 is expressed in the cortex and epidermis of the mature root.

4.4.7 Glutamate Receptors (GLR)

Arabidopsis genome has 20 *AtGLR* genes (compared with only 11 in humans), which can be grouped into three clades (Chiu et al. 2002). All of the 20 genes are expressed in roots. Homologues of GLRs exist in cyanobacteria (GluR0) where they function as glutamate-gated K^+ channels (Chen et al. 1999). Glutamate receptors (GLR's) are non-selective cation channels involved in Ca^{2+} influx and are differentially activated by amino acids especially glutamic acid and glycine (Stephens et al. 2008; Kudla et al. 2010; Price et al. 2012).

4.4.8 Potassium Transport in Leaves

Information on properties and regulation of membrane potassium transport in leaves is limited. Most of the studies on mechanism of K^+ transport have been confined to roots since it is the primary route of K uptake. Leaf tissues consist of epidermis, mesophyll, guard cells and vascular system along with their intracellular organelles. There is considerable variation in K^+ ion concentration in cells of leaf tissues. Epidermal cells rely heavily on inorganic ions (primarily K^+) for osmotic adjustment since they are virtually unable to produce organic solutes. In mesophyll cells, while K^+ is the dominant osmoticum, 20–30% of osmolytes are organic solutes. To maintain cytosolic K^+ homeostasis in mesophyll cells for protection and maintenance of optimal photosynthetic activity, the concentration of K^+ in epidermal cells may decline to very low levels. A significant difference was also measured between K^+ uptake of epidermal cells at the growing (leaf base) and at the fully extended (leaf tip) regions of corn leaves (Shabala 2003).

4.4.8.1 K^+ Channels in Stomatal Guard Cells

There are two major types of K^+ channels present at the plasma membrane of guard cells:

1. Voltage-dependent K^+-selective inward rectifying (*KIR*).
2. Outward rectifying (*KOR*) channels (Pilot et al. 2001; Schroeder et al. 2001; Szyroki et al. 2001; Zimmermann et al. 2001).

KIR channels such as AKT1, AKT2/3, AtKC1 and KAT2 mediate stomatal opening and are activated by membrane hyperpolarisation. KOR channels such as GORK, a voltage-gated outwardly rectifying K^+ channel of the guard cell membrane (Hosy et al. 2003) in the *Arabidopsis* genome, mediate stomatal closure and are opened by voltages more positive than E_K. Guard cells also possess a wide range of either depolarisation- or hyperpolarisation-activated non-selective cation channels (*NSCC*) (Demicchik et al. 2002).

4.4.8.2 K^+ Channels in Mesophyll Cell Plasma Membranes

Mesophyll cell plasma membranes of leaves contain both KIRs and KORs in addition to NSCCs. Active K^+ transporters such as HAK/KT/KUP and HKT types are also present.

4.4.8.3 K^+ Channels in Epidermis

Subsidiary cells of maize (Majore et al. 2002) were found to contain two time-dependent Ca^{2+}-regulated K^+-selective channels (KIRs and KORs). Such channels were also reported for barley epidermis. Expressions of AKT2 genes (Chérel et al. 2002) and the HAK/KT/KUP K^+ transporters (Su et al. 2002) have been attributed to epidermal cells.

4.4.8.4 K^+ Channels in Vascular Tissues

The most abundant K^+ channels in the phloem tissue are AKT3 (Marten et al. 1999; Cherel et al. 2002) and their homologues (Golldack et al. 2003), which mediate potassium influx and efflux in the phloem loading process. Another major type of K^+ channel detected in minor veins is KAT2 (Pilot et al. 2001), involved in K^+ loading into the phloem sap. High-affinity K^+ transporters, McHAKs (Su et al. 2002) and the HKT1 transporters are also present in the vascular tissue in leaves (Schachtman 2000).

4.4.8.5 K^+ Channels in Vacuole

The most abundant K^+-permeable channels present in tonoplast are slow-activating (*SV*) and fast-activating (*FV*) vacuolar channels. The SV channel is permeable to both mono- and divalent cations and is activated by cytosolic Ca^{2+} and positive vacuolar voltage. The FV channel is selective for monovalent cations only, activated by positive voltages, and may be blocked by divalent cations (Allen and Sanders 1997). Both SV and FV channels are ubiquitous in plant tissues, including mesophyll and guard cell vacuoles. Vacuolar two-pore K^+ channels (TPKs) play an important role in maintaining K^+ homeostasis. Rice genome encodes two TPK isoforms, TPKa localised in the tonoplast of large lytic vacuole (LV) and TPKb in the tonoplast of smaller protein storage vacuoles (PSV) that contain membrane transporters (Isayenkov et al. 2011).

4.4.8.6 K^+ Channels in Chloroplast

The transport barrier in the chloroplast is the inner membrane, which contains transporters for a selected numbers of low molecular weight substrates. The outer membrane contains specific pore-forming proteins and is permeable to substances with molecular weight of several kDa (Pottosin 1992). Most of these 'pores' are also able to conduct ions (Neuhaus and Wagner 2000).

Massive light-driven transport of H^+ into the thylakoid lumen is electrically balanced by the counter flow of other ions (Hinnah and Wagner 1998). This process is mediated by weakly voltage-dependent cation-selective channels, equally permeable to K^+ and Mg^{2+} (Pottosin and Schonknecht 1996). Several types of cation-permeable channels have been found at thylakoid membranes of different species (Pottosin 1992; Pottosin and Schonknecht 1996; Hinnah and Wagner 1998). All of them belong to the NSCC class. Channel conductance varied greatly from 60 pS (Pottosin and Schonknecht 1996) to very high values (non-selective porin-like maxi channel with 1,016 pS conductance (Pottosin 1992). Most of these channels show bimodal gating (Pottosin 1992). However, some channels showed only moderate voltage dependence (Pottosin and Schonknecht 1996), suggesting that additional mechanisms to regulate the thylakoid cation channel activity might be involved. It has been recently reported (Carraretto et al. 2013) that a thylakoid-located two-pore K^+ channel TPK3 modulates the composition of proton motive force (pmf) through ion counterbalancing to convert photochemical energy into physiological functions. In *Arabidopsis*, the channel is found in the thylakoid stromal lamellae.

4.4.8.7 K^+ Channels in Mitochondria

Petrussa et al. (2001) reported that plant mitochondria possess a K^+ selective, voltage-dependent channel, which is opened by cyclosporin, regulated by the redox state and inhibited by nucleotides. The ATP-inhibited plant mitochondrial K^+ channel (PmitoKATP) discovered in durum wheat mitochondria catalyses the electrophoretic K^+ uniport through the inner mitochondrial membrane. PmitoKATP restricts ROS production by lowering mitochondrial membrane potential $\Delta\Psi$. At moderate oxidative stress, this happens without impairing ATP synthesis (Pastore et al. 2013).

4.4.9 Effect of K^+ Uptake on Drought Resistance

Potassium has been reported to improve drought resistance of crops by reducing leaf osmotic potential, increasing turgor, bound water content and water use efficiency (Li 2006). It controls activity of superoxide dismutase (SOD) and mitigates possible injury from active oxygen derived from drought stress to plasma membrane. Potassium increases proline content of leaves and suppresses malondialdehyde (MDA) content (induced by drought) to strengthen drought resistance of crops. Potassium maintains the balance of internal hormone level of CTK, ABA and ethylene.

ABA has been known to regulate K^+ channel activity in guard cells (Assmann and Shimazaki 1999; Luan 2002). Regulation of K^+ channels may be mediated by either ABA-induced increase in cytosolic free Ca^{2+}, which activates KIR (Assmann and Wang 2001; Schroeder et al. 2001; Kohler et al. 2003), or via an ABA-induced Ca^{2+}-independent pathway (Blatt 2000). The latter is probably mediated by the pH effect on activity of FV and SV vacuolar channels (Assmann and Wang 2001). Also, not only gating properties and/or open probability of K^+ channels but their expression *per se* might be affected by ABA (Pilot et al. 2003). Finally, mechano-sensory (*SAS*) channel involvement cannot be ruled out (Cosgrove and Hedrich 1991).

4.4.10 K^+ Transporters and Salt Tolerance

Although plants have an absolute requirement for K^+ and Na^+ is toxic for many biological reactions in the cytoplasm, this does not apply to vacuolar processes (Flowers and Läuchli 1983; Subbarao et al. 2003). Na^+ can undertake osmotic functions, reducing the total K^+ requirements and improving growth when the lack of K^+ is a limiting factor (see Sect. 18.3.1 and 18.4.2).

HKT transporters (high-affinity K^+ transporter), which mediate Na^+-specific transport or Na^+/K^+ transport, play a key role in regulation of Na^+ homeostasis (Rodriguez-Navarro and Rubio 2006; Munns and Tester 2008). There is only one HKT gene in *Arabidopsis thaliana* (Uozumi et al. 2000) and eight genes in rice (Horie et al. 2001; Garciadeblas et al. 2003). *HKT* transporters are divided into two main subfamilies (Platten et al. 2006). Members of subfamily 1 have a serine residue in the first pore loop of the protein, which is replaced by glycine in most members of subfamily 2. The division is also associated with differences in Na^+ and K^+ selectivity (Horie et al. 2001; Maser et al. 2002; Garciadeblas et al. 2003).

Gene members of subfamily 1 are all Na^+-specific transporters. Some of them are expressed in cells in the stele rather than the root cortex and regulate root-to-shoot transport of Na^+ by removing Na^+ from the xylem sap as it flows to the shoot.

Members of subfamily 2 are Na^+/K^+ cotransporters or Na^+ and K^+ uniporters, except *OsHKT2;2* (*OsHKT2*). Some of them are specifically expressed in the root cortex and may serve to scavenge Na^+ under conditions of K^+ deficiency and so provide ionic homeostasis. Under saline conditions, the expression of those genes may be downregulated. OsHKT2;1 mediated the transport of Na^+ into roots of K^+-starved plants and enhanced their growth but was downregulated when plants were exposed to 30 mM NaCl (Horie et al. 2007). TaHKT2;1 (TaHKT1) and HvHKT2;1 (HvHKT1) in wheat and barley roots mediate Na^+ uptake into roots of K^+-starved plants (Laurie et al. 2002; Haro et al. 2005).

References

Ache P, Becker D, Ivashikina N, Dietrich P, Roelfsema MRG, Hedrich R (2000) GORK, a delayed outward rectifier expressed in guard cells of *Arabidopsis thali-*

ana, is a K^+-selective, K^+-sensing ion channel. FEBS Lett 486:93–98

Ahn SJ, Shin R, Schachtman DP (2004) Expression of *KT/KUP* genes in *Arabidopsis* and the role of root hairs in K^+ uptake. Plant Physiol 134:1135–1145

Allen GJ, Sanders D (1997) Vacuolar ion channels of higher plants. Adv Bot Res 25:218–252

Arazi T, Kaplan B, Fromm H (2000) A high-affinity calmodulin-binding site in a tobacco plasma-membrane channel protein coincides with a characteristic element of cyclic nucleotide-binding domains. Plant Mol Biol 42:591–601

Assmann SM, Shimazaki K (1999) The multi-sensory guard cell: stomatal responses to blue light and abscisic acid. Plant Physiol 119:809–815

Assmann SM, Wang XQ (2001) From milliseconds to millions of years: guard cells and environmental responses. Curr Opin Plant Biol 4:421–428

Balague C, Lin B, Alcon C, Flottes G, Malmstrom S, Kohler C, Neuhaus G, Pelletier G, Gaymard F, Roby D (2003) HLM1, an essential signaling component in the hypersensitive response, is a member of the cyclic nucleotide-gated channel ion channel family. Plant Cell 15:365–379

Banuelos MA, Garciadeblas B, Cubero B, Rodriguez-Navarro A (2002) Inventory and functional characterization of the HAK potassium transporters of rice. Plant Physiol 130:784–795

Blatt MR (2000) Cellular signalling and volume control in stomatal movements of plants. Annu Rev Cell Dev Biol 16:221–241

Bridges D, Fraser M, Moorhead G (2005) Cyclic nucleotide binding proteins in the *Arabidopsis thaliana* and *Oryza sativa* genomes. BMC 6:6

Carraretto L, Formentin E, Teardo E, Checchetto V, Tomizioli M et al (2013) A thylakoid-located two pore K^+ channel controls photosynthetic light utilization in plants. Science 342(6154):114–118

Cellier F, Conejero G, Ricaud L, Luu DT, Lepetit M, Gosti F, Casse F (2004) Characterization of AtCHX17, a member of the cation/H^+ exchangers, CHX family, from *Arabidopsis thaliana* suggests a role in K^+ homeostasis. Plant J 39:834–846

Chen GQ, Cui CH, Mayer ML, Gouaux E (1999) Functional characterization of a potassium-selective prokaryotic glutamate receptor. Nature 402:817–821

Chérel I (2004) Regulation of K^+ channel activities in plants: from physiological to molecular aspects. J Exp Bot 55:337–351

Cherel I, Michard E, Platet N, Mouline K, Alcon C, Sentenac H, Thibaud JB (2002) Physical and functional interaction of the Arabidopsis K^+ channel AKT2 and phosphatase AtPP2CA. Plant Cell 14:1133–1146

Chiu JC, Brenner ED, DeSalle R, Nitabach MN, Holmes TC, Coruzzi GM (2002) Phylogenetic and expression analysis of the glutamate receptor-like gene family in *Arabidopsis thaliana*. Mol Biol Evol 19:1066–1082

Cosgrove DJ, Hedrich R (1991) Stretch-activated chloride, potassium, and calcium channels coexisting in plasma membranes of guard cells of *Vicia faba*. Planta 186:143–153

Czempinski K, Frachisse JM, Maurel C, Barbier-Brygoo H, Mueller-Roeber B (2002) Vacuolar membrane localization of the *Arabidopsis* 'two-pore' K+ channel KCO1. Plant J 29:809–820

Davies C, Shin R, Liu W, Thomas MR, Schachtman DP (2006) Transporters expressed during grape berry (*Vitis vinifera* L.) development are associated with an increase in berry size and berry potassium accumulation. J Exp Bot 57:3209–3216

Deeken R, Geiger D, Fromm J, Koroleva O, Ache P, Langenfeld-Heyser R, Sauer N, May ST, Hedrich R (2002) Loss of the AKT2/3 potassium channel affects sugar loading into the phloem of *Arabidopsis*. Planta 216:334–344

Demidchik V, Davenport RJ, Tester M (2002) Nonselective cation channels in plants. Annu Rev Plant Biol 53:67–107

Dennison KL, Robertson WR, Lewis BD, Hirsch RE, Sussman MR, Spalding EP (2001) Functions of AKT1 and AKT2 potassium channels determined by studies of single and double mutants of *Arabidopsis*. Plant Physiol 127:1012–1019

Flowers TJ, Läuchli A (1983) Sodium versus potassium: substitution and compartmentation. In: Läuchli A, Bieleski RL (eds) Inorganic plant nutrition, vol 15B. Springer, Berlin, pp 651–681

Garciadeblas B, Benito B, Rodriguez-Navarro A (2002) Molecular cloning and functional expression in bacteria of the potassium transporters CnHAK1 and CnHAK2 of the seagrass *Cymodocea nodosa*. Plant Mol Biol 50:623–633

Garciadeblás B, Senn ME, Bañuelos MA, Rodríguez-Navarro A (2003) Sodium transport and HKT transporters: the rice model. Plant J 34:788–801

Gassman W, Rubio F, Schroeder JI (1996) Alkali cation selectivity of the wheat root high-affinity potassium transporter HKT1. Plant J 10:869–952

Gaymard F, Cerutti M, Horeau C, Lemaillet G, Urbach S, Ravallec M, Devauchelle G, Sentenac H, Thibaud J-B (1996) The baculovirus/insect cell system as an alternative to *Xenopus* oocytes, first characterization of the AKT1 K^+ channel from *Arabidopsis thaliana*. J Biol Chem 271:22863–22870

Gaymard F, Pilot G, Lacombe B, Bouchez D, Bruneau D, Boucherez J, Michaux-Ferrière N, Thibaud J-B, Sentenac H (1998) Identification and disruption of a plant shaker-like outward channel involved in K^+ release into the xylem sap. Cell 94:647–655

Gierth M, Mäser P (2007) Potassium transporters in plants – involvement in K^+ acquisition, redistribution and homeostasis. FEBS Lett 581(12): 2348–2356

Gierth M, Maser P, Schroeder JI (2005) The potassium transporter AtHAK5 functions in K^+ deprivation-induced high-affinity K^+ uptake and AKT1 K^+ channel contribution to K^+ uptake kinetics in *Arabidopsis* roots. Plant Physiol 137:1105–1114

Golldack D, Quigley F, Michalowski CB, Kamasani UR, Bohnert HJ (2003) Salinity stress-tolerant and -sensitive rice (*Oryza sativa* L.) regulate *AKT1*-type potassium channel transcripts differently. Plant Mol Biol 51:71–81

Grabov A (2007) Plant KT/KUP/HAK potassium transporters: single family – multiple functions. Ann Bot 99(6):1035–1041

Hampton CR, Bowen HC, Broadley MR, Hammond JP, Mead A, Payne KA et al (2004) Cesium toxicity in *Arabidopsis*. Plant Physiol 136:3824–3837

Haro R, Banuelos MA, Senn ME, Berrero-Gil J, Rodríguez-Navarro A (2005) HKT1 mediates sodium uniport in roots: pitfalls in the expression of HKT1 in yeast. Plant Physiol 139:1495–1506

Hinnah SC, Wagner R (1998) Thylakoid membranes contain a high-conductance channel. Eur J Biochem 253:606–613

Hong JP, Takeshi Y, Kondou Y, Schachtman DP, Matsui M, Shin R (2013) Identification and characterization of transcription factors regulating arabidopsis *HAK5*. Plant Cell Physiol 54(9):1478–1490

Horie T, Yoshida K, Nakayama H, Yamada K, Oiki S, Shinmyo A (2001) Two types of HKT transporters with different properties of Na^+ and K^+ transporter in *Oryza sativa*. Plant J 27:129–138

Horie T, Costa A, Kim TH, Han MJ, Horie R, Leung H-Y, Miyao A, Hirochika H, An G, Schroeder JI (2007) Rice OsHKT2;1 transporter mediates large Na^+ influx component into K^+-starved roots for growth. EMBO J 26:3003–3014

Hoshi T (1995) Regulation of voltage dependence of the KAT1 channel by intracellular factors. J Gen Physiol 105:309–328

Hosy E, Vavasseur A, Mouline K, Dreyer I, Gaymard F, Poree F, Boucherez J, Lebaudy A, Bouchez D, Very A-A et al (2003) The *Arabidopsis* outward K^+ channel GORK is involved in regulation of stomatal movements and plant transpiration. Proc Natl Acad Sci U S A 100:5549–5554

Hua BG, Mercier RW, Leng Q, Berkowitz GA (2003) Plants do it differently. A new basis for potassium/sodium selectivity in the pore of an ion channel. Plant Physiol 132:1353–1361

Isayenkov S, Isner JH, Maathuis FJM (2011) Rice two-pore K^+ channels are expressed in different types of vacuoles. Plant Cell 23(2):756–768

Kim EJ, Kwak JM, Uozumi N, Schroeder JI (1998) *AtKUP1*: an *Arabidopsis* gene encoding high-affinity potassium transport activity. Plant Cell 10:51–62

Köhler C, Neuhaus G (2000) Characterisation of calmodulin binding to cyclic nucleotide-gated ion channels from *Arabidopsis thaliana*. FEBS Lett 471:133–136

Köhler C, Merkle T, Neuhaus G (1999) Characterisation of a novel gene family of putative cyclic nucleotide- and calmodulin-regulated ion channels in *Arabidopsis thaliana*. Plant J 18:97–104

Kohler B, Hills A, Blatt MR (2003) Control of guard cell ion channels by hydrogen peroxide and abscisic acid indicates their action through alternate signaling pathways. Plant Physiol 131:385–388

Kudla J, Batistic O, Hashimoto K (2010) Calcium signals: the lead currency of plant information processing. Plant Cell 22:541–563

Lacombe B, Pilot G, Gaymard F, Sentenac H, Thibaud JB (2000) pH control of the plant outwardly rectifying potassium channel SKOR. FEBS Lett 466:351–354

Laurie S, Feeney KA, Maathuis FJM, Heard PJ, Brown SJ, Leigh RA (2002) A role for HKT1 in sodium uptake by wheat roots. Plant J 32:139–149

Leng Q, Mercier RW, Hua BG, Fromm H, Berkowitz GA (2002) Electrophysiological analysis of cloned cyclic nucleotide-gated ion channels. Plant Physiol 128:400–410

Li F (2006) Potassium and water interaction, international workshop on soil potassium and K-fertiliser interaction, Agricultural College, Guangxi University, Nanning Guangxi-530005, November 14

Luan S (2002) Signaling drought in guard cells. Plant Cell Environ 25:229–237

Majore I, Wilhelm B, Marten I (2002) Identification of K^+ channels in the plasma membrane of maize subsidiary cells. Plant Cell Physiol 43:844–852

Marshner H (1995) Mineral nutrition of higher plants. Academic, London, pp 313–323

Marten I, Hoth S, Deeken R, Ache P, Ketchum KA, Hoshi T, Hedrich R (1999) AKT3, a phloem-localized K^+ channel, is blocked by protons. Proc Natl Acad Sci U S A 96:7581–7586

Maser P, Thomine S, Schroeder JI, Ward JM, Hirschi K, Sze H et al (2001) Phylogenetic relationships within cation transporter families of *Arabidopsis*. Plant Physiol 126:1646–1667

Mäser P, Eckelman B, Vaidyanathan R et al (2002) Altered shoot/root Na^+ distribution and bifurcating salt sensitivity in Arabidopsis by genetic disruption of the Na^+ transporter AtHKT1. FEBS Lett 531:157–161

Mouline K, Véry A-A, Gaymard F, Boucherez J, Pilot G, Devic M, Bouchez D, Thibaud JB, Sentenac H (2002) Pollen tube development and competitive ability are impaired by disruption of a Shaker K^+ channel in *Arabidopsis*. Genes Dev 16:339–350

Munns R, Tester M (2008) Mechanisms of salinity tolerance. Annu Rev Plant Biol 59:65–681

Nakamura RL, McKendree WL Jr, Hirsch RE, Sedbrook JC, Gaber RF, Sussman MR (1995) Expression of an *Arabidopsis* potassium channel gene in guard cells. Plant Physiol 109:371–374

Neuhaus HE, Wagner R (2000) Solute pores, ion channels, and metabolite transporters in the outer and inner envelope membranes of higher plant plastids. Biochim Biophys Acta 1465:307–323

Pastore D, Soccio M, Laus MN, Trono D (2013) The uniqueness of the plant mitochondrial potassium channel. BMP Rep 46(8):391–397

Petrussa E, Casolo V, Braidot E, Chiandussi E, Macri F, Vianello A (2001) Cyclosporin A induces the opening of a potassium-selective channel in higher plant mitochondria. J Bioenerg Biomembr 33:107–117

Pilot G, Lacombe B, Gaymard F, Cherel I, Boucherez J, Thibaud JB, Sentenac H (2001) Guard cell inward K^+ channel activity in *Arabidopsis* involves expression of the twin channel subunits KAT1 and KAT2. J Biol Chem 276:3215–3221

Pilot G, Gaymard F, Mouline K, Cherel I, Sentenac H (2003) Regulated expression of *Arabidopsis* Shaker K^+ channel genes involved in K^+ uptake and distribution in the plant. Plant Mol Biol 51:773–787

Platten JD et al (2006) Nomenclature for HKT transporters, key determinants of plant salinity tolerance. Trends Plant Sci 11:372–374

Pottosin II (1992) Single channel recording in the chloroplast envelope. FEBS Lett 308:87–90

Pottosin II, Schöonknecht G (1996) Ion channel permeable for divalent and monovalent cations in native spinach thylakoid membranes. J Membr Biol 152:223–233

Price MB, Jelesko J, Okumoto S (2012) Glutamate receptor homologs in plants: functions and evolutionary origins. Front Plant Sci 3:235

Rodriguez-Navarro A, Rubio F (2006) High-affinity potassium and sodium transport systems in plants. J Exp Bot 57:1149–1160

Ruan YL, Llewellyn DJ, Furbank RT (2001) The control of single-celled cotton fiber elongation by developmentally reversible gating of plasmodesmata and coordinated expression of sucrose and K^+ transporters and expansion. Plant Cell 13:47–60

Rubio F, Gassmann W, Schroeder JI (1995) Sodium-driven potassium uptake by the plant potassium transporter HKT1 and mutations conferring salt tolerance. Science 270:1660–1663

Rubio F, Santa-María GE, Rodríguez-Navarro A (2000) Cloning of *Arabidopsis* and barley cDNAs encoding HAK potassium transporters in root and shoot cells. Physiol Plant 109:34–43

Santa-Maria GE, Rubio F, Dubcovsky J, Rodriguez-Navarro A (1997) The *HAK1* gene of barley is a member of a large gene family and encodes a high-affinity potassium transporter. Plant Cell 9:2281–2289

Schachtman DP (2000) Molecular insights into the structure and function of plant K^+ transport mechanisms. Biochim Biophys Acta 1465:127–139

Schonknecht G, Spoormaker P, Steinmeyer R, Bruggeman L, Ache P, Dutta R, Reintanz B, Godde M, Hedrich R, Palme K (2002) KCO1 is a component of the slow-vacuolar (SV) ion channel. FEBS Lett 511:28–32

Schroeder JI, Allen GJ, Hugouvieux V, Kwak JM, Waner D (2001) Guard cell signal transduction. Annu Rev Plant Physiol Plant Mol Biol 52:627–658

Senn ME, Rubio F, Banuelos MA, Rodriguez-Navarro A (2001) Comparative functional features of plant potassium HvHAK1 and HvHAK2 transporters. J Biol Chem 276:44563–44569

Shabala S (2003) Regulation of potassium transport in leaves: from molecular to tissue Level. Ann Bot 92:627–634

Stephens NR, Qi Z, Spalding EP (2008) Glutamate receptor subtypes evidenced by differences in desensitization and dependence on the GLR3.3 and GLR3.4 genes. Plant Physiol 146:529–538

Su H, Golldack D, Zhao CS, Bohnert HJ (2002) The expression of HAK-type K^+ transporters is regulated in response to salinity stress in common ice plant. Plant Physiol 129:1482–1493

Subbarao GV, Ito O, Berry WL, Wheeler RM (2003) Sodium: a functional plant nutrient. Crit Rev Plant Sci 22:391–416

Szyroki A, Ivashikina N, Dietrich P, Roelfsema MRG, Ache P, Reintanz B, Deeken R, Godde M, Felle H, Steinmeyer R et al (2001) KAT1 is not essential for stomatal opening. Proc Natl Acad Sci U S A 98:2917–2921

Talke IN, Blaudez D, Maathuis FJM, Sanders D (2003) CNGCs: prime targets of plant cyclic nucleotide signalling? Trends Plant Sci 8:286–293

Uozumi N, Kim EJ, Rubio F, Yamaguchi T, Muto S, Tsuboi A, Bakker EP, Nakamura T, Schroeder JL (2000) The *Arabidopsis HKT1* gene homolog mediates inward Na^+ currents in *Xenopus laevis* oocytes and Na^+ uptake in *Saccharomyces cerevisiae*. Plant Physiol 122:1249–1259

Véry A-A, Sentenac H (2002) Cation channels in the *Arabidopsis* plasma membrane. Trends Plant Sci 7:168–175

Véry A-A, Sentenac H (2003) Molecular mechanisms and regulation of K^+ transport in higher plants. Annu Rev Plant Biol 54:575–603

Walker DJ, Leigh RA, Miller AJ (1996) Potassium homeostasis in vacuolated plant cells. Proc Natl Acad Sci U S A 93:10510–10514

Wang YH, Garvin DF, Kochian LV (2002) Rapid induction of regulatory and transporter genes in response to phosphorus, potassium, and iron deficiencies in tomato roots: evidence for cross talk and root/rhizosphere-mediated signals. Plant Physiol 130:1361–1370

Wang M, Zheng Q, Shen Q, Guo S (2013) The critical role of potassium in plant stress response. Int J Mol Sci 14:7370–7390. doi:10.3390/ijms14047370

White P, Karley A (2010) Potassium. In: Hell R, Mendel RR (eds) Cell biology of metals and nutrients. Springer, Berlin/Heidelberg, pp 199–224

Zimmermann S, Hartje S, Ehrhardt T, Plesch G, Mueller-Roeber B (2001) The K^+ channel SKT1 is co-expressed with KST1 in potato guard cells – both channels can co-assemble via their conserved K-T domains. Plant J 28:517–527

5 Calcium (Ca) Uptake

Abstract

Healthy plants growing under conditions of adequate calcium (Ca^{2+}) supply have a calcium content of 0.1–5 % of their shoot dry weight. A steady supply of Ca^{2+} is required for normal plant growth. Cytoplasmic concentration of Ca^{2+} needs to be strictly regulated at nanomolar (nM) range, though Ca^{2+} concentration in μM to mM ranges occurs in cell wall and plasma membrane externally and vacuole, endoplasmic reticulum, plastids and mitochondria internally.

Movement of Ca^{2+} is slow and its distribution unequal within the plants. The older leaves contain more Ca^{2+} than the younger ones. Since Ca^{2+} at higher concentration is cytotoxic, its movement through phloem is strictly regulated. Calcium is involved in regulating various fundamental processes such as cytoplasmic streaming, thigmotropism, gravitropism, cell division, cell elongation, cell differentiation, cell polarity, photomorphogenesis and plant defence and stress responses. Calcium also functions as a sensing and signalling molecule. Various abiotic stresses, such as cold, heat, salinity, drought, osmotic and oxidative stresses, physical stimuli – touch and swaying of the plants by wind – etc., cause transient perturbations of cytosolic Ca^{2+} concentration, which are restored to basal levels within minutes. Calcium homeostasis in cytoplasm is achieved through regulation of influx/efflux of Ca^{2+} ion by (i) calcium channels, (ii) Ca^{2+}/H^+ antiporters and (iii) Ca-ATPases.

5.1 Occurrence of Calcium and Soil Reactions

Calcium (Ca) is the fifth most abundant element and constitutes about 3.5 % of the earth's crust. Most of the soils are moderate to rich in Ca^{2+} (0.7–1.5 %), except strongly acidic tropical soils (0.1–0.3 %) and calcareous soils in arid and semiarid regions (1–30 %) (Pasricha and Sarkar 2002). The alkaline soils are rich in exchangeable Na^+ but deficient in Ca^{2+}. Primarily Ca^{2+} is present as $CaCO_3$ possibly along with dolomite [$CaMg(CO_3)_2$] and gypsum ($CaSO_4 \cdot 2H_2O$). There is an equilibrium among solution Ca^{2+}, exchangeable Ca^{2+} and mineral Ca^{2+}.

G.N. Mitra, *Regulation of Nutrient Uptake by Plants: A Biochemical and Molecular Approach*, DOI 10.1007/978-81-322-2334-4_5, © Springer India 2015

$$\text{Mineral } Ca^{2+} \rightarrow \text{Solution } Ca^{2+} \rightleftarrows \text{Exchangeable } Ca^{2+}$$

Slow dissolution of mineral Ca^{2+} enriches solution Ca^{2+} which is in equilibrium with exchangeable Ca^{2+} through sorption and desorption.

5.2 Ca Content of Plants

Healthy plants growing under conditions of adequate Ca supply have a calcium content of 0.1–5 % of their shoot dry weight (White 2003). A steady supply of 1–10 mM Ca is required for normal plant growth (Gilroy et al. 1993).

Plants have been classified according to their Ca requirement as 'calcifuges', which grow in acid soils with low Ca, and 'calcicoles', which grow in calcareous soils with high calcium. The capacity to tolerate toxic concentrations of Fe, Al and Mn determines the natural flora of acid soils, and the tolerance to deficiency of P and Fe determines the flora of calcareous soils. Calcicoles such as *Crassulaceae*, *Brassicaceae* and *Fabaceae* have a relatively higher shoot concentration of Ca. Calcifuges such as *Apiales* and *Asterales*, which are potassium plants, have low Ca concentration in their shoots with a higher K/Ca ratio (White 2003).

5.3 Functions of Calcium in Plants

Calcium is an essential element for plant nutrition and has a role in various plant processes. It is a structural component of cell. It is directly involved in various physiological and biochemical reactions and functions as a sensing and signalling molecule.

5.3.1 Calcium as a Cell Wall Constituent and Its Role in Root Cation Exchange Capacity

Calcium is a constituent of cell wall. Acidic pectin residues (galacturonic acid) are secreted as methyl esters during cell wall formation. These are subsequently de-esterified by pectin methyl esterase enzyme liberating free carboxyl groups, which bind Ca^{2+}. Rigidity of cell wall depends upon its Ca content and higher Ca content imparts more rigidity. There could be interactions between Ca and molecules other than pectin that could contribute to cell wall structure and extensibility (Hepler 2005).

Plant root CEC is located in its apoplast and is attributed to the carboxylic groups of galacturonic acid of cell wall pectins in the middle lamella (Sattelmacher 2001). Calcium content of shoot $[Ca]_{shoot}$ has been correlated with root cation exchange capacity (CEC). Variation in root CEC of monocots has been found to be parallel to pectin content of shoot cell walls (White 2001). Cell wall provides an enormous reservoir of Ca^{2+} (10 μM–10 mM) in contrast to cytosol (100–200 nM). High concentration of Ca^{2+} (0.1–1.0 mM) is required on the outer surface of plasma membrane to maintain its structural and functional integrity (Gilroy et al. 1993; Hepler 2005). Calcium probably binds to the phospholipids of lipid bilayer of plasma membrane and stabilises it. Within the cytoplasm, vacuole, which often constitutes 90–95 % of volume of cell, contains Ca^{2+} in mM concentration. Endoplasmic reticulum, mitochondria and plastids also have Ca^{2+} concentration higher than cytoplasm.

5.3.2 Movement of Calcium Within the Plant

Movement of Ca^{2+} is slow and its distribution unequal within the plants. The older leaves contain more Ca^{2+} than the younger ones. Since Ca^{2+} at higher concentration is cytotoxic, its movement through phloem is strictly regulated. Most of the Ca^{2+} transported through xylem is sequestered and locally deposited. Concentration of Ca^{2+} in xylem sap has been reported to be around 300 μM and 16.5 mM (White et al. 1992; De Silva et al. 1998) depending upon external Ca^{2+} abundance. The excessive Ca^{2+} within cytoplasm, which move to vacuole, is deposited as insoluble salts of oxalic, phosphoric and phytic acids (Borchert 1990).

5.3.3 Involvement of Calcium in Fundamental Processes in Plants

Calcium is involved in regulating various fundamental processes such as cytoplasmic streaming, thigmotropism, gravitropism, cell division, cell elongation, cell differentiation, cell polarity, photomorphogenesis and plant defence and stress responses (Song et al. 2008). It has been reported that both in flowering and nonflowering plants, cytoplasmic streaming is permitted at a low Ca^{2+} concentration of 0.1 μM, but an elevated concentration of 1 μM inhibits the process (Hepler 2005). Cytoplasmic concentration of Ca^{2+} needs to be strictly regulated at nanomolar (nM) range (100–200 nM), though it is enclosed by cell wall and plasma membrane externally and vacuole, endoplasmic reticulum, plastids and mitochondria internally with Ca^{2+} concentration in μM to mM ranges.

5.3.4 Abiotic Stress and Calcium Signature

Various abiotic stresses, such as cold, heat, salinity, drought, osmotic and oxidative stresses, physical stimuli – touch and swaying of the plants by wind – etc., cause transient perturbations of cytosolic Ca^{2+} concentration, which are restored to basal levels within minutes (White 2003; Reddy et al. 2011). According to Monshausen et al. (2009), mechanical stimuli such as touch and bending stimulate distinct pattern of Ca^{2+} responses in the roots of *Arabidopsis*. There is monophasic elevation of cytosolic Ca^{2+} at the touch site, whereas bending involves biphasic elevation of cytosolic Ca^{2+} in the cells on the convex side of the roots. Transient perturbations of cytosolic Ca^{2+} concentrations also occur in response to hormones. All such changes are triggered by cellular second messengers such as NAADP, IP3, IP6, Sphingosine-1-phosphate and cADPR (Navazio et al. 2000; Lemtiri-Chlieh et al. 2003; Kudla et al. 2010). A term 'Ca^{2+}-signature' is used to define the pattern of perturbation in cytosolic Ca^{2+} concentration in its intensity, amplitude and duration caused by physiological, developmental or environmental changes (Webb et al. 1996; White 2003; Kudla et al. 2010).

5.4 Mechanism of Calcium Uptake by Plants

Calcium homeostasis in cytoplasm is achieved through regulation of influx/efflux of Ca^{2+} ion by (i) calcium channels, (ii) Ca^{2+}/H^+ antiporters and (iii) Ca-ATPases.

5.4.1 Influx of Ca^{2+}

Influx of Ca^{2+} into cytoplasm occurs primarily through Calcium channels.

5.4.2 Calcium Channels

Calcium permeable channels are found in all plants and have been classified according to their voltage dependence as (i) hyperpolarisation-activated cation channels (HACC), (ii) depolarisation-activated cation channels (DACC) and (iii) voltage-independent cation channels (VICC). There are also outward-rectifying cation (KORC or NORC) channels (de Boer 1999) and mechano-sensitive (stretch-activated) and second messenger-activated Ca^{2+} channels (White 1998; Leng et al. 1999).

5.4.2.1 Hyperpolarisation-Activated Cation Channels (HACCs)

Hyperpolarisation-activated Ca^{2+} channels (HACCs) are found only in cells of the elongation zone such as growing root apex, a region with a high demand for Ca^{2+} uptake for cell division and cell expansion as well as in cells from the endodermis (Kiegle et al. 2000; White 2003).

HACCs have been identified in onion epidermal cells (Pickard and Ding 1993), suspension-cultured tomato cells (Blumwald et al. 1998), leaf mesophyll cells (Stoelzle et al. 2003) and stomatal guard cells (Perfus-Barbeoch et al. 2002).

These channels are permeable to many divalent cations such as Ba^{2+}, Ca^{2+}, Mg^{2+}, Mn^{2+}, Cd^{2+} and Zn^{2+}. HACCs are mechano-sensitive and regulated by cytosolic Ca^{2+} concentration.

5.4.2.2 Depolarisation-Activated Cation Channels (DACCs)

Plasma membrane of plant cells contains several types of DACCs. These are permeable to both mono- and divalent cations, which include toxic cations in addition to Ca^{2+}. The outwardly rectifying KORC channels for K^+ found in plasma membrane are also Ca^{2+} permeable DACCs. These channels catalyse a large K^+ efflux with a small Ca^{2+} influx, which might increase cytosolic Ca^{2+} concentration to coordinate ion transport, metabolism and gene expression (White 2003). DACCs contribute to short and transient influx of Ca^{2+} in response to various external stimuli such as chilling or microbial interaction (Kudla et al. 2010).

5.4.2.3 Voltage-Independent Cation Channels (VICCs)

VICCs are present in the plasma membrane of plant cells. They differ in their cation selectivity, voltage dependence and pharmacology. They are possibly the only Ca^{2+} permeable channels open at the resting potential of most of the plant cells (White 2003). They are permeable to both mono- and divalent cations.

5.4.2.4 Ligand-Gated Channels (Cyclic Nucleotide-Gated Channels, CNGCs)

Cyclic nucleotide-gated channels (CNGCs) are ligand-gated channels located in the plasma membrane. There are 20 CNGC genes identified in *Arabidopsis*. CNGCs are activated, when they bind to the cyclic nucleotides cAMP or cGMP. This biding site partially overlaps with the biding domain of calmodulin, the Ca^{2+}-binding protein. The CNGCs are inactivated, when Ca^{2+}/calmodulin binds to this domain. As a result, Ca^{2+} itself modulates its influx mediated by CNGCs (Kudla et al. 2010). Several CNGCs (CNGC4, CNGC 11, CNGC 12) are activated in response to attack by pathogens (Balague et al. 2003; Yoshioka et al. 2006; Urquhart et al. 2007). Apart from response to pathogen attack, CNGCs are involved in several other functions such as CNGC3 and CNGC10 are involved in maintaining Na^+/K^+ balance during salt stress. CNGC18 is asymmetrically localised in the tip region of growing pollen tube and regulates pollen tip growth (Kudla et al. 2010). CNGC5 and CNGC6 genes encode unique cGMP-activated non-selective Ca^{2+} permeable cation channels in the plasma membrane of *Arabidopsis* guard cells (Wang et al. 2014).

5.4.2.5 Glutamate Receptors (GLRs)

Glutamate receptors (GLRs) are non-selective cation channels involved in Ca^{2+} influx. *Arabidopsis* has 20 genes that code for GLRs, which increase cytosolic Ca^{2+} and are differentially activated by amino acids especially glutamic acid and glycine (Stephens et al. 2008; Kudla et al. 2010; Price et al. 2012). Apart from their role in Ca^{2+} nutrition, GLRs are also involved in Ca^{2+}-mediated response to cold stress (Meyerhoff et al. 2005) and Al-toxicity (Sivaguru et al. 2003). AtGLR1.1 regulates expression of enzymes involved in C and N metabolism and ABA biosynthesis in *Arabidopsis* (Kang and Turano 2003; Kang et al. 2004). Disruption of OsGLR3.1 gene in rice causes reduced growth of primary and adventitious roots especially in early seedling stage due to reduced mitotic activity of the root apical meristem (Li et al. 2006).

5.4.3 Efflux of Ca^{2+}

Calcium is present in mM concentration in vacuole, whereas its concentration in cytosol is in nanomolar range. Efflux of Ca^{2+} from cytosol is carried out by (i) Ca/H^+ antiporters, which mediate a high-affinity low turnover efflux, and (ii) P-type Ca-ATPases, which mediate a low-affinity high-capacity efflux of Ca^{2+}. Antiporters reduce signal-mediated influx of Ca^{2+} concentration by a few micromolar, whereas ATPases maintain the low resting concentration of Ca^{2+} (Hirschi et al. 1996).

5.4.3.1 Ca^{2+}/H^+ Antiporters

Calcium proton antiporters, also known as calcium exchangers (CAXs), are a group of proteins coded by six genes present in *Arabidopsis*. They regulate homeostasis of Ca^{2+} and other divalent cations such as Mn^{2+}, Zn^{2+}, Cd^{2+}, Hg^{2+} and Ni^{2+} (Zhao et al. 2008; Kudla et al. 2010). A steep pH gradient exists across the vacuolar membrane, the tonoplast. While its cytosolic side maintains the physiologic pH, the vacuole has a significantly lower pH of 4–5. A pH gradient is established across tonoplast by proton pumps such as H^+- ATPase or H^+- pyrophosphatase. The CAX transporters take advantage of this pH gradient to move cations from cytoplasm to vacuole in exchange for H^+, which is present abundantly inside it (Kamiya and Maeshima 2004). The CAXs have been put into three groups based on their phylogenetic relationship. All of the plant CAXs are grouped as type I CAXs. They have approximately 400 amino acids and 11 transmembrane domains (TMs). The polypeptide is divided into two half proteins by a short negatively charged loop between TM6 and TM7 termed as 'acidic motif' (Ivey et al. 1993). Type II CAXs are found in animals and also in fungi and protozoa. Type III CAXs are found exclusively in bacteria (Shigaki et al. 2006).

CAX1 from *Arabidopsis* is low-affinity and high-capacity Ca^{2+} transporter localised in the vacuolar membrane. Its activity is regulated by an N-terminal auto-inhibitory domain (36 amino acids). *Arabidopsis* has 12 Ca^{2+}/H^+ antiporters (CAX1–CAX11 and MHX). They have a 9-amino-acid Ca^{2+} domain which exists in the hydrophilic loop between TM1 and TM2 (Song et al. 2008). CAXs similar to CAX1 of *Arabidopsis* are also found in other plants, bacteria and fungi. VCAX1 involved in Ca^{2+} homeostasis has been found in the tonoplast of mung bean (Ueoka-Nakanishi et al. 2000). 'OsCAX1a' present in tonoplast of rice has been reported to be involved in transport of Ca^{2+} and Mn^{2+}. GmCAX1 present in plasma membrane of soybean is involved in transport of Na^+, K^+ and Li^+ (Luo et al. 2005). OsCAX3 is present in the plasma membrane of rice and involved in Ca^{2+} homeostasis (Qi et al. 2005). Zhao et al. (2009) reported interaction between CAX1, Ca^{2+} and CAX3, which form a 'hetero-CAX' with unique transporter function in response to stress and during flowering and seed germination. Autoinhibition of CAX1 could be overcome through such hetero-CAX formation.

5.4.3.2 P-Type Ca^{2+}ATPase

P-type Ca^{2+}ATPases are part of the P-type ATPase superfamily of ion pumps energised by hydrolysis of ATP and specific to a large number of cations (they form a phosphorylated intermediate and hence are named P-types). There are two classes of P-type Ca^{2+}ATPases, (i) P2A- or ER-type Ca^{2+}ATPases named ECAs (4 members in *Arabidopsis*) and (ii) P2B Ca^{2+}ATPases (10 members in *Arabidopsis*, all of which contain an auto-inhibitory N-terminal domain, hence named as ACAs). P2A-type Ca^{2+}ATPases are located (Kudla et al. 2010) in the ER (ECA1, Liang et al. 1997), the Golgi (ECA3, Mills et al. 2008) and endosomes (ECA3, Li et al. 2008). The P2B Ca^{2+}ATPases are localised in the ER (ACA2, Harper et al. 1998), vacuole (ACA4 and ACA11, Geisler et al. 2000; Lee et al. 2007) and plasma membrane (ACA8, ACA9 and ACA10, Bonza et al. 2000; Schiott et al. 2004 and George et al. 2008) and at the plastid envelop (ACA1, Huang et al. 1993).

Plant P-type ATPases have 8–12 transmembrane domains with N and C terminals exposed to cytoplasm and a large cytoplasmic segment, which contains the phosphorylation and ATP binding sites (Axelsen and Palmgren 2001).

5.4.4 Ca^{2+} Sensing and Signalling

5.4.4.1 EF Hands

A large set of calcium-binding proteins in plants acts as cellular Ca^{2+} sensors and as first information translation point (Luan et al. 2002; Batistič and Kudla 2004; McCormack et al. 2005; Kim et al. 2007; Kudla et al. 2010). These proteins have one or more highly conserved Ca^{2+} binding helix-turn-helix structures known as EF hands, which bind Ca^{2+} with high affinity (Strynadka and James 1989; White 2003). Pairs of EF hands may interact through antiparallel β-sheets, which cooperatively bind Ca^{2+}.

5.4.4.2 Types of EF Hands

There are one to six EF hands in Ca^{2+} sensors (Reddy et al. 2011). The EF hand containing sensors are of two types, (i) sensor relays and (ii) sensor responders (Sanders et al. 2002).

Sensor Relays

Sensor relays such as calmodulins (CaMs), CaM-like proteins (CMLs) and calcineurin B-like proteins (CBLs) undergo calcium-induced conformational change (sensing) that is relayed to an interacting partner, which responds with changes in its enzyme activity or structure (e.g. calmodulin stimulation of an ACA pump activity).

Sensor Responders

Sensor responders undergo a calcium-induced conformational change that alters protein's own activity or structure such as Ca^{2+}-dependent protein kinases (CDPKs), Ca^{2+} and Ca^{2+} CaM-dependent protein kinases (CCaMKs), some DNA- or lipid-binding proteins and a few enzymes (Harper and Harmon 2005; Reddy et al. 2011).

5.4.4.3 Calcium-Binding Proteins

Calmodulin (CaM)

Calmodulin (CaM) is a Ca^{2+}-binding protein found in apoplast, cytosol, ER and nucleus of plant cells. CaM concentration in cytosol is about 5–40 μM (Zielinski 1998). CaM is involved in Ca^{2+}-dependent responses to light, gravity, mechanical stress, phytohormones, pathogens, osmotic stress, salinity, heavy metals, xenobiotics, anoxia, oxidative stress, heat shock and chilling (Zielinski 1998; Snedden and Fromm 2001; Reddy 2001; Rudd and Franklin-Tong 2001; Fasano et al. 2002; White 2003). CaM is a small acidic protein (17 kDa), highly conserved, with two globular domains each containing two EF hands connected by a flexible α-helical spacer (Zielinski 1998; Reddy 2001; Snedden and Fromm 2001; Luan et al. 2002; White 2003). Several CaM genes encode identical proteins and other genes encode its isoforms (Zielinski 1998; Snedden and Fromm 2001). Seven genes in *Arabidopsis* encode CaM isoforms. CAM2/3/5/6 differ from CAM7 by substitution of one amino acid, whereas CAM1/4 by four amino acids (McCormack et al. 2005). CAM7 (not CAM 2/3/5) has been reported to act as a transcriptional regulator, which directly interacts with promoters of several light inducible genes (Kudla et al. 2010).

CaM-Like Proteins (CMLs)

Plants contain CaM-like proteins, which have limited homology to CaM (less than 75 % homology to canonical CaM isoforms). They have one to six EF hands (Luan et al. 2002; Zielinski 2002; White 2003).

In *Arabidopsis* these proteins consist of CaBP 22 (Ling and Zielinski 1993), TCH2, TCH3 (Braam et al. 1997), AtCP1 (Jang et al. 1998), centrins (Cordeiro et al. 1998), NADPH oxidase (Torres et al. 1998), homologues of rice ABA-inducible EFA27 (Frandsen et al. 1996) and Ca^{2+}-binding protein phosphatases such as ABI1 and ABI2 (Leung et al. 1997).

Calcineurin B-Like Proteins (CBLs)

CBLs have three EF hands (Luan et al. 2002). There are at least 10 *AtCBL* genes, which code for Ca^{2+} sensor proteins involved in salt tolerance (Luan et al. 2002; Xiong et al. 2002). There is induction of expression of *AtCBL1* gene in response to drought, cold, wounding and salinity (Kudla et al. 1999; Piao et al. 2001). There is accumulation of transcripts of AtCBL1 and AtCBL2 in response to illumination (Nozawa et al. 2001).

Calcium-Dependent Protein Kinases (CDPKs)

CDPKs can be grouped into four classes: (i) Ca^{2+}-dependent protein kinases (CDPKs), (ii) CDPK-related proteins (CRKs), (iii) CaM-dependent protein kinases (CaMKs) and (iv) chimeric Ca^{2+} CaM-dependent protein kinases (CCaMKs). CDPKs are ubiquitous in plants.

There are at least 34 genes in *Arabidopsis* genome, which encode CDPKs. A similar number of genes encode CDPKs in other plants (Harmon et al. 2001; Cheng et al. 2002; White 2003). They generally have four EF hands at their C-terminus, which bind Ca^{2+} to activate their serine/threonine kinase activity. Individual CDPKs differ in their affinities for Ca^{2+} (Lee et al. 1998). CDPKs are not integral part of membrane proteins. They are however associated with ER, cytoskeleton, nucleus and plasma membrane. CDPKs convert cytosolic Ca^{2+} signals into biochemical and genetic responses through phosphorylation of different target proteins, including membrane solute transporters (Ca^{2+} ATPases AtACA2), ion and water channels, NADPH oxidases, proteases and DNA-binding proteins (Reddy 2001; Rudd and Franklin-Tong 2001; Cheng et al. 2002; Sanders et al. 2002). Specific CDPKs are induced in different plants as a response to various types of stresses, such as cold, drought, salinity, anoxia and mechanical intrusion, wounding and pathogen elicitors (Saijo et al. 2000; Anil et al. 2000; Romeis et al. 2001; Chico et al. 2002; Lee et al. 2003). There are other protein kinases such as CCaMKs with CaM-binding domains and three EF hands. They require Ca^{2+} for auto-phosphorylation but Ca^{2+} and CaM for substrate phosphorylation. CCaMKs are present in legume, maize, tobacco and other plants but not in *Arabidopsis* (DeFalco et al. 2010).

Ca^{2+}-Binding Proteins Without EF Hands

Ca^{2+}-binding proteins without EF hands include annexins, calreticulin, calsequestrin, calnexin and BiP. These proteins are involved in Ca^{2+} homeostasis, protein folding and post-translational modifications (Crofts and Denecke 1998; Michalak et al. 1998).

5.4.5 Ca^{2+}-Regulated Gene Expression and Abiotic Stress Responses

It is evident from several global studies that reprogramming of transcriptome is an important part of stress signalling and adaptation (Reddy et al. 2011). Transcriptomic changes are brought about by changes in gene expression, which are regulated by transcription factors (Brivanlou and Darnell 2002). It has been reported from several studies that perturbation in cellular or nuclear Ca^{2+} levels modulates gene expression (Braam 1992; van Der Luit et al. 1999; Kaplan et al. 2006; Reddy et al. 2011).

5.4.5.1 Mechanisms of Gene Expression by Signal-Induced Cellular Ca^{2+} Level

The different mechanisms of gene expression by signal-induced cellular Ca^{2+} level have been described by Reddy et al. (2011) as follows:

(i) Activated Ca^{2+} sensors (Ca^{2+} CaM and Ca^{2+} CML) may directly bind to *cis* elements in the promoter of specific gene and induce or repress their expression.
(ii) Activated Ca^{2+} sensors may bind to DNA-binding proteins and activate or inactivate them resulting in expression or repression of gene expression.
(iii) Cellular-elevated Ca^{2+} may activate Ca^{2+}-regulated protein kinases (CDPKs, CBKs [CaM-binding protein kinases] or CCaMK [Ca^{2+} CaM-binding protein kinases]) and phosphatases, which in turn phosphorylate/dephosphorylate specific DNA-binding region.

Increase in levels of extracellular Ca^{2+} results in increase in expression of several genes including those involved in encoding Ca^{2+} sensors. Expression of some genes in response to heat or cold shock also depends on external Ca^{2+} concentration (Braam 1992; Polisensky and Braam 1996). Bioinformatic analysis of *Arabidopsis* genome indicates the presence of 230 Ca^{2+}-responsive genes, of which 162 are upregulated and 68 downregulated. A significant occurrence of two consensus ABRE (abscisic acid-responsive element) *cis* elements (CACGTG [T/C/G]) and its coupling element ([C/A] ACGCG [T/C/G]) has been found (Kaplan et al. 2006). It has been observed from kinetic studies that Ca^{2+}-responsive genes reach their maximum expression within 30 minutes in response to a stimulus (Kaplan et al. 2006).

5.4.5.2 Ca^{2+}- and CaM-Binding Transcription Factors

More than 2,000 proteins (>7 % of total proteome) of *Arabidopsis* genome have been identified as possible DNA-binding TFs. These are classified into 58 families according to their DNA-binding domains and other conserved motifs (Zhang et al. 2011). Functions of many of these are yet to be discovered. About half of them are found in plants (Riechmann et al. 2000). Among them about 90 calcium-binding proteins (CBPs) are grouped into 10 families; all members of some families are CaM binding (CAMTAs), whereas members of other families (WRKY, Myb, etc.) interact with CaMs or CMLs.

Calmodulin-Binding Transcription Factors (CAMTA Proteins)

Calmodulin-binding transcription factors (CAMTA proteins) may partly convert Ca^{2+} signatures into transcriptional response (Kudla et al. 2010). A family of six members of CAMTAs is found in *Arabidopsis*. All of them share conserved structural domains including an N-terminal CG-1 domain, which binds to DNA cis element (CAMTA-binding sites) and also abscisic acid-responsive elements (ABREs). The C-terminal-binding domain of CAMTA interacts with calmodulin (Finkler et al. 2007).

CAMTA transcript levels are induced by cold and heat treatment (CAMTA1, CAMTA3 to CAMTA6) and salinity (CAMTA1 to CAMTA4 and CAMTA6) (Yang and Poovaiah 2002). Individual CAMTAs are involved in multiple signal transduction pathways and stress responses.

MYB Family

Several members of MYB family TFs have been found to bind Ca^{2+}/CAM (Popescu et al. 2007). MYB TFs have structurally conserved MYB domain, which contains up to four imperfect repeats (R) of about 52 amino acids. Plant MYB are grouped into four classes according to number of repeats (R) as 4R-MYB, 3RMYB, R2R3MYB (double repeats) and MYB-related (containing single or partial repeats). It has been reported that a soybean (*Glycine max*) CaM, Gm-Cam4, mediates Ca^{2+} signalling response by activating R2R3-MYB2 TF (Yoo et al. 2005).

WRKY Family

Members of WRKY TF family are also activated by Ca^{2+}/CaM. WRKYs have a conserved DNA-binding domain (WRKYGQK) and an atypical Zn^{2+} finger structure. WRKY TFs bind specifically to W-box DNA *cis* element (C/T) TGAC (C/T) (Eulgem and Somssich 2007). Several WRKYs specifically of IId subfamily have been found to bind different isoforms of CAM in *Arabidopsis* genome (Popescu et al. 2007) (WRKYs are grouped into three families, I, II and III. The group II members are further subdivided into five subgroups IIa, IIb, IIc, IId and IIe) (Reddy et al. 2011).

Other Ca^{2+}-/CAM-Binding TFs

Other Ca^{2+}-/CAM-binding TFs found in plants include basic leucine zipper (bZIP) TFs, MADS box TFs, four scarecrow like TFs, two NAM TFs, etc. (Reddy et al. 2011).

At least two directly *Ca^{2+}-binding TFs* have been identified, which do not involve CaM or CMLs (Reddy et al. 2011). These are NIG1 (NaCl INDUCED GENE) TF and AtCAM7 TF in *Arabidopsis*.

5.4.5.3 Ca^{2+}-Regulated Gene Expression in Response to Some Specific Abiotic Stress

Drought

Physiological response of plants to water stress conditions includes increasing efficiency of water uptake from soil, conserving water within the cells and reducing transpiration loss by regulating closure of stomata (Yang et al. 2010). It is reported that more than 95 % of water translocated through plants exit through the stomatal pores, which are also involved in uptake of CO_2 for use in photosynthesis. Cytosolic Ca^{2+} regulates closure of stomata by two mechanisms: (i) short-term Ca^{2+}-reactive closure and (ii) long-term Ca^{2+}-programmed closure (Allen et al. 2001; Sanders et al. 2002).

(i) Short-term Ca^{2+}-reactive closures are rapid reactions induced by cytosolic Ca^{2+}, when it exceeds a threshold limit.
(ii) Long-term Ca^{2+}-programmed closure, which involves prevention of stomatal reopening, is controlled by specific Ca^{2+} signature: Ca^{2+} oscillation within a defined range of amplitude, frequency, duration and overall transient number (Kudla et al. 2010).

Exogenous Ca^{2+} has been reported to enhance drought resistance, inhibit synthesis of activating oxides, protect the structure of plasma membrane, maintain normal photosynthesis and regulate the metabolism of plant hormones. Cellular Ca^{2+} as a second messenger transmits drought signal and induces physiological response to water stress (Zhang et al. 2001; Tuberosa et al. 2007; Song et al. 2008). Ca^{2+}/CaM messenger system is reported to be involved in controlling stress resistance of rice seedlings; blocking messenger transduction, drought resistance and salt tolerance; and decreasing cold resistance (Zong et al. 2000). Ca^{2+} treatment of rice seedlings increases protection against membrane lipid peroxidation, stabilises membranes and increases their drought resistance (Lu et al. 1993).

Microarray analysis of *Arabidopsis* genome shows that several hundred genes are expressed in a specific pattern due to water deficiency in plants (Seki et al. 2002; Yamaguchi-Shinozaki and Shinozaki 2006; Reddy et al. 2011). Such expressions are induced by many Ca^{2+}-binding proteins (protein kinases/phosphatases) and TFs (AREBs and DREBs), chaperones and molecules involved in osmo-protectant metabolism (Reddy et al. 2011).

The synthesis of phytohormone ABA is induced under water stress conditions. The increased levels of ABA signal closure of guard cells and induce expression of drought stress-related genes. These genes encode proteins, which provide dehydration tolerance to plants (Reddy et al. 2011). ABA may regulate ABA-responsive genes through cellular Ca^{2+} changes (Kaplan et al. 2006). It is reported that in the presence of Ca^{2+}, the overexpression of TaTPC1 (which functions in Ca^{2+} import in wheat cytosol) accelerates stomatal closing (Wang et al. 2005).

Cold

Ca^{2+}-permeable channel proteins have been reported to be primary temperature sensors in plants and are involved in plant response to cold stress (Plieth et al. 1999). It has been observed in alfalfa, barley and *Arabidopsis* that Ca^{2+} influx acts as signal transduction element for gene expression at low temperature (Plieth et al. 1999; Busconi et al. 2001). Cold acclimation by temperate plants involves changes in gene expression (Fowler and Thomashow 2002; Kreps et al. 2002; Reddy et al. 2011). A large number of genes of CBF regulon are induced during the process of cold acclimation. These genes are activated by transcription factors, C-repeat-binding factors and CBF 1, 2 and 3 also called DREB 1B, 1C and 1A, respectively (Riechmann et al. 2000; Maruyama et al. 2004; Sakamoto et al. 2004; Vogel et al. 2005; Reddy et al. 2011). The induction of KIN1 a member of CBF regulon due to cold requires a rapid increase of cytosolic Ca^{2+} (Monroy et al. 1997). A number of cold-responsive genes contain CAMTA-binding sequence CGCG and may be regulated transcriptionally by CAMTA proteins on exposure to cold (Doherty et al. 2009).

Heat

Plants in response to higher temperature synthesise heat shock proteins (HSPs), a number of which have been characterised. Their transcription is tightly regulated by TFs. Elevation of cellular Ca^{2+} due to heat changes expression of several genes including Ca^{2+} sensors (Braam 1992; Zhang et al. 2009). Overexpression of a CaM-binding phosphatase (PP7) in *Arabidopsis* has been found to increase expression of heat shock proteins and provide thermotolerance. A CaM-binding protein kinase (CBK) in *Arabidopsis* phosphorylates heat shock TF (At HSFA1a) and regulates transcription of HSPs, which provide thermotolerance (Liu et al. 2007). CAMTA1 is also involved in heat shock response (Galon et al. 2010).

Salt

Low pH and high salinity cause greater damage to plants under Ca^{2+}-deficient conditions. External and apoplastic Ca^{2+} directly alleviate symptoms produced by ion stresses and mineral toxicities, such as proton, Al^{3+} and Cl^- toxicities, and help maintain a favourable K^+/Na^+ balance in the plants under conditions of salt stress (Plieth 2005; Song et al. 2008).

A large number of genes are activated on exposure to salinity, including ion channels, receptors, signalling molecules and genes involved in producing compatible molecules such as osmo-protectants, glycine betaine and proline (Tuteja 2007; Reddy et al. 2011). The salt stress-mediated Ca^{2+} signatures are decoded by 'salt overly sensitive' (SOS) pathway. Under saline conditions SOS1, a plasma membrane-localised Na^+/H^+ antiporter exports Na^+ to the apoplast. The SOS3 (CBL4)/SOS2 (CIPK24) complex modulates the expression of SOS1 and regulates ion homeostasis (Chinnusamy et al. 2004; Mahajan et al. 2008; Reddy et al. 2011).

Saline stress and other abiotic and biotic stress upregulate a number of CAMTA family TFs (Galon et al. 2010). Salt-induced Ca^{2+} signalling has also been found to activate MYB2 TF, which is an upstream regulator of a number of salt- and dehydration-responsive genes (Yoo et al. 2005). A soybean CaM isoform induced by salt stress is Gm-CaM4. Overexpression of Gm-CaM4 induces constitutive expression of salt- and dehydration-responsive genes, including proline-synthesising enzyme P5CS1 (Δ-1-pyrroline-5-carboxylate synthetase-1), which facilitates proline accumulation and provides protection against salt stress (Yoo et al. 2005).

GTL1 (GT-2 lIKE-1) TF a CaM-binding member of GTL family downregulates drought resistance. Water stress represses expression of GTL1.

Mechanical Stimuli

Mechanical stimuli such as touch and wind induce elevation of cytosolic Ca^{2+} concentration (Braam 2005). Different types of mechanical stimuli induce distinct type of Ca^{2+} response in *Arabidopsis* roots. Touch stimuli induce monophasic elevation of cytosolic Ca^{2+} concentration at the touch site. Bending induces biphasic transient elevation of cytosolic Ca^{2+} concentration on the convex (stretching) side (Fig. 5.1).

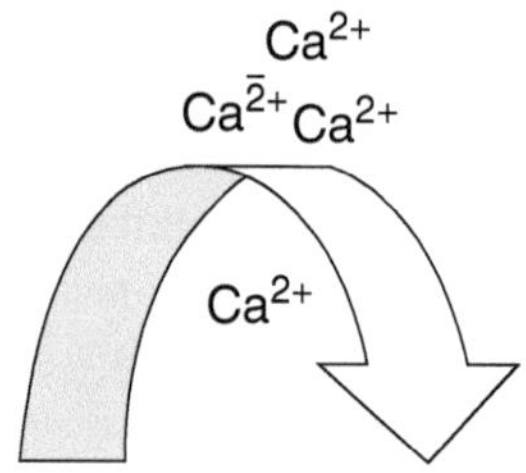

Fig. 5.1 High cytosolic Ca^{2+} concentration on the convex side

Such responses are essential for the apoplastic alkalisation and expression of membrane-localised NADPH oxidase enzyme, RBOH C, which has been shown to contribute to ROS production related to root hair elongation (Monshausen et al. 2009).

Mechanical stimuli induce expression of several CaM and CaM-related genes (Braam et al. 1997; van Der Luit et al. 1999; Walley and Dehesh 2010; Reddy et al. 2011). Mechanical stress-induced transcriptomic study and bioinformatic analysis of data identified an over-represented cis element 'CGCGTT' termed as rapid stress-response element (RSRE) in the promoter region of rapid wound-responsive genes (Walley et al. 2007). This cis element contains the CAMT core cis element 'CGCG'. This indicates that CAMTAs are probably involved in stress response to wounding (Walley et al. 2007; Walley and Dehesh 2010; Reddy et al. 2011).

5.4.6 Biotic Stress

5.4.6.1 Ca Signature and Early Perception of Pathogen Attack

PAMP Molecules

It is now well established that perception of pathogen or conserved components of microbial cells [pathogen-associated molecular pattern (PAMP) molecules] induces influx of Ca^{2+} across plasma membrane resulting in increase in levels of cytosolic and/or nuclear Ca^{2+} levels. This constitutes an early signalling mechanism of pathogen attack (Lecourieux et al. 2002, 2006; Hu et al. 2004; Ali et al. 2007; Ma et al. 2007, 2008). Such changes in Ca^{2+} signatures also occur during root nodulation by symbiotic nitrogen-fixing bacteria and for fungal mycorrhizal associations with roots, to acquire P (Shaw and Long 2003; Lecourieux et al. 2006; Kosuta et al. 2008).

Elicitors

Ca^{2+} signatures are triggered by a number of elicitors, which consist of a variety of compounds such as proteins (cryptogein, Avr2, Avr4, Avr5, Avr9, Pep13), oligogalacturonides, chitosans, β-heptaglucosans, lipopolysaccharides, xylanases and BcPG1 (Garcia-Brugger et al. 2006). Elicitors are a group of diverse compounds, which are constituents of pathogens or secreted by them, or are released by cell walls of plant or pathogen by hydrolytic enzymes from the pathogen or plant (Garcia-Brugger et al. 2006).

Elicitor perception is rapidly followed by Ca^{2+} influx and intracellular Ca^{2+} signal resulting in protein kinase (PK) activation. Elicitors are readily recognised by receptor-like kinases (RLKs) located in cytoplasm or plasma membrane (Garcia-Brugger et al. 2006).

Pathogen-induced cytoplasmic Ca^{2+} influx takes place through various channels, such as plasma membrane-localised cyclic nucleotide-gated channels (CNGC), a family of CaM and cyclic nucleotide binding ion channel and possibly other channels, pumps and transporters (Ali et al. 2007; Ma et al. 2007, 2008, 2009; Reddy et al. 2011). It has been established that the same elicitors that induce Ca^{2+} signatures also induce defence-related genes at the transcriptional level (DeFalco et al. 2010; Reddy et al. 2011).

5.4.6.2 CaMs and Plant Pathogen Signalling

It has been reported that CaM and/or a CML is involved in plant pathogen signalling and innate immune response (Heo et al. 1999; Chiasson et al. 2005; Takabatake et al. 2007). In soybean constitutive expression of CaM (SCaM-4 and SCaM-5) enhances resistance to a broad spectrum of virulent and avirulent pathogens (Heo et al. 1999). According to Takabatake et al. (2007), 13 CaM genes of tobacco fall into three groups according to their amino acid homology. Wound-inducible type I isoforms NtCaM1 and NtCaM2 are moderately induced by tobacco mosaic virus (TMV)-mediated hypersensitive reaction (HR). Type II isoforms NtCaM3–NtCaM12 show little response, but type III isoform NtCaM13 is highly induced by TMV-mediated HR. They suggest that type III isoforms are probably involved in providing basal defence against necrotrophic pathogens in tobacco. Chiasson et al. (2005) report that CML43 gene in *Arabidopsis* and APR134 gene of tomato-encoding CaM/CML proteins are important mediators of Ca^{2+}-dependent signals during the plant immune response to bacterial pathogens. These studies indicate that CaMs play a critical role in transducing pathogen-related Ca^{2+} signature to downstream component of plant defence signalling (Reddy et al. 2011).

5.4.6.3 CaM Interacting Transcription Factor (TF) Families Involved in Plant Defence Against Pathogen Infection

CaM interacting transcription factor (TF) families such as CAMTA, WRKY, bZIP (TGA) and CBP60s are involved in plant defence against pathogen infection. The plant-specific CBP60s have been isolated from maize (Reddy et al. 1993), tobacco (Lu and Harrington 1994), *Arabidopsis* (Reddy et al. 2002) and bean (*Phaseolus vulgaris*; Ali et al. 2003). CBP60s are involved in modulating expression of defence genes against pathogen attack. It has been reported that different types of pathogens activate distinct defence pathway involving specific regulators (Ton et al. 2002). Necrotrophic pathogens appear to activate ethylene (ET) and jasmonic acid (JA)-dependent defence response. Biotrophic pathogens also induce salicylic acid (SA)-dependent defence response (Thomma et al. 2001). These pathways mutually inhibit each other, which indicate that there is crosstalk among them. Plants can adapt a specific pathway depending on the type of pathogen attack (Reymond and Farmer 1998; Spoel et al. 2003).

CBP60s are differentially expressed in response to biotic stress and elicitors. According to Zhang et al. (2010), CBP60g (a CaM interacting TF) and SARD1 (salicylic acid-deficient resistance1) represent plant-specific family of DNA-binding proteins. Both proteins are induced in response to pathogen infection and bind to the promoter region of ICS1 (isochorismate synthase1) involved in the synthesis of salicylic acid and control SA synthesis at transcriptional level.

Calmodulin-binding transcription factors (CAMTAs) as stated earlier are involved in plant response to abiotic and biotic stress. All CAMTAs are nuclear localised (Galon et al. 2010) and are induced upon wounding (Yang and Poovaiah 2002). Phytohormones and secondary messengers, which mediate plant response to biotic and abiotic stress, also regulate expression of CAMTAs. Abscisic acid regulates expression of (CAMTA2 and CAMTA4 to CAMTA6) methyl jasmonate (CAMTA1, 3 and 4), ethylene (CAMTA1, 3 and 4), H_2O_2 (CAMTA2-6), salicylic acid (CAMTA2 and CAMTA4-6) and auxin (CAMTA1) (Yang and Poovaiah 2002; Galon et al. 2010; Reddy et al. 2011).

WRKY TFs either positively or negatively regulate plant immunity to pathogens (Pandey and Somssich 2009). The CaM-binding TFs are also positive or negative regulators of plant defence. Such positive and negative regulations appear to be dependent upon stages of disease and lifestyle of pathogen. Majority of positive regulation of WRKYs become active at the early stages of pathogen attack and the negative regulation of WRKYs at the later stages of infection (Reddy et al. 2011).

5.4.6.4 CDPKs and MAMP Signalling

Specific CDPKs (Ca-dependent protein kinases) play a critical role in the initial MAMP (microbe-associated molecular pattern) signalling. According to Boudsocq et al. (2010) (using a functional genomic screen and genome-wide gene expression profiling) four CDPKs (CDPK4, 5, 6 and 11) are Ca^{2+}-sensor protein kinases critical for transcriptional reprogramming in plant innate immune signalling. CDPKs and MAPK (mitogen-activated protein kinase) cascades act differentially in four MAMP-mediated regulatory programmes to control early genes involved in the synthesis of defence peptides and metabolites, cell wall modifications and redox signalling. Transcriptome profile comparison suggests that CDPKs are the convergence point of signalling triggered by most MAMPs. PAMP (pathogen-associated molecular pattern) perception induces CDPKs by regulating Ca^{2+}-influx channels (Ma et al. 2009; Kwaaitaal et al. 2011; Rasmussen et al. 2012). It is reported that Ca^{2+}-ATPases regulate Ca^{2+} efflux and control innate immune defences (Zhu et al. 2010).

Some of the Ca^{2+}- and CaM-binding proteins are not TFs but are involved in plant defence against pathogens. A calmodulin-binding MLO protein located in the plasma membrane of barley plays a sensor role in modulating plant defence against powdery mildew.

The involvement of Ca^{2+}-signalling pathway in plant defence against pathogens is complex and comprises of several cascades of interrelated reactions, which result in either induction or repression of specific genes, which are involved in coding defence-related proteins.

References

Ali GS, Reddy VS, Lindgren PB, Jakobek JL, Reddy AS (2003) Differential expression of genes encoding calmodulin-binding proteins in response to bacterial pathogens and inducers of defense responses. Plant Mol Biol 51:803–815

Ali R, Ma W, Lemtiri-Chlieh F, Tsaltas D, Leng Q, von Bodman S, Berkowitz GA (2007) Death don't have no mercy and neither does calcium: *Arabidopsis* CYCLIC NUCLEOTIDE GATED CHANNEL2 and innate immunity. Plant Cell 19:1081–1095

Allen GJ, Chu SP, Harrington CL, Schumacher K, Hoffmann T, Tang YY, Grill E, Schroeder JI (2001) A defined range of guard cell calcium oscillation parameters encodes stomatal movements. Nature 411:1053–1057

Anil VS, Harmon AC, Sankara Rao K (2000) Spatio-temporal accumulation and activity of calcium-dependent protein kinases during embryogenesis, seed development, and germination in sandalwood. Plant Physiol 122:1035–1043

Axelsen KB, Palmgren MG (2001) Inventory of the superfamily of P-type ion pumps in Arabidopsis. Plant Physiol 126:696–706

Balague C, Lin B, Alcon C, Flottes G, Malmstrom S, Kohler C, Neuhaus G, Pelletier G, Gaymard F, Roby D (2003) HLM1, an essential signaling component in the hypersensitive response, is a member of the cyclic nucleotide-gated channel ion channel family. Plant Cell 15:365–379

Batistič O, Kudla J (2004) Integration and channeling of calcium signaling through the CBL calcium sensor/CIPK protein kinase network. Planta 219:915–924

Blumwald E, Aharon GS, Lam BC-H (1998) Early signal transduction pathways in plant-pathogen interactions. Trends Plant Sci 3:342–346

Bonza MC, Morandini P, Luoni L, Geisler M, Palmgren MG, De Michelis MI (2000) At-ACA8 encodes a plasma membrane localized calcium-ATPase of *Arabidopsis* with a calmodulin-binding domain at the N terminus. Plant Physiol 123:1495–1506

Borchert R (1990) Ca^{2+} as a developmental signal in the formation of Ca-oxalate crystal spacing patterns during leaf development in *Carva ovate*. Planta 182:339–347

Boudsocq M, Willmann MR, McCormack M, Lee H, Shan L, He P, Bush J, Cheng SH, Sheen J (2010) Differential innate immune signalling via Ca($^{2+}$) sensor protein kinases. Nature 464(7287):418–422

Braam J (1992) Regulated expression of the calmodulin-related TCH genes in cultured *Arabidopsis* cells: induction by calcium and heat shock. Proc Natl Acad Sci U S A 89:3213–3216

Braam J (2005) In touch: plant responses to mechanical stimuli. New Phytol 165:373–389

Braam J, Sistrunk ML, Polisensky DH, Xu W, Purugganan MM, Antosiewicz DM, Campbell P, Johnson KA (1997) Plant responses to environmental stress: regulation and functions of the *Arabidopsis* TCH genes. Planta 203:S35–S41

Brivanlou AH, Darnell JE Jr (2002) Signal transduction and the control of gene expression. Science 295:813–818

Busconi M, Bosco CD, Crosatti C, Baldi P, Marè C, Grossi M, Mastrangelo AM, Rizza F, Cattivelli L, Stanca AM (2001) The cold-regulated genes are involved in the physiological response of barley to cold environment. ICL Agri Sci 14:17–27

Cheng S-H, Willmann MR, Chen H-C, Sheen J (2002) Calcium signaling through protein kinases. The *Arabidopsis* calcium-dependent protein kinase gene family. Plant Physiol 129:469–485

Chiasson D, Ekengren SK, Martin GB, Dobney SL, Snedden WA (2005) Calmodulin-like proteins from *Arabidopsis* and tomato are involved in host defense against *Pseudomonas syringae* pv. Tomato. Plant Mol Biol 58:887–897

Chico JM, Raíces M, Téllez-Iñón MT, Ulloa RM (2002) A calcium-dependent protein kinase is systemically induced upon wounding in tomato plants. Plant Physiol 128:256–270

Chinnusamy V, Schumaker K, Zhu JK (2004) Molecular genetics perspectives on cross-talk and specificity in abiotic stress signalling in plants. J Exp Bot 55:225–236

Cordeiro MCR, Piqueras R, de Oliveira DE, Castresana C (1998) Characterization of early induced genes in *Arabidopsis thaliana* responding to bacterial inoculation: identification of centrin and of a novel protein with two regions related to kinase domains. FEBS Lett 434:387–393

Crofts AJ, Denecke J (1998) Calreticulin and calnexin in plants. Trends Plant Sci 3:396–399

De Boer AH (1999) Potassium translocation into the root xylem. Plant Biol 1:36–45

DeFalco TA, Bender KW, Snedden WA (2010) Breaking the code: Ca^{2+} sensors in plant signaling. Biochem J 425:27–40

De Silva DLR, Hetherington AM, Mansfield TA (1998) The regulation of apoplastic calcium in relation to intracellular signaling in stomatal guard cells. Z Pflanzenernähr Bodenkd 161:533–539

Doherty CJ, Van Buskirk HA, Myers SJ, Thomashow MF (2009) Roles for *Arabidopsis* CAMTA transcription factors in cold-regulated gene expression and freezing tolerance. Plant Cell 21:972–984

Eulgem T, Somssich IE (2007) Networks of WRKY transcription factors in defense signaling. Curr Opin Plant Biol 10:366–371

Fasano JM, Massa GD, Gilroy S (2002) Ionic signaling in plant responses to gravity and touch. J Plant Growth Regul 21:71–88

Finkler A, Ashery-Padan R, Fromm H (2007) CAMTAs: calmodulin-binding transcription activators from plants to human. FEBS Lett 581:3893–3898

Fowler S, Thomashow MF (2002) *Arabidopsis* transcriptome profiling indicates that multiple regulatory pathways are activated during cold acclimation in addition to the CBF cold response pathway. Plant Cell 14:1675–1690

Frandsen G, Müller-Uri F, Nielsen M, Mundy J, Skriver K (1996) Novel plant Ca^{2+}-binding protein expressed in response to abscisic acid and osmotic stress. J Biol Chem 271:343–348

Galon Y, Aloni R, Nachmias D, Snir O, Feldmesser E, Scrase-Field S, Boyce JM, Bouché N, Knight MR, Fromm H (2010) Calmodulin-binding transcription activator 1 mediates auxin signaling and responds to stresses in *Arabidopsis*. Planta 232:165–178

Garcia-Brugger A, Lamotte O, Vandelle E, Bourque S, Lecourieux D, Poinssot B, Wendehenne D, Pugin A (2006) Early signaling events induced by elicitors of plant defenses. Mol Plant Microbe Interact 19(7):711–724

Geisler M, Frangne N, Gomes E, Martinoia E, Palmgren MG (2000) The ACA4 gene of *Arabidopsis* encodes a vacuolar membrane calcium pump that improves salt tolerance in yeast. Plant Physiol 124:1814–1827

George L, Romanowsky SM, Harper JF, Sharrock RA (2008) The ACA10 Ca2+–ATPase regulates adult vegetative development and inflorescence architecture in *Arabidopsis*. Plant Physiol 146:716–728

Gilroy S, Bethke PC, Jones RL (1993) Calcium homeostasis in plants. J Cell Sci 106:453–462

Harmon AC, Gribskov M, Gubrium E, Harper JF (2001) The CDPK superfamily of protein kinases. New Phytol 151:175–183

Harper JF, Harmon A (2005) Plants, symbiosis and parasites: a calcium signaling connection. Nat Rev Mol Cell Biol 6:555–566

Harper JF, Hong B, Hwang I, Guo HQ, Stoddard R, Huang JF, Palmgren MG, Sze H (1998) A novel calmodulin regulated Ca2+–ATPase (ACA2) from *Arabidopsis* with an N-terminal autoinhibitory domain. J Biol Chem 273:1099–1106

Heo WD, Lee SH, Kim MC, Kim JC, Chung WS, Chun HJ, Lee KJ, Park CY, Park HC, Choi JY et al (1999) Involvement of specific calmodulin isoforms in salicylic acid-independent activation of plant disease resistance responses. Proc Natl Acad Sci U S A 96:766–771

Hepler PK (2005) Calcium: a central regulator of plant growth and development. Plant Cell 17(8):2142–2155

Hirschi KD, Zhen RG, Cunningham KW, Rea PA, Fink GR (1996) CAX1, an H+/Ca2+ antiporter from *Arabidopsis*. Proc Natl Acad Sci U S A 93:8782–8786

Hu XY, Neill SJ, Cai WM, Tang ZC (2004) Induction of defense gene expression by oligogalacturonic acid requires increases in both cytosolic calcium and hydrogen peroxide in *Arabidopsis thaliana*. Cell Res 14:234–240

Huang L, Berkelman T, Franklin AE, Hoffman NE (1993) Characterization of a gene encoding a Ca^{2+}-ATPase-like protein in the plastid envelope. Proc Natl Acad Sci U S A 90:10066–10070

Ivey DM, Guffanti AA, Zemsky J, Pinner E, Karpel R, Padan E, Schuldiner S, Krulwich TA (1993) Cloning and characterization of a putative Ca2+/H+ antiporter gene from *Escherichia coli* upon functional complementation of Na+/H+ antiporter-deficient strains by the over-expressed gene. J Biol Chem 268:11296–11303

Jang HJ, Pih KT, Kang SG, Lim JH, Jin JB, Piao HL, Hwang I (1998) Molecular cloning of a novel Ca^{2+}-binding protein that is induced by NaCl stress. Plant Membr Biol 37:839–847

Kamiya T, Maeshima M (2004) Residues in internal repeats of the rice cation/H+ exchanger are involved in the transport and selection of cations. J Biol Chem 279:812–819

Kang J, Turano FJ (2003) The putative glutamate receptor 1.1 (AtGLR1.1) functions as a regulator of carbon and nitrogen metabolism in *Arabidopsis thaliana*. Proc Natl Acad Sci U S A 100:6872–6877

Kang J, Mehta S, Turano FJ (2004) The putative glutamate receptor 1.1 (AtGLR1.1) in *Arabidopsis thaliana* regulates abscisic acid biosynthesis and signaling to control development and water loss. Plant Cell Physiol 45:1380–1389

Kwaaitaal M, Huisman R, Maintz J, Reinstädler A, Panstruga R (2011) Ionotropic glutamate receptor (iGluR)-like channels mediate MAMP-induced calcium influx in *Arabidopsis thaliana*. Biochem J 440:355–365

Kaplan B, Davydov O, Knight H, Galon Y, Knight MR, Fluhr R, Fromm H (2006) Rapid transcriptome changes induced by cytosolic Ca2+ transients reveal ABRE-related sequences as Ca2+–responsive cis elements in *Arabidopsis*. Plant Cell 18:2733–2748

Kiegle E, Moore CA, Haseloff J, Tester MA, Knight MR (2000) Cell-type-specific calcium responses to drought, salt and cold in the *Arabidopsis* root. Plant J 23:267–278

Kim BG, Waadt R, Cheong YH, Pandey GK, Dominguez-Solis JR, Schultke S, Lee SC, Kudla J, Luan S (2007) The calcium sensor CBL10 mediates salt tolerance by regulating ion homeostasis in *Arabidopsis*. Plant J 52:473–484

Kosuta S, Hazledine S, Sun J, Miwa H, Morris RJ, Downie JA, Oldroyd GE (2008) Differential and chaotic calcium signatures in the symbiosis signaling pathway of legumes. Proc Natl Acad Sci U S A 105:9823–9828

Kreps JA, Wu Y, Chang HS, Zhu T, Wang X, Harper JF (2002) Transcriptome changes for *Arabidopsis* in response to salt, osmotic, and cold stress. Plant Physiol 130:2129–2141

Kudla J, Xu Q, Harter K, Gruissem W, Luan S (1999) Genes for calcineurin B-like proteins in *Arabidopsis* are differentially regulated by stress signals. Proc Natl Acad Sci U S A 96:4718–4723

Kudla J, Batistic O, Hashimoto K (2010) Calcium signals: the lead currency of plant information processing. Plant Cell 22:541–563

Lecourieux D, Mazars C, Pauly N, Ranjeva R, Pugin A (2002) Analysis and effects of cytosolic free calcium increases in response to elicitors in *Nicotiana plumbaginifolia* cells. Plant Cell 14:2627–2641

Lecourieux D, Ranjeva R, Pugin A (2006) Calcium in plant defense-signaling pathways. New Phytol 171:249–269

Lee JY, Yoo BC, Harmon AC (1998) Kinetic and calcium-binding properties of three calcium-dependent protein kinase isoenzymes from soybean. Biochemistry 37:6801–6809

Lee SM, Kim HS, Han HJ, Moon BC, Kim CY, Harper JF, Chung WS (2007) Identification of a calmodulin-regulated autoinhibited Ca2+–ATPase (ACA11) that is localized to vacuole membranes in Arabidopsis. FEBS Lett 581:3943–3949

Lee SS, Cho HS, Yoon GM, Ahn J-W, Kim H-H, Pai H-S (2003) Interaction of NtCDPK1 calcium-dependent protein kinase with NtRpn3 regulatory subunit of the 26S proteasome in *Nicotiana tabacum*. Plant J 33:825–840

Lemtiri-Chlieh F, MacRobbie EA, Webb AA, Manison NF, Brownlee C, Skepper JN, Chen J, Prestwich GD, Brearley CA (2003) Inositol hexakisphosphate mobilizes an endomembrane store of calcium in guard cells. Proc Natl Acad Sci U S A 100:10091–10095

Leng Q, Mercier RW, Yao WZ, Berkowitz GA (1999) Cloning and first functional characterization of a plant cyclic nucleotide-gated cation channel. Plant Physiol 121:753–761

Leung J, Merlot S, Giraudat J (1997) The *arabidopsis* ABSCISIC ACID-INSENSITIVE2 (ABI2) and ABI1 genes encode homologous protein phosphatases 2C involved in abscisic acid signal transduction. Plant Cell 9:759–771

Li J, Zhu S, Song X, Shen Y, Chen H, Yu J, Yi K, Liu Y, Karplus VJ, Wu P, Deng XW (2006) A rice glutamate receptor-like gene is critical for the division and survival of individual cells in the root apical meristem. Plant Cell 18:340–349

Li X, Chanroj S, Wu Z, Romanowsky SM, Harper JF, Sze H (2008) A distinct endosomal Ca2+/Mn2+ pump affects root growth through the secretory process. Plant Physiol 147:1675–1689

Liang F, Cunningham KW, Harper JF, Sze H (1997) ECA1 complements yeast mutants defective in Ca2+ pumps and encodes an endoplasmic reticulum-type Ca2+-ATPase in *Arabidopsis thaliana*. Proc Natl Acad Sci U S A 94:8579–8584

Ling V, Zielinski RE (1993) Isolation of an *Arabidopsis* cDNA sequence encoding a 22 kDa calcium-binding protein (CaBP-22) related to calmodulin. Plant Mol Biol 22:207–214

Liu HT, Li GL, Chang H, Sun DY, Zhou RG, Li B (2007) Calmodulin-binding protein phosphatase PP7 is involved in thermotolerance in *Arabidopsis*. Plant Cell Environ 30:156–164

Lu Y, Harrington HM (1994) Isolation of tobacco cDNA clones encoding calmodulin-binding proteins and characterization of a known calmodulin-binding domain. Plant Physiol Biochem 32:413–422

Lu SY, Li YC, Guo ZF, Li BS, Li MQ (1993) Enhancement of drought resistance of rice seedlings by calcium. Chin J Rice 13:161–164

Luan S, Kudla J, Rodriguez-Concepcion M, Yalovsky S, Gruissem W (2002) Calmodulins and calcineurin B-like proteins: calcium sensors for specific signal response coupling in plants. Plant Cell 14(suppl):S389–S400

Luo GZ, Wang HW, Huang J, Tian AG, Wang YJ, Zhang JS, Chen SY (2005) A putative plasma membrane cation/proton antiporter from soybean confers salt tolerance in *Arabidopsis*. Plant Mol Biol 59:809–820

Ma W, Yoshioka K, Berkowitz GA (2007) Cyclic nucleotide gated channels and Ca2+-mediated signal transduction during plant innate immune response to pathogens. Plant Signal Behav 2:548–550

Ma W, Smigel A, Tsai Y-C, Braam J, Berkowitz GA (2008) Innate immunity signaling: cytosolic Ca2+ elevation is linked to downstream nitric oxide generation through the action of calmodulin or a calmodulin-like protein. Plant Physiol 148(2):818–828

Ma W, Qi Z, Smigel A, Walker RK, Verma R, Berkowitz GA (2009) Ca2+, cAMP, and transduction of non-self perception during plant immune responses. Proc Natl Acad Sci U S A 106:20995–21000

Mahajan S, Pandey GK, Tuteja N (2008) Calcium- and salt-stress signaling in plants: shedding light on SOS pathway. Arch Biochem Biophys 471:146–158

Maruyama K, Sakuma Y, Kasuga M, Ito Y, Seki M, Goda H, Shimada Y, Yoshida S, Shinozaki K, Yamaguchi-Shinozaki K (2004) Identification of cold-inducible downstream genes of the *Arabidopsis* DREB1A/CBF3 transcriptional factor using two microarray systems. Plant J 38:982–993

McCormack E, Tsai YC, Braam J (2005) Handling calcium signaling: *Arabidopsis* CaMs and CMLs. Trends Plant Sci 10:383–389

Meyerhoff O, Muller K, Roelfsema MR, Latz A, Lacombe B, Hedrich R, Dietrich P, Becker D (2005) AtGLR3.4, a glutamate receptor channel-like gene is sensitive to touch and cold. Planta 222:418–427

Michalak M, Mariani P, Opas M (1998) Calreticulin, a multifunctional Ca^{2+} binding chaperone of the endoplasmic reticulum. Biochem Cell Biol 76:779–785

Mills RF, Doherty ML, Lopez-Marques RL, Weimar T, Dupree P, Palmgren MG, Pittman JK, Williams LE (2008) ECA3, a Golgi-localized P2A-type ATPase, plays a crucial role in manganese nutrition in *Arabidopsis*. Plant Physiol 146:116–128

Monroy AF, Labbé E, Dhindsa RS (1997) Low temperature perception in plants: effects of cold on protein phosphorylation in cell-free extracts. FEBS Lett 410:206–209

Monshausen GB, Bibikova TN, Weisenseel MH, Gilroy S (2009) Ca2+ regulates reactive oxygen species production and pH during mechano-sensing in *Arabidopsis* roots. Plant Cell 21:2341–2356

Navazio L, Bewell MA, Siddiqua A, Dickinson GD, Galione A, Sanders D (2000) Calcium release from the endoplasmic reticulum of higher plants elicited by the NADP metabolite nicotinic acid adenine dinucleotide phosphate. Proc Natl Acad Sci U S A 97:8693–8698

Nozawa A, Koizumi N, Sano H (2001) An *Arabidopsis* SNF1-related protein kinase, AtSR1, interacts with a calcium-binding protein, AtCBL2, of which transcripts respond to light. Plant Cell Physiol 42:976–981

Pandey SP, Somssich IE (2009) The role of WRKY transcription factors in plant immunity. Plant Physiol 150:1648–1655

Pasricha NS, Sarkar AK (2002) Secondary nutrients. In: Fundamentals of soil science. Indian Society of Soil Science, Division of Soil Science and Agricultural Chemistry, New Delhi

Perfus-Barbeoch L, Leonhardt N, Vavasseur A, Forestier C (2002) Heavy metal toxicity: cadmium permeates through calcium channels and disturbs the plant water status. Plant J 32:539–548

Piao HL, Lim JH, Kim SJ, Cheong G-W, Hwang I (2001) Constitutive over-expression of AtGSK1 induces NaCl stress responses in the absence of NaCl stress and results in enhanced NaCl tolerance in *Arabidopsis*. Plant J 27:305–314

Pickard BG, Ding JP (1993) The mechanosensory calcium-selective ion channel: key component of a plasmalemmal control centre? Aust J Plant Physiol 20:439–459

Plieth C (2005) Calcium: just another regulator in the machinery of life? Ann Bot 96:1–8

Plieth C, Hansen UP, Knight H, Knight MR (1999) Temperature sensing by plants: the primary characteristics of signal perception and calcium response. Plant J 18:491–497

Polisensky DH, Braam J (1996) Cold-shock regulation of the *Arabidopsis* TCH genes and the effects of modulating intracellular calcium levels. Plant Physiol 111:1271–1279

Popescu SC, Popescu GV, Bachan S, Zhang Z, Seay M, Gerstein M, Snyder M, Dinesh-Kumar SP (2007) Differential binding of calmodulin-related proteins to their targets revealed through high-density *Arabidopsis* protein microarrays. Proc Natl Acad Sci U S A 104:4730–4735

Price MB, Jelesko J, Okumoto S (2012) Glutamate receptor homologs in plants: functions and evolutionary origins. Front Plant Sci 3:235

Qi BS, Li CG, Chen YM, Lu PL, Hao FS, Shen GM, Chen J, Wang XC (2005) Functional analysis of rice Ca^{2+}/H^{+} antiporter OsCAX3 in yeast and its subcellular localization in plant. Prog Biochem Biophys 32:876–881 (in Chinese)

Rasmussen MW, Roux M, Petersen M, Mundy J (2012) MAP kinase cascades in *Arabidopsis* innate immunity. Front Plant Sci 3:169

Reddy AS (2001) Calcium: silver bullet in signaling. Plant Sci 160:381–404

Reddy ASN, Takezawa D, Fromm H, Poovaiah BW (1993) Isolation and characterization of two cDNAs that encode for calmodulin-binding proteins from corn root tips. Plant Sci 94:109–117

Reddy AS, Day IS, Narasimhulu SB, Safadi F, Reddy VS, Golovkin M, Harnly MJ (2002) Isolation and characterization of a novel calmodulin-binding protein from potato. J Biol Chem 277:4206–4214

Reddy ASN, Ali GS, Celesnik H, Day IS (2011) Coping with stresses: roles of calcium- and calcium/calmodulin-regulated gene expression. Plant Cell J 23(6):2010–2032

Reymond P, Farmer EE (1998) Jasmonate and salicylate as global signals for defense gene expression. Curr Opin Plant Biol 1:404–411

Riechmann JL et al (2000) *Arabidopsis* transcription factors: genome-wide comparative analysis among eukaryotes. Science 290:2105–2110

Romeis T, Ludwig AA, Martin R, Jones JDG (2001) Calcium-dependent protein kinases play an essential role in a plant defence response. EMBO J 20:5556–5567

Rudd JJ, Franklin-Tong VE (2001) Unraveling response-specificity in Ca^{2+} signaling pathways in plant cells. New Phytol 151:7–33

Saijo Y, Hata S, Kyozuka J, Shimamoto K, Izui K (2000) Over-expression of a single Ca^{2+}-dependent protein kinase confers both cold and salt/drought tolerance on rice plants. Plant J 23:319–327

Sakamoto H, Maruyama K, Sakuma Y, Meshi T, Iwabuchi M, Shinozaki K, Yamaguchi-Shinozaki K (2004) *Arabidopsis* Cys2/His2-type zinc-finger proteins function as transcription repressors under drought, cold, and high-salinity stress conditions. Plant Physiol 136:2734–2746

Sanders D, Pelloux J, Brownlee C, Harper JF (2002) Calcium at the crossroads of signaling. Plant Cell 14:S401–S417

Sattelmacher B (2001) The apoplast and its significance for plant mineral nutrition. New Phytol 149:167–192

Schiott M, Romanowsky SM, Baekgaard L, Jakobsen MK, Palmgren MG, Harper JF (2004) A plant plasma membrane Ca2+ pump is required for normal pollen tube growth and fertilization. Proc Natl Acad Sci U S A 101:9502–9507

Seki M et al (2002) Monitoring the expression profiles of 7000 *Arabidopsis* genes under drought, cold and high-salinity stresses using a full-length cDNA microarray. Plant J 31:279–292

Shaw SL, Long SR (2003) Nod factor elicits two separable calcium responses in *Medicago truncatula* root hair cells. Plant Physiol 131:976–984

Shigaki T, Rees I, Nakhleh I, Hirschi KD (2006) Identification of three distinct phylogenetic groups of CAX cation/proton antiporters. J Mol Evol 63:815–825

Sivaguru M, Pike S, Gassmann W, Baskin TI (2003) Aluminum rapidly depolymerizes cortical microtubules and depolarizes the plasma membrane: evidence that these responses are mediated by a glutamate receptor. Plant Cell Physiol 44:667–675

Snedden W, Fromm H (2001) Calmodulin as a versatile calcium signal transducer in plants. New Phytol 151:35–66

Song WY, Zhang ZB, Shao HB, Guo XL, Cao HX, Zhao HB, Fu ZY, Hu XJ (2008) Relationship between calcium decoding elements and plant abiotic-stress resistance. Int J Biol Sci 4(2):116–125

Spoel SH, Koornneef A, Claessens SM, Korzelius JP, Van Pelt JA, Mueller MJ, Buchala AJ, Metraux JP, Brown R, Kazan K, Van Loon LC, Dong X, Pieterse CM (2003) NPR1 modulates cross-talk between salicylate- and jasmonate-dependent defense pathways through a novel function in the cytosol. Plant Cell 15:760–770

Stephens NR, Qi Z, Spalding EP (2008) Glutamate receptor subtypes evidenced by differences in desensitization and dependence on the GLR3.3 and GLR3.4 genes. Plant Physiol 146:529–538

Stoelzle S, Kagawa T, Wada M, Hedrich R, Dietrich P (2003) Blue light activates calcium-permeable channels in *Arabidopsis* mesophyll cells via the phototropin signaling pathway. Proc Natl Acad Sci U S A 100:1456–1461

Strynadka NCJ, James MNG (1989) Crystal structures of the helix–loop–helix calcium-binding proteins. Annu Rev Biochem 58:951–998

Takabatake R, Karita E, Seo S, Mitsuhara I, Kuchitsu K, Ohashi Y (2007) Pathogen-induced calmodulin isoforms in basal resistance against bacterial and fungal pathogens in tobacco. Plant Cell Physiol 48:414–423

Thomma BP, Penninckx IA, Broekaert WF, Cammue BP (2001) The complexity of disease signaling in *Arabidopsis*. Curr Opin Immunol 13:63–68

Ton J, Van Pelt JA, Van Loon LC, Pieterse CM (2002) Differential effectiveness of salicylate-dependent and jasmonate/ethylene-dependent induced resistance in *Arabidopsis*. Mol Plant-Microbe Interact 15:27–34

Torres MA, Onouchi H, Hamada S, Machida C, Hammond-Kosack KE, Jones JDG (1998) Six *Arabidopsis thaliana* homologues of the human respiratory burst oxidase ($gp91^{phox}$). Plant J 14:365–370

Tuberosa R, Giuliani S, Parry MAJ, Araus JL (2007) Improving water use efficiency in Mediterranean agriculture: what limits the adoption of new technologies? Ann Appl Biol 2:157–162

Tuteja N (2007) Mechanisms of high salinity tolerance in plants. Methods Enzymol 428:419–438

Ueoka-Nakanishi H, Tsuchiya T, Sasaki M, Nakanishi Y, Cunningham KW, Maeshima M (2000) Functional expression of mung bean Ca2+/H+ antiporter in yeast and its intracellular localization in the hypocotyl and tobacco cells. Eur J Biochem 267:3090–3098

Urquhart W, Gunawardena AH, Moeder W, Ali R, Berkowitz GA, Yoshioka K (2007) The chimeric cyclic nucleotide-gated ion channel ATCNGC11/12 constitutively induces programmed cell death in a Ca2+ dependent manner. Plant Mol Biol 65:747–761

van Der Luit AH, Olivari C, Haley A, Knight MR, Trewavas AJ (1999) Distinct calcium signaling pathways regulate calmodulin gene expression in tobacco. Plant Physiol 121:705–714

Vogel JT, Zarka DG, Van Buskirk HA, Fowler SG, Thomashow MF (2005) Roles of the CBF2 and ZAT12 transcription factors in configuring the low temperature transcriptome of *Arabidopsis*. Plant J 41:195–211

Walley JW, Dehesh K (2010) Molecular mechanisms regulating rapid stress signaling networks in *Arabidopsis*. J Integr Plant Biol 52:354–359

Walley JW, Coughlan S, Hudson ME, Covington MF, Kaspi R, Banu G, Harmer SL, Dehesh K (2007) Mechanical stress induces biotic and abiotic stress responses via a novel cis-element. PLoS Genet 3:1800–1812

Wang YJ, Yu JN, Chen T, Zhang ZG, Hao YJ, Zhang JS, Chen SY (2005) Functional analysis of a putative Ca21 channel gene TaTPC1 from wheat. J Exp Bot 56:3051–3060

Wang Y-F, Munemasa S, Nishimura N, Ren HM, Robert N Han M et al (2014) Identification of cyclic GMP-activated non-selective Ca^{2+} permeable cation channels and associated CNGC5 and CNGC6 genes in *Arabidopsis* guard cells. www.plantphysiol.org

Webb AAR, Mc Ainsh MR, Taylor JE, Hetherington AM (1996) Calcium ions as intracellular second messengers in higher plants. Adv Bot Res 22:45–96

White PJ (1998) Calcium channels in the plasma membrane of root cells. Ann Bot 81:173–183

White PJ (2001) The pathways of calcium movement to the xylem. J Exp Bot 52:891–899

White PJ (2003) Calcium in plants. Ann Bot 92(4):487–511. doi:10.1093/aob/mcg164

White PJ, Banfield J, Diaz M (1992) Unidirectional Ca^{2+} fluxes in roots of rye (*Secale cereale* L.). A comparison of excised roots with roots of intact plants. J Exp Bot 43:1061–1074

Xiong L, Schumaker KS, Zhu J-K (2002) Cell signaling during cold, drought, and salt stress. Plant Cell 14:S165–S183

Yamaguchi-Shinozaki K, Shinozaki K (2006) Transcriptional regulatory networks in cellular responses and tolerance to dehydration and cold stresses. Annu Rev Plant Biol 57:781–803

Yang T, Poovaiah BW (2002) A calmodulin-binding/CGCG box DNA-binding protein family involved in multiple signaling pathways in plants. J Biol Chem 277:45049–45058

Yang S, Vanderbeld B, Wan J, Huang Y (2010) Narrowing down the targets: towards successful genetic engineering of drought-tolerant crops. Mol Plant 3:469–490

Yoo JH, Park CY, Kim JC et al (2005) Direct interaction of a divergent CaM isoform and the transcription factor, MYB2, enhances salt tolerance in *Arabidopsis*. J Biol Chem 280:3697–3706

Yoshioka K, Moeder W, Kang HG, Kachroo P, Masmoudi K, Berkowitz G, Klessig DF (2006) The chimeric *Arabidopsis* CYCLIC NUCLEOTIDE-GATED ION CHANNEL11/12 activates multiple pathogen resistance responses. Plant Cell 18:747–763

Zhang WJ, Zhang J, Liu F, Li GM, Guan JF (2001) The relationship between Ca^{2+} and drought-resistance in plants. Chin Bull Bo 18:473–478

Zhang W, Zhou RG, Gao YJ, Zheng SZ, Xu P, Zhang SQ, Sun DY (2009) Molecular and genetic evidence for the key role of AtCaM3 in heat-shock signal transduction in *Arabidopsis*. Plant Physiol 149:1773–1784

Zhang Y, Xu S, Ding P, Wang D, Cheng YT, He J, Gao M, Xu F, Li Y, Zhu Z, Li X, Zhang Y (2010) Control of salicylic acid synthesis and systemic acquired resistance by two members of a plant-specific family of transcription factors. Proc Natl Acad Sci U S A 107:18220–18225

Zhang H, Jin J, Tang L, Zhao Y, Gu X, Gao G, Luo J (2011) Plant TFDB 2.0: update and improvement of the comprehensive plant transcription factor database. Nucleic Acids Res 48:D1114–D1117

Zhao J, Barkla BJ, Marshall J, Pittman JK, Hirschi KD (2008) The *Arabidopsis* cax3 mutants display altered salt tolerance, pH sensitivity and reduced plasma membrane H+–ATPase activity. Planta 227:659–669

Zhao J, Shigaki Y, Mei H, Guo Y-Q, Cheng N-H, Hirischi KD (2009) Interaction between *Arabidopsis* Ca^{2+}/H^{+} exchangers CAX1 and CAX3. J Biol Chem 284:4605–4615

Zhu X, Caplan J, Mamillapalli P, Czymmek K, Dinesh-Kumar SP (2010) Function of endoplasmic reticulum calcium ATPase in innate immunity-mediated programmed cell death. EMBO J 29:1007–1018

Zielinski RE (1998) Calmodulin and calmodulin-binding proteins in plants. Annu Rev Plant Physiol Plant Mol Biol 49:697–725

Zielinski RE (2002) Characterization of three new members of the *Arabidopsis thaliana ca*lmodulin gene family: conserved and highly diverged members of the gene family functionally complement a yeast calmodulin null. Planta 214:446–455

Zong H, Liu EE, Guo ZF, Li MQ (2000) Enhancement of drought resistance of rice seedlings. J South China Agri Univ 21:63–65

6 Magnesium (Mg) Uptake

Abstract

Magnesium (Mg^{2+}) is the most abundant free divalent cation in the cytosol of the plants. The free Mg^{2+} level in the cytosol is strictly regulated due to its role in photosynthesis and on membrane ionic currents. About 90 % of Mg^{2+} is complexed with cytoplasmic ATP.

Mg^{2+} acts as cofactor of many enzymes, such as RNA polymerase, ATPases, protein kinases, phosphatases, carboxylases and glutathione synthetase. It is required for aggregation of ribosomes and is the central atom of chlorophyll molecule. The proteins involved in transport of Mg^{2+} across biological membranes have unique structures. Al^{3+} tolerance of plants could be improved by upregulation of genes of AtMGT family.

6.1 Occurrence of Mg and Soil Reactions

Mg^{2+} is unique in its chemical properties among the biologically active divalent cations. It has the smallest ionic radius, highest charge density and largest hydrated radius. Mg^{2+} often interacts with other molecules maintaining its hydration sphere. There is a 400-fold difference between volumes of hydrated and non-hydrated states (Li et al. 2001; Geberta et al. 2009).

Earth's crust contains about 1.93 % of Mg. The Mg content of soil may vary from 0.1 % coarse-textured humid soils to 4 % in fine-textured soils from arid or semiarid region. The sources of Mg in soil are the Mg-bearing minerals, such as dolomite (12.2 % MgO), biotite (2–20 % MgO), augite (15.7 % MgO) and a number of other minerals. Weathering of these minerals releases Mg^{2+} ions to the soil solution. Coarse-textured humid soils show high degree of Mg deficiency. Mg^{2+}ions unlike Ca^{2+} are more susceptible to leaching since they are not as strongly in adsorbed to clay minerals or organic matter due to their large hydrated radius. In the average values for exchangeable Mg^{2+} is less than 0.5 milli equivalent/100 g soil as compared to about 2.0 meq for exchangeable Ca^{2+}.

6.2 Mg Content of Plants

Mg^{2+} concentration in crops varies from 0.1 to 0.4 %. The critical limit of Mg^{2+} in dry banana leaves has been reported to be 0.3 % and of coconut 0.2 % (14th fond) (Mitra 2006).

Magnesium (Mg^{2+}) is the most abundant free divalent cation in the cytosol of the plants. The

G.N. Mitra, *Regulation of Nutrient Uptake by Plants: A Biochemical and Molecular Approach*,
DOI 10.1007/978-81-322-2334-4_6,

free Mg^{2+} level in the cytosol, however, is strictly regulated due to its role in photosynthesis and on membrane ionic currents (Shaul 2002). The concentration of Mg^{2+} in the metabolic pool of leaf cells (cytoplasm and chloroplast) is reported to be 2–10 mM (Leigh and Wyn Jones 1986). Free Mg^{2+} concentration is considerably less since Mg^{2+} forms complexes with various molecules such as ATP. About 90 % of Mg^{2+} has been found to be complexed with cytoplasmic ATP in the mung bean root tip, and the concentration of free $Mg^{2+}$is only 0.4 mM as compared to total Mg^{2+} concentration of 3.9 mM in this tissue (Yazaki et al. 1988). Vacuole is the main organelle, which is involved in Mg^{2+} homeostasis in the cytosol and chloroplast (Marschner 1995). Total Mg^{2+} concentration in barley mesophyll vacuoles is 3–7 mM (Dietz et al. 1992).

6.3 Functions of Mg in Plants

Mg^{2+} acts as cofactor of many enzymes, such as RNA polymerase, ATPases, protein kinases, phosphatases, carboxylases and glutathione synthetase. It is required for aggregation of ribosomes and is the central atom of chlorophyll molecule. Small variation in Mg^{2+} level in the cytosol and chloroplast strongly affects key photosynthetic enzymes (Shaul 2002). During the process of chlorophyll formation, insertion of Mg^{2+} into the porphyrin structure is catalysed by Mg^{2+}-chelatase (Walker and Weinstein 1991; Papenbrock et al. 2000). Chlorophyll breakdown is caused by Mg^{2}-dechelatase with the formation of pheophytin (Langmeier et al. 1993). Mg^{2+} substitution in vivo in the chlorophyll by heavy metals (Hg^{2+}, Cu^{2+}, Cd^{2+}, Ni^{2+}, Zn^{2+}, Pb^{2+}) under conditions of heavy metal stress impairs photosynthesis (Kupper et al. 1996, 1998). Mg^{2+} is involved in both light and dark reactions of photosynthesis. Mg^{2+}-deficient leaves are therefore highly photosensitive (Shaul 2002). Mg^{2+} deficiency affects root growth of the plants and hence nutrient and water uptake (Marschner 1995). Mg^{2+} is also involved in Ca^{2+}-based signal transduction processes (Baumann et al. 1991). Mg^{2+} deprivation elicits rapid Ca^{2+} uptake and activates Ca^{2+}/calcineurin signalling (Wiesenberger et al. 2007). Low magnesium concentrations may become a limiting factor for functional intracellular communication (Geberta et al. 2009).

6.4 Mechanism of Mg Uptake by Plants

A number of transporter proteins have been identified, which are involved in transport of Mg^{2+} across biological membranes. These include CorA and CorA homologues found in different living organisms including higher plants.

6.4.1 CorA (Cobalt-Resistant Phenotype of Bacterial Mutants)

The proteins involved in transport of Mg^{2+} across biological membranes have unique structures (Shaul 2002; Gardner 2003; Moomaw and Maguire 2008; Geberta et al. 2009). The bacterial membrane transport proteins for Mg^{2+}, CorA (named from cobalt-resistant phenotype of bacterial mutants), have been studied well (Kehres et al. 1998; Moncrief and Maguire 1999; Niegowski and Eshaghi 2007). The crystal structure of CorA shows that the protein is a homopentamer. Each monomer has two closely placed C-terminal transmembrane domains (TMs), the first of which ends invariably with GMN (Gly-Meth-Asn) motif of tripeptide. The first TM domains of each subunit protein line the pores of the channels, located on the plasma membrane. The five N terminals form a large cone-shaped funnel within the cytosol (Eshaghi et al. 2006; Lunin et al. 2006; Payandeh and Pai 2006). The mechanism of Mg^{2+} transport involves the binding of the fully hydrated cation to an extracellular binding loop, which connects the TM domains. No electrostatic interactions are involved in passage of the cation through the membrane. However, one of the cytosolic domains carries extremely high concentration of positive charge and the other negative charge. This appears to help control of Mg^{2+} flux along with an intracellular Mg^{2+} bound between domains of each monomer (Maguire 2006). CorA appears to be a constitutive gene since it is not transcriptionally regulated.

6.4.2 CorA Homologue Proteins and AtMGT Family of Mg^{2+} Transporter Proteins

CorA homologue proteins have been found in all living organisms. In yeasts Mrs2p (named after impaired mitochondrial RNA splicing phenotype of mutants) protein is a CorA homologue. It is located in the inner mitochondrial membrane (Kolisek et al. 2003; Weghuber et al. 2006; Schindl et al. 2007). ALR protein (named after aluminium-resistant phenotype of mutant) located in the plasma membrane of yeasts is a homologue of CorA/MRS2 proteins (Liu et al. 2002; Lee and Gardener 2006; Wachek et al. 2006).

CorA homologues have been identified in plants. *Arabidopsis* has ten members of this gene family initially named as AtMRS2 (Schock et al. 2000) and subsequently AtMGT (Li et al. 2001) for Mg^{2+} transport. CorA-MRS2-ALR superfamily of Mg^{2+} transporter proteins is present in different living organisms, but they have low sequence similarity. All of these proteins, however, have the conserved GMN (Gly-Met-Asn) motif at the end of the first of two conserved transmembrane domains near the C terminus. Mutation of the GMN motif is reported to abolish Mg^{2+} transport. Naturally occurring variants GVN and GIN are associated with transport of other divalent cations such as Zn^{2+} and Cd^{2+}. This whole class of proteins has been named as 2-TM-GxN type within the so-called metal ion transporter (MIT) superfamily (Knoop et al. 2005).

According to Li et al. (2001), the AtMGT family of Mg^{2+} transporter proteins coded by *Arabidopsis* genome constitute Mg^{2+} transporter of higher plants as well. AtMGT1 to AtMGT9 are closely related. AtMGT10 is the most divergent of the plant family. AtMGT10 has been renamed as AtMRS2-11 (Drummond et al. 2006) and is reported to be involved in Mg^{2+} transport in chloroplast/plastids. There is tissue-specific expression of MRS2/MGT family of genes in plants. Six of them expressed in root tissues are probably involved in Mg^{2+} supply and distribution after uptake from soil (Geberta et al. 2009). AtMGT1 protein is located in the plasma membrane of *Arabidopsis*. AtMGT1 has the highest affinity for Mg^{2+}, but is capable of transporting other divalent cations (Ni^{2+}, Co^{2+}, Fe^{2+}, Mn^{2+} and Cu^{2+}) at a considerably higher concentration beyond the normal physiological range (Li et al. 2001). AtMGT1, AtMGT7 and AtMGT9 show higher expression in roots, but their probable involvement in Mg^{2+} uptake is not well understood (Chen et al. 2012). AtMGT7 is localised in the endoplasmic reticulum and AtMGT9 highly expressed in mature anthers, leaves and young roots (Chen et al. 2009). AtMGT5 in *Arabidopsis* operates as a dual function transporter in a concentration-dependent manner. It functions as a Mg^{2+} importer at micromolar levels but facilitates efflux at millimolar range. AtMGT5 protein is localised in the mitochondria. It mediates Mg^{2+} transport between cytosol and mitochondria. AtMGT5 gene is exclusively expressed in anthers at the early stages of flower development (Li et al. 2008). There are nine Mg^{2+} transporter proteins encoded by rice genome, which are homologues of AtMRS2/MGT gene family.

Proteome analysis of cellular compartments indicates that MRS2-1/MGT2 is localised in the tonoplast (Carter et al. 2004). MRS2-5/MGT3 is localised either in the tonoplast (Whiteman et al. 2008) or plasma membrane (Alexanderson et al. 2004). MHX, the Mg^{2+}/H^{+} exchanger which is not related to MRS2/MGT gene family, is located in the tonoplast (Shaul et al. 1999). MRS2-6/MGT5 is localised in the mitochondria (Li et al. 2008), MRS2-4/MGT6 and MRS2-11/MGT10 located in chloroplast (Froehlich et al. 2003; Drumond et al. 2006; Geberta et al. 2009).

6.4.3 Role of Mg^{2+} in Alleviation of Al^{3+} Toxicity

Al^{3+} toxicity is an agricultural problem particularly in acid soils. Soluble Al^{3+} inhibits root growth at micromolar concentration and affects nutrient and water uptake (Kochian et al. 2004; Ma 2007; Delhaize et al. 2012). Earlier experiments have reported the role of Mg^{2+} in alleviating Al^{3+} toxicity. It has been observed that grasses and cereals treated with Al^{3+} show Mg^{2+} deficiency (Tan et al. 1991), and application of higher levels of Mg^{2+} can alleviate Al^{3+} toxicity (Tan et al. 1991; Matsumoto 2000). It has also been shown that Al^{3+} inhibits Mg^{2+} uptake by roots (Rengel and Robinson 1989). Alleviation of Al^{3+} toxicity by application of Mg^{2+} has been observed in a number of crop plants such as sorghum (*Sorghum bicolor*; Tan et al. 1992), soybean (*Glycine max*; Silva et al. 2001a), wheat (*Triticum aestivum*; Ryan et al. 1994), rice (*Oryza sativa*; Watanabe and Okada 2005) and rice bean (*Vigna umbellata*; Yang et al. 2007). There is however a difference among crop species with respect to alleviation of Al^{3+} toxicity by application of Mg^{2+}. In some crops such as soybean and rice bean, Mg^{2+} at micromolar concentration is required to alleviate Al^{3+} toxicity (Silva et al. 2001a; Yang et al. 2007), but for rice and wheat, millimolar concentrations are required (Ryan et al. 1997; Watanabe and Okada 2005). The hydrated radius of Mg^{2+} and Al^{3+} is similar (Bose et al. 2011). At millimolar concentration Mg^{2+} can effectively compete with Al^{3+} for the same binding sites of the roots. Enhanced excretion of organic acids is also a likely mechanism in Mg^{2+}-mediated alleviation of Al^{3+} toxicity. Addition of 50 μg of Mg^{2+} has been found to enhance citrate concentration in root tip of soybean. When toxic level of Al^{3+} is added, it excretes citrate, which forms a nontoxic citrate-Al complex in the rhizosphere. This leads to increased levels of Al^{3+} tolerance (Silva et al. 2001b).

Li et al. (2001) suggested that Al^{3+} tolerance of plants could be improved by upregulation of genes of AtMGT family. Overexpression of *Arabidopsis* Mg^{2+} transporter gene AtMGT1 in *Nicotiana benthamiana* has been observed to confer higher Al^{3+} tolerance and is associated with increased Mg^{2+} uptake (Deng et al. 2006).

6.4.4 Mechanism of Al Tolerance by Rice

Rice is the most Al^{3+}-tolerant crop among the cereals. This is due to presence of multiple Al-tolerance genes involved in detoxification of Al^{3+} at different cellular levels regulated by a transcription factor ART1 (Al^{3+} resistance transcription factor 1) (Tsutsui et al. 2011). ART1 is a Cys2-His2-type Zn finger TF and is constitutively expressed in roots (Yamaji et al. 2009). ART1 regulates expression of 31 genes downstream through a *cis*-acting element, GGN (T/g/a/C)V(C/A/g)S(C/G). This element was found in the promoter region of 29 genes out of 31 ART1-regulated genes (Tsutsui et al. 2011). Some of the downstream ART1-regulated genes include STAR1/STAR2 (sensitive to Al^{3+} rhizotoxicity1/2), Nrat1 (Nramp Al^{3+} transporter1), OsFRDL4 (rice ferric reductase defective like4) and OsALS1 (rice aluminium sensitive1) (Chen et al. 2012). OsFRDL4 is involved in citrate transport from roots to rhizosphere, where secreted citrate binds Al^{3+} to form a nontoxic compound (Yokosho et al. 2011).

It has been recently reported that an ART1-regulated gene OsMGT1 encodes a Mg^{2+} transporter located in the rice roots and shoots in the absence of Al^{3+}. The expression of this gene is upregulated only in roots rapidly and specifically under Al^{3+} stress to increase Mg^{2+} uptake and concentration in the root cell sap, which confers Al^{3+} tolerance to rice plant (Chen et al. 2012). OsMGT1 transporter is localised in the plasma membrane. It is a high-affinity transporter for Mg^{2+}. OsMGT1 is probably involved in Mg^{2+} uptake from soil since Mg^{2+} concentration in most soils is around 20–200 μM (Epstein 1972; Chen et al. 2012).

OsMGT1 (Os01g0869200) cloned from rice contains six exons and five introns and encodes a peptide containing 418 amino acids. OsMGT1 transporter is a membrane-bound protein with two transmembrane domains near the C terminus. It has 63–81 % amino acid similarity with AtMGT family of *Arabidopsis* Mg^{2+} transporters and has the Gly-Met-Asn-conserved motif at the end of the first transmembrane domain (Chen et al. 2012).

References

Alexandersson E, Saalbach G, Larsson C, Kjellbom P (2004) *Arabidopsis* plasma membrane proteomics identifies components of transport, signal transduction and membrane trafficking. Plant Cell Physiol 45:1543–1556

Baumann O, Walz B, Somlyo AV, Somlyo AP (1991) Electron probe microanalysis of calcium release and magnesium uptake by endoplasmic reticulum in bee photoreceptors. Proc Natl Acad Sci U S A 88:741–744

Bose J, Babourina O, Rengel Z (2011) Role of magnesium in alleviation of aluminium toxicity in plants. J Exp Bot 62:2251–2264

Carter C, Pan S, Zouhar J, Avila EL, Girke T, Raikhel NV (2004) The vegetative vacuole proteome of *Arabidopsis thaliana* reveals predicted and unexpected proteins. Plant Cell 16:3285–3303

Chen J, Li LG, Liu ZH, Yuan YJ, Guo LL, Mao DD, Tian LF, Chen LB, Luan S, Li DP (2009) Magnesium transporter AtMGT9 is essential for pollen development in *Arabidopsis*. Cell Res 19:887–898

Chen ZC, Yamaji N, Motoyama R, Nagamura Y, Ma JF (2012) Up-regulation of a magnesium transporter gene OsMGT1 is required for conferring aluminum tolerance in rice plant. Physiology 159(4):1624–1633

Delhaize E, Ma JF, Ryan PR (2012) Transcriptional regulation of aluminium tolerance genes. Trends Plant Sci 17:341–348

Deng W, Luo K, Li D, Zheng X, Wei X, Smith W, Thammina C, Lu L, Li Y, Pei Y (2006) Overexpression of an Arabidopsis magnesium transport gene, AtMGT1, in Nicotiana benthamiana confers Al tolerance. J Exp Bot 57:4235–4243

Dietz KJ, Schramm M, Lang B, Lanzl-Schramm A, Durr C, Martinoia E (1992) Characterization of the epidermis from barley primary leaves II The role of the epidermis in ion compartmentation. Planta 187:431–437

Drummond RSM, Tutone A, Li YC, Gardner RC (2006) A putative magnesium transporter AtMRS2-11 is localized to the plant chloroplast envelope membrane system. Plant Sci 170:78–89

Epstein E (1972) Mineral nutrition of plants: principles and perspectives. Wiley, New York, pp 128–131

Eshaghi S, Niegowski D, Kohl A, Molina DM, Lesley SA, Nordlund P (2006) Crystal structure of a divalent metal ion transporter CorA at 2.9 angstrom resolution. Science 313:354–357

Froehlich JE, Wilkerson CG, Ray WK, McAndrew RS, Osteryoung KW, Gage DA, Phinney BS (2003) Proteomic study of the *Arabidopsis thaliana* chloroplastic envelope membrane utilizing alternatives to traditional two-dimensional electrophoresis. J Proteome Res 2:413–425

Gardner RC (2003) Genes for magnesium transport. Curr Opin Plant Biol 6:263–267

Geberta M, Meschenmosera K, Svidováb S, Weghuberb J, Schweyenb R, Eiflera K, Lenza H, Weyanda K, Knoopa V (2009) A root-expressed magnesium transporter of the MRS2/MGT gene family in *Arabidopsis thaliana* allows for growth in Low-Mg2+ environments. Plant Cell 21(12):4018–4030

Kehres DG, Lawyer CH, Maguire ME (1998) The CorA magnesium transporter gene family. Microb Comp Genomics 3:151–169

Knoop V, Groth-Malonek M, Gebert M, Eifler K, Weyand K (2005) Transport of magnesium and other divalent cations: evolution of the 2-TM-GxN proteins in the MIT superfamily. Mol Genet Genomics 274:205–216

Kochian LV, Hoekenga OA, Pineros MA (2004) How do crop plants tolerate acid soils? Mechanisms of aluminum tolerance and phosphorous efficiency. Annu Rev Plant Biol 55:459–493

Kolisek M, Zsurka G, Samaj J, Weghuber J, Schweyen RJ, Schweigel M (2003) Mrs2p is an essential component of the major electrophoretic Mg^{2+} influx system in mitochondria. EMBO J 22:1235–1244

Kupper H, Kupper F, Spiller M (1996) Environmental relevance of heavy metal-substituted chlorophylls using the example of water plants. J Exp Bot 47:259–266

Kupper H, Kupper F, Spiller M (1998) *In situ* detection of heavy metal substituted chlorophylls in water plants. Photosynth Res 58:123–133

Langmeier M, Ginsburg S, Matile P (1993) Chlorophyll breakdown in senescent leaves – demonstration of Mg-dechelatase activity. Physiol Plant 89:347–353

Lee JM, Gardner RC (2006) Residues of the yeast ALR1 proteins that are critical for magnesium uptake. Curr Genet 49:7–20

Leigh RA, Wyn-Jones RG (1986) Cellular compartmentation in plant nutrition: the selective cytoplasm and the promiscuous vacuole. In: Tinker B, Lauchli A (eds) Advances in plant nutrition 2. Praeger Scientific, New York, pp 249–279

Li L, Tutone AF, Drummond RSM, Gardner RC, Luan S (2001) A novel family of magnesium transport genes in *Arabidopsis*. Plant Cell 13:2761–2775

Li LG, Sokolov LN, Yang YH, Li DP, Ting J, Pandy GK, Luan S (2008) A mitochondrial magnesium transporter functions in *Arabidopsis* pollen development. Mol Plant 1:675–685

Liu GJ, Martin DK, Gardner RC, Ryan PR (2002) Large Mg^{2+}-dependent currents are associated with the increased expression of ALR1 in *Saccharomyces cerevisiae*. FEMS Microbiol Lett 213:231–237

Lunin VV, Dobrovetsky E, Khutoreskaya G, Zhang R, Joachimiak A, Doyle DA, Bochkarev A, Maguire ME, Edwards AM, Koth CM (2006) Crystal structure of the CorA Mg^{2+} transporter. Nature 440:833–837

Ma JF (2007) Syndrome of aluminum toxicity and diversity of aluminum resistance in higher plants. Int Rev Cytol 264:225–252

Maguire ME (2006) Magnesium transporters: properties, regulation and structure. Front Biosci 1(11):3149–3163

Marschner H (1995) Mineral nutrition of higher plants. Academic, London/San Diego

Matsumoto H (2000) Cell biology of aluminum toxicity and tolerance in higher plants. Int Rev Cytol 200:1–46

Mitra, G. N. (2006) Nutrient Management of crops in soils of Orissa. IFFCO, India.

Moncrief MB, Maguire ME (1999) Magnesium transport in prokaryotes. J Biol Inorg Chem 4:523–527

Moomaw AS, Maguire ME (2008) The unique nature of Mg^{2+} channels. Physiology (Bethesda) 23:275–285

Niegowski D, Eshaghi S (2007) The CorA family: structure and function revisited. Cell Mol Life Sci 64:2564–2574

Papenbrock J, Mock HP, Tanaka R, Kruse E, Grimm B (2000) Role of magnesium chelatase activity in the early steps of the tetrapyrrole biosynthetic pathway. Plant Physiol 122:1161–1169

Payandeh J, Pai EF (2006) A structural basis for Mg^{2+} homeostasis and the CorA translocation cycle. EMBO J 25(16):3762–3773

Rengel Z, Robinson DL (1989) Competitive aluminum ion inhibition of net magnesium ion uptake by intact *Lolium multiflorum* roots. Plant Physiol 91: 1407–1413

Ryan PR, Kinraide TB, Kochian LV (1994) Al^{3+}-Ca^{2+} interactions in aluminum rhizotoxicity. Planta 192:98–102

Ryan PR, Reid RJ, Smith FA (1997) Direct evaluation of the Ca2+- displacement hypothesis for Al toxicity. Plant Physiol 113:1351–1357

Schindl R, Weghuber J, Romanin C, Schweyen RJ (2007) Mrs2p forms a high conductance Mg^{2+} selective channel in mitochondria. Biophys J 93:3872–3883

Schock I, Gregan J, Steinhauser S, Schweyen R, Brennicke A, Knoop V (2000) A member of a novel Arabidopsis thaliana gene family of candidate Mg2+ ion transporters complements a yeast mitochondrial group II intron-splicing mutant. Plant J 24:489–501

Shaul O (2002) Magnesium transport and function in plants: the tip of the iceberg. Bio Met 15:309–323

Shaul O, Hilgemann DW, Almeida-Engler J, Van M, Inzé M, Galili G (1999) Cloning and characterization of a novel Mg^{2+}/H^{+} exchanger. EMBO J 18:3973–3980

Silva IR, Smyth TJ, Carter TE, Rufty TW (2001a) Altered aluminum root elongation inhibition in soybean genotypes in the presence of magnesium. Plant Soil 230:223–230

Silva IR, Smyth TJ, Israel DW, Raper CD, Rufty TW (2001b) Magnesium ameliorates aluminum rhizotoxicity in soybean by increasing citric acid production and exudation by roots. Plant Cell Physiol 42:546–554

Tan K, Keltjens WG, Findenegg GR (1991) Role of magnesium in combination with liming in alleviating acid-soil stress with the aluminum-sensitive sorghum genotype CV323. Plant Soil 136:65–72

Tan K, Keltjens WG, Findenegg GR (1992) Aluminium toxicity with sorghum genotypes in nutrient solutions and its amelioration by magnesium. J Plant Nutr Soil Sci 155:81–86

Tsutsui T, Yamaji N, Ma JF (2011) Identification of a *Cis-acting* element of ART1, a C_2H_2-type zinc-finger transcription factor for aluminum tolerance in rice. Plant Physiol 156(2):925–931

Wachek M, Aichinger MC, Stadler JA, Schweyen RJ, Graschopf A (2006) Oligomerization of the Mg^{2+}-transport proteins Alr1p and Alr2p in yeast plasma membrane. FEBS J 273:4236–4249

Walker CJ, Weinstein JD (1991) Further characterization of magnesium chelatase in isolated developing cucumber chloroplasts – substrate-specificity, regulation, intactness, and ATP requirements. Plant Physiol 95:1189–1196

Watanabe T, Okada K (2005) Interactive effects of Al, Ca and other cations on root elongation of rice cultivars under low pH. Ann Bot (Lond) 95:379–385

Weghuber J, Dieterich F, Froschauer EM, Svidovà S, Schweyen RJ (2006) Mutational analysis of functional domains in Mrs2p, the mitochondrial Mg^{2+} channel protein of *Saccharomyces cerevisiae*. FEBS J 273:1198–1209

Whiteman SA, Serazetdinova L, Jones AM, Sanders D, Rathjen J, Peck SC, Maathuis FJ (2008) Identification of novel proteins and phosphorylation sites in a tonoplast enriched membrane fraction of *Arabidopsis thaliana*. Proteomics 8:3536–3547

Wiesenberger G, Steinleitner K, Malli R, Graier WF, Vormann J, Schweyen RJ, Stadler JA (2007) Mg^{2+} deprivation elicits rapid Ca^{2+} uptake and activates Ca^{2+}/calcineurin signaling in *Saccharomyces cerevisiae*. Eukaryot Cell 6:592–599

Yamaji N, Huang CF, Nagao S, Yano M, Sato Y, Nagamura Y, Ma JF (2009) A zinc finger transcription factor ART1 regulates multiple genes implicated in aluminum tolerance in rice. Plant Cell 21:3339–3349

Yang JL, You JF, Li YY, Wu P, Zheng SJ (2007) Magnesium enhances aluminum-induced citrate secretion in rice bean roots (*Vigna umbellata*) by restoring plasma membrane H^{+}-ATPase activity. Plant Cell Physiol 48:66–73

Yazaki Y, Asukawagawa N, Ishikawa Y, Ohta E, Sakata M (1988) Estimation of cytoplasmic free Mg^{2+} levels and phosphorylation potentials in mung bean root tips by *in vivo* ^{31}P NMR spectroscopy. Plant Cell Physiol 29:919–924

Yokosho K, Yamaji N, Ma JF (2011) An Al-inducible MATE gene is involved in external detoxification of Al in rice. Plant J 68:1061–1069

7 Sulphur (S) Uptake

Abstract

Globally soils of many countries are deficient in S. They need application of S-containing fertilisers to meet the crop demand and to maintain their yield and quality parameters. Sulphur plays an important ecological role in defence against herbivores and pathogens. S-containing defence compounds are widely distributed in plants, such as antimicrobial peptides called defensins and thionins, elemental S and glucosinolates.

There are five groups of gene families encoding sulphate transporters in plants.

7.1 Occurrence of Sulphur and Soil Reactions

Sulphur (S) content of the earth's crust averages about 0.06–0.10 %. Soils derive S from the S-bearing metal sulphide minerals. Due to weathering, S is oxidised to SO_4^{2-} and converted to soluble and insoluble salts. The SO_4^{2-} content of sea water is about 2,700 ppm and fresh water 0.5–50 ppm. About 90 % of total S in soils is present in the organic form. The readily available forms of S in soils are solution and adsorbed SO_4^{2-}.

Atmosphere which is enriched by S-containing gases primarily SO_2 from industrial emissions constitutes another source of soil S. It is estimated that 30–40 % of S in top soil at Rothamsted, UK, is derived from atmospheric deposition (Zhao et al. 2001). This source is getting depleted with greater environmental awareness and the industries adopting clean technology to reduce SO_2 emission (McGrath et al. 1996). There is further depletion of S in soils due to use of relatively cheaper S-free fertilisers for crop production. Application of S-fertilisers in optimum doses does not have any residual effect since clay minerals do not bind sulphate and it is leached out of soil. Annual applications of S as sulphate >50 kg S ha^{-1} for more than 150 years in the Broadbalk experiments in Rothamsted have not resulted in any build-up of S in the soil (Zhao et al. 2001).

Globally soils of many countries are deficient in S and need application S-containing fertilisers to meet the crop demand for maintaining their yield and quality parameters (McGrath et al. 1996). S deficiency is observed in many types of soils in India, especially in coarse-textured alluvial soils (entisols and inceptisols), red and lateritic soils (Alfisols) and Vertisols. The critical limit of $CaCl_2$ (0.15 %) extractable sulphur in soils is about 10 ppm.

G.N. Mitra, *Regulation of Nutrient Uptake by Plants: A Biochemical and Molecular Approach*,
DOI 10.1007/978-81-322-2334-4_7,

7.2 Sulphur Content of Plants

Sulphur (S) is an essential plant nutrient and is considered as the fourth major nutrient after N, P and K (TSI 2008). It is also of importance in human and animal nutrition. The total S content of plant tissues has been reported to be 0.5–1.5 % of the dry weight of the plants (Zhao et al. 1993; Marschner 1995; Burandt et al. 2001). Among the crops, *Brassica* species such as rapeseed mustard, cabbage, turnips, etc., have the highest requirement of S, followed by legumes (berseem and clover) and oilseeds (groundnut and soybean) (Pasricha and Sarkar 2002).

7.3 Functions of S in Plants

Sulphur is a constituent of various organic plant constituents. Sulphur is a constituent of amino acids cysteine and methionine, which are involved in maintaining protein structure and conformation. It is a constituent of coenzymes and prosthetic groups such as lipoic acid, coenzyme A, thiamine, etc. Sulphur compounds are involved in response to abiotic and biotic stress, such as glutathione in the detoxification of active oxygen species. Sulphur plays an important ecological role in defence against herbivores and pathogens. S-containing defence compounds are widely distributed in plants, such as antimicrobial peptides called defensins and thionins (Broekaert et al. 1995), elemental S (Williams and Cooper 2003), glucosinolates of the Brassicales and a number of compounds with limited occurrence in specific plant families. Glucosinolates have been well studied and contribute to flavour and anticancer properties of certain vegetables such as broccoli, cabbage, cauliflower, mustard and radish (Falk et al. 2007).

Defensins Defensins are cysteine-rich small and stable polypeptides, found in plants. They constitute an innate defence system of plants against attack by pathogens. They have antifungal, antibacterial, proteinase inhibitory and insect amylase inhibitory properties. Transgenic plants overexpressing defensins are strongly resistant to attack by fungal pathogens (Stoz et al. 2009).

The CO_2 assimilation rate, activity of Rubisco enzyme and abundance of protein are affected in young wheat plants due to S deficiency (Gilbert et al. 1997). There is a general inhibition of de novo functioning of photosynthetic apparatus under S-deficient conditions (Hawkesford 2000).

7.3.1 Effects of S on Yield and Quality of Crops

Field experiments at Rothamsted show that yield loss due to S deficiency in oilseed rape can be up to 70 % and in cereals up to 50 % (Zhao et al. 2001). Oil seed crops generally have a higher requirement of S as compared to other crops. S deficiency also affects quality of crops. Under limiting S availability, wheat grains accumulate low sulphate storage proteins such as ω-gliadin and high molecular weight subunits of glutenin at the expense of S-rich proteins. Such changes in protein composition affect dough rheology. Bread-making quality of wheat is closely correlated with S content of grain rather than N content (Zhao et al. 1999). Adequate S supply has been reported to increase both yield and malting quality of barley. S application significantly increases concentration of S-methyl methionine (the precursor of dimethyl sulphide) in kilned mart, which affects beer flavour (Zhao et al. 2006).

According to The Sulphur Institute (2008), application of S in S-deficient soils of India has been found to increase crop yields of rice by 17 %, wheat by 25 %, groundnut by 32 %, mustard by 30 %, soybean by 25 %, sunflower by 20 %, linseed by 16 %, pigeon pea by 22 % and green gram by 20 %. The oil content of oil seeds is reported to increase due to S application as follows: sunflower, 3.8 %; linseed, 6 %; soybean, 9.2 %; mustard, 9.2 %; and groundnut, 11.3 %. There are also reports of increase in protein content of grains, more in oilseeds than pulses, and of cysteine and methionine contents of grain protein due to S application (Tandon 1991). Quality of tea has been reported to improve due to S application (TSI 2008) and morphine, codeine and thebaine content of opium (Subrahmanyam et al. 1991). Plant glucosinolate content of Brassicales has been reported to increase from 25 % to more than 50-fold depending upon the plant species, amount of S fertiliser used and type of treatment (Falk et al. 2007).

7.4 Mechanism of Sulphur Uptake by Plants

7.4.1 Forms of Sulphur Taken Up and Its Mobilisation Within the Plant

Sulphate (SO_4^{2-}) is the major form of inorganic-S taken up directly from soil and transported in xylem (Falk et al. 2007). The cytoplasmic concentration of sulphate remains more or less constant. The excess sulphate is stored in the vacuole. The mobilisation of vacuolar pool of S is reported to be a slow process in roots and mature leaves especially in oil seed rape (Blake-Kalff et al. 1998; Hawkesford 2000). The inefficiency of mobilisation of S reserve is probably the reason for high S requirement of oilseed rape (Hawkesford 2000). In soybean higher redistribution of S occurs, when N limitation causes proteolysis (Sunarpi and Anderson 1997). There appears to be remobilisation of S from flag leaf to the grain in wheat, when there is adequate supply of S (Hawkesford 2000). In general shoot growth is more significantly affected than root growth in response to S availability (Marschener 1995). Under prolonged S deprivation, the partitioning of S between shoot and root is in favour of root growth (Buchner et al. 2004a).

7.4.2 Constituents of Sulphur Pool

The primary constituents of S pools are sulphate and S in protein fractions. Other smaller pools include amino acids, methionine and cysteine; the tri-peptide glutathione, sulpholipids; and the secondary compounds, glucosinolates, found in Brassicaceae (Hawkesford 2000).

7.4.3 Pathway for Assimilation of Sulphur in Plants

While plants take up S as sulphate (SO_4^{2-}), it is reduced to sulphide (S^{2-}) before it is assimilated into organic constituents of plants. Sulphate is first acted upon by ATP-sulphurylase (ATPS) to form adenosine-5′-phosphosulphate (APS). In plastids of plants, APS is reduced by APS reductase to sulphite. This is further reduced to sulphide by sulphite reductase (SiR). Sulphide is then incorporated into amino acid skeleton of O-acetyl serine (OAS) to form cysteine, catalysed by OAS (thiol) lyase (OALS) (Logan et al. 1996; Davidiana and Koprivab 2010) (Fig. 7.1). Other forms of S, such as H_2S from pedosphere, are absorbed through foliar absorption and converted directly into cysteine (Stuiver and De Kok 2001; Buchner et al. 2004a)

7.4.4 Plant Sulphate Transporters

Sulphate taken up by roots has to move through several inter- and intracellular membranes to get

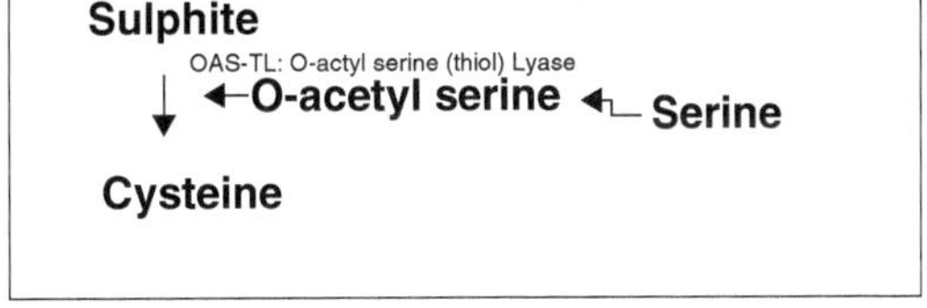

Fig. 7.1 Assimilation of sulphate into organic compounds (*ATPS* ATP-sulphurylase, *APS* adenosine-5′-phosphate sulphate, *APR* APS reductase, *SiR* sulphite reductase, *SAT* serine actyltransferase, *OAS-TL* O-acetylserine (thiol) lyase, *APK* APS kinase, *PAPS* 3′-phophoadenosine 5′- phosphosulphate)

into the cell and the organelles within the cell. It has to move from cell to cell through plasmodesmata to reach the distant leaf chloroplast, which is the principal site for reduction of sulphate to sulphide and its consequent assimilation into plant metabolism (Davidiana and Koprivab 2010). This involves several sulphate transport steps and coordinated gene regulation-encoding proteins involved in sulphate uptake, transport and assimilation. Plasma membrane sulphate transport is probably a pH-dependent proton-coupled cotransport involving $3H^+$/sulphate stoichiometry (Hawkesford et al. 1993; Smith et al. 1995). The sulphate transporter protein expressed in the plasma membrane of root cells consist of a single polypeptide chain of around 70–74 kD. A large number of sulphate transporter genes have been identified from *Arabidopsis,* rice and other plants (Smith et al. 1995; Smith et al. 1997; Takahashi et al. 1996, 1997, 1999, 2000; Yoshimoto et al. 2003; Howarth et al. 2003). At Rothamsted, sulphate transporters have been cloned from a number of agriculturally important crops such as wheat, barley, maize, oil seed rape, potato and tomato (Zhao et al. 2001).

7.4.5 Gene Family Encoding Sulphate Transporters

The gene family encoding sulphate transporter proteins in plants has been described by Hawkesford (2003). In *Arabidopsis* the gene family consists of 14 isoforms, which can be subdivided into five groups. Wheat, *Brassica oleracea* and rice have similar gene groups of sulphate transporters and probably are close homologues with similar functions (Buchner et al. 2004a, b, c, 2010; Shinmachi et al. 2010). Alignment and phylogenetic analysis of the first four groups of *Arabidopsis* and rice sulphate transporter proteins indicate that all have 12 transmembrane-spanning domains and a STAS domain at the carboxy terminal (Aravind and Koonin 2000). The fifth group more diverse but closely related with two smaller proteins lacks the STAS domain (Hawkesford 2003). The five groups with their locations and function are as follows:

Group 1 The transporters are high-affinity transporters and located in the plasma membrane.

Group 2 The transporters are low-affinity transporters also located in the plasma membrane.

Group 3 The transporters are of unknown function and may be associated with heterodimer association (Kataoka et al. 2004a).

Group 4 The transporters are involved in efflux of sulphate across tonoplast of vacuole into cytoplasm (Kataoka et al. 2004b).

Group 5 A member of group 5 sulphate transporter, Sultr 5;2, is probably an intracellular transporter involved in Mo (molybdenum) metabolism in *Arabidopsis* and is named as mot1 (Tomatsu et al. 2007; Baxter et al. 2008). The uptakes of Mo and Se (selenium) are probably through sulphate uptake pathway (Shinmachi et al. 2010).

7.4.6 Expression of Different Groups of Sulphate Transporters in Plants

7.4.6.1 Arabidopsis

Group 1 sulphate transporters from *Arabidopsis*, AtSultr1;1 and AtSultr1;2, are expressed primarily in epidermis and cortex of root tissues. The two sulphate transporters appear to be differentially regulated (Buchner et al. 2004b). One of the transporters, AtSultr1;2, mediates sulphate uptake both under sulphate-deficient and sufficient conditions and is insensitive to external sulphate concentration. The second transporter, AtSultr1;1, is expressed under sulphate-deficient conditions but almost absent when sulphate concentration is high. AtSultr1;2 is the major facilitator of sulphate uptake by plants (Takahashi et al. 2000; Shibagaki et al. 2002; Yoshimoto et al. 2002). Sultr1;3 is located in the sieve elements-companion cell element of the phloem and mediates the source to sink translocation of sulphate in plants (Yoshimoto et al. 2003).

Group 2 transporters of *Arabidopsis* are low-affinity sulphate transporters. AtSultr2;1and AtSultr2;2 transporters are localised in xylem parenchyma cells and pericycle of roots (Takahashi et al. 1997, 2000; Yoshimoto et al. 2003; Davidiana and Koprivab 2010) and may facilitate long-distance sulphate transport from root to shoot. The long-distance transport of sulphate through xylem is an important step in the distribution of sulphate in the plant tissues (Kataoka et al. 2004a, b).

Several members of Group 3 sulphate transporters are expressed in seeds in different stages of seed development and control sulphate translocation within developing seeds (Sirko et al. 2009; Davidiana and Koprivab 2010) A member of Group 3 sulphate transporter Sultr3;5 identified from *Arabidopsis* is localised in xylem parenchyma cells and pericycle along with Sultr2;1. The open reading frame of Sultr3;5 encodes a polypeptide of 634 amino acids with 12 transmembrane domains between the cytosolic N and C terminals (Kataoka et al. 2004a, b). Expression of Sultr3;5 alone does not have any sulphate uptake function in yeast, which indicates that it is a nonfunctional-type transporter by itself. Sultr3;5 expressed constitutively in roots reinforces the sulphate transport activity of Sultr2;1, the low-affinity sulphate transporter. The co-localisation Sultr3;5 with Sultr2;1 in xylem parenchyma cells and pericycle and their co-expression has a synergistic effect on sulphate retrieval from apoplast to xylem parenchyma cells in the vasculature of *Arabidopsis* roots (Kataoka et al. 2004a, b). This facilitates enhanced sulphate transport from roots to shoots.

Two members of Sultr4 subfamily, AtSultr4;1 and AtSultr4;2, are involved in transport of sulphate from vacuole across tonoplast to cytosol (Kataoka et al. 2004b). Expression of Sultr4;2 is more responsive to sulphate deprivation than Sultr4;1.

7.4.6.2 Wheat

The expression of different sulphate transporters in wheat with and without S has been reported from field experiment on winter wheat in Broadbalk fields, Rothamsted, UK (Shimanchi et al. 2010). These studies indicate that Sultr1;1, Sultr1;3 and Sultr2;1 are located in the plasma membrane. Sultr1;1, a high-affinity sulphate transporter, has higher expression in all tissues of wheat collected from (−S) plots. Sultr1;3, a high-affinity sulphate transporter expressed in phloem of *Arabidopsis,* is expressed in all tissues of wheat irrespective of S treatment. Sultr2;1 a sulphate transporter expressed in the vascular tissue, the central cylinder and root caps of *Arabidopsis* is expressed in all tissues of wheat except grain. Sultr1;1 and Sultr 1;3 are expressed in grains, not Sultr2;1. The sulphate transporter, Sultr4;1, which is involved in efflux of sulphate from vacuole to cytoplasm in *Arabidopsis,* is expressed in all tissues of wheat. Sultr5;1, which belongs to the SulP family (Hawlkesford 2003) without any transport function, is expressed in all tissues of wheat. Sultr5;2, which is involved in Mo accumulation in *Arabidopsis,* is expressed in all tissues of wheat.

7.4.6.3 Brassica

Sulphur deprivation increases the sulphate uptake capacity of the roots of *Brassica oleracea* seedlings. There is a concurrent increase in the expression of genes encoding specific sulphate transporters in the roots and other parts of the plant (Buchner et al. 2004a, b). A complete gene family corresponding to 12 different sulphate transporters has been isolated from *Brassica oleracea and Brassica napus* species. Based on sequence analysis, the *Brassica* sulphate transporter genes have been classified into 4 different groups according to their tissue specificity and sulphate ion availability. The sulphate transporter genes of Groups 1, 2 and 4 are induced or upregulated in response to sulphate deficiency. The expression of genes of Group 3 sulphate transporters is not affected by sulphate status (Buchner et al. 2004a). A comparison of sequence of the coding region of mRNA of the *Brassica* sulphate transporters and phylogenetic analysis with corresponding sulphate transporters of *Arabidopsis* indicates that three transporters BSultr1;1, 1;2 and 1;3 belong to Group 1. One sulphate transporter, BSultr2;1, belongs to Group 2. There are five sulphate transporters in Grou-3, BSultr3;1, 3;2. 3;3. 3;4 and 3;5. Group 4 sulphate transporter family consists of two transporters, BSultr4;1 and BSultr4;2. There is 84–89 % similarity between mRNA-coding regions of sulphate transporters of *Arabidopsis* and *Brassica* (Buchner et al. 2004a). While close phylogenetic relationship and high sequence similarities between sulphate transporters of *Arabidopsis* and *Brassica* sp. suggest their similar or identical functions, differences are found especially under sulphur-sufficient conditions. All of the sulphate transporters are not expressed in the roots of *B. oleracea.* Abundant expression of BSultr1;2 in roots indicates that it is primarily responsible for sulphate uptake (Takahashi et al. 2000; Shibagaki et al. 2002). Unlike *Arabidopsis*, out of two Group 2 low-affinity sulphate transporters, only one BSultr2;2 is expressed in the vascular tissues of *Brassica oleracea.* Expression BSutr1;3 in *Brassica* roots is low (Buchner et al. 2004a).

7.4.6.4 Rice

Godwin et al. (2003) isolated two sulphate transporter genes, OsSultr1;1 and OsSultr4;1, from a genomic library and the coding regions of their corresponding cDNAs generated by RT-PCR. OsSultr1;1 is localised in the roots and its expression strongly induced by S deficiency. OsSultr4;1 is expressed both in roots and shoots and appears to be significantly different from OsSultr1;1. Rice has been reported to have 14 isoforms of sulphate transporters (Kumar et al. 2011). OsSultr2;1 from rice with close phylogenetic relationship with Group 2 *Arabidopsis* sulphate transporters has been identified, but its role in sulphate transport is yet to be conclusively established.

7.4.6.5 Other Plants

Several homologues of Sutr3 subfamily are suggested to be involved in sulphate transport and delivery to the developing embryo of chick pea (*Cicer arietinum*) (Tabe et al. 2003). A symbiosome-specific Sultr3 sulphate transporter (SST1) is essential for development of functional nodules in *Lotus japonicus* (Krusell et al. 2005). ZmSultr1;1 a high-affinity sulphate transporter gene expressed in roots of maize has been functionally characterised (Nocito et al. 2006).

7.4.7 Regulation of Sulphate Uptake

7.4.7.1 Regulation by Other Nutrients

It is generally known that uptake and assimilation of sulphate is regulated by nutrient status of plants. Regulatory pathways are well organised to maintain a balance among uptake, assimilation and storage of sulphate in plants. Sulphur uptake is closely coordinated with nitrogen and carbon metabolism. There is an induction of the genes of high-affinity sulphate transporters due to addition of sucrose (Maruyama-Nakashita et al. 2004b). Nitrogen deficiency strongly reduces sulphur uptake and consequently a significant reduction in accumulation of transcripts of high-affinity sulphate transporters AtSultr1;1 and AtSultr1;2 (Maruyama-Nakashita et al. 2004b). However, S deficiency does not decrease total N content although there is an increase in O-acetyl serine (OAS), which is the precursor for synthesis of cysteine and has a role in regulation of sulphate uptake and reduction (Hawkesford 2000).

7.4.7.2 Regulation by OAS (O-Acetyl Serine)

It is reported that addition of OAS to plants with adequate supply of sulphate leads to increase in mRNA levels of sulphate transporters, sulphate uptake rates and tissue content of glutathione and cysteine. While sulphate, cysteine and glutathione act as negative regulator of sulphate transporter gene expression, OAS overrides such effects and acts as a positive regulator (Smith et al. 1997). The set of genes regulated under S-deficient conditions differ considerably between leaves and roots (Hirai et al. 2003).

7.4.7.3 SURE (Sulphur-Responsive Element)

Upregulation of AtSultr 1;1 in *Arabidopsis* under S-deficient conditions requires protein phosphatase as an upstream regulatory factor (Maruyama-Nakashita et al. 2004a). The promoter region of

AtSultr1;1 has been found to contain a 16 bp sulphur-responsive element (SURE), which includes an auxin-responsive factor (ARF) binding sequence (GAGACA). Within the conserved ARF binding site, there is a 5 bp core element (GAGAC), which regulates expression of a set of genes required for adaptation of plants to sulphur-deprived conditions (Maruyama-Nakashita et al. 2005). In wheat a gene specifically responsive to S deficiency has been found to contain a six base pair binding sequence of *cis*-acting sulphur-responsive element motif. The position of this motif on the promoter of wheat sulphur-deficiency-induced-1 (sdi1) gene is similar to the position of SURE in *Arabidopsis* promoter (Howarth et al. 2009).

7.4.7.4 SLIM1 (Sulphur Limitation 1)

A transcriptional regulator sulphur limitation1 (SLIM1) has been reported to upregulate AtSultr1;1, AtSultr1;2 and AtSultr4;1 gene expression in response to S deprivation in *Arabidopsis*. SLIM1 also called EIL3 has functional identity with the transcription factors ETHYLENE-INSENSITIVE-LIKE (EIL) family. No other member of EIL family is a sulphur limitation regulator (Maruyama-Nakashita et al. 2006). SLIM1 does not need SURE element for its regulatory activity. While SURE element is present in the promoter of AtSultr1;1, it is absent in the promoter of AtSultr1;2, which is the major sulphate transporter in the roots. AtSultr1;2 is probably controlled by SLIM1 under S-deficient conditions. A strong upregulation of AtSultr1;1 by SURE element under S deficiency reinforces sulphate uptake by AtSutr1;2 (Davidiana and Koprivab 2010).

7.4.7.5 miRNA (microRNA)

It has been reported that different microRNAs (miRNA) are involved in nutrient stress signal transduction pathways and nutrient homeostasis in plants (Kuo and Chiou 2011) (see Sect. 3.4.9 and Box 3.1). Expression of miR395 is significantly upregulated during S deficiency. Two families of genes involved in sulphate metabolism are targeted by miR395:

1. The APS gene-coding ATP-sulphurylase isoforms: ATPS1, ATPS3 and ATPS4
2. The genes of low-affinity sulphate transporters, Sultr2;1, which are located in the xylem parenchyma cells of roots and shoots

Sultr2;1 is cleaved by miR395 (Liang et al. 2010). Distribution of S is impaired from older to younger leaves in miR395 overexpressing plants (Liang et al. 2010). According to Kawashima et al. (2009), miR395 loci are expressed in the vascular system of leaves, roots and root tips under S-deficient conditions. Translocation of miR395 from leaves to roots through phloem is not necessary under S-deficient conditions. Induction of miR395 is controlled by the transcription factor SLIM1 involved in S-assimilation pathway (Kawashima et al. 2009).

References

Aravind L, Koonin EV (2000) The STAS domain: a link between anion transporters and antisigma-factor antagonists. Curr Biol 10:53–55

Baxter I, Muthukumar B, Park HC, Buchner P, Lahner B, Danku J, Zhao K, Lee J, Hawkesford MJ, Guerinot ML et al (2008) Variation in molybdenum content across broadly distributed populations of *Arabidopsis thaliana* is controlled by a mitochondrial molybdenum transporter (MOT1). PLoS Genet 4:1–13

Blake-Kalff MMA, Harrison KR, Hawkesford MJ, Zhao FJ, McGrath SP (1998) Allocation of sulfur within oilseed rape (*Brassica napus* L.) leaves in response to sulfur-deficiency. Physiol Plant 118:1337–1344

Broekaert WF, Terras FRG, Cammue BPA, Osborn RW (1995) Plant defensins: novel antimicrobial peptides as components of the host defense system. Plant Physiol 108:1353–1358

Buchner P, Elisabeth C, Stuiver E, Westerman S, Wirtz M, Hell R, Hawkesford MJ, De Kok LJ (2004a) Regulation of sulfate uptake and expression of sulfate transporter genes in *Brassica oleracea* as affected by atmospheric H_2S and pedospheric sulfate nutrition. Plant Physiol 136:3396–3408

Buchner P, Prosser IM, Hawkesford MJ (2004b) Phylogeny and expression of paralogous and orthologous sulphate transporter genes in diploid and hexaploid wheat. Genome 47:526–534

Buchner P, Takahashi H, Hawkesford MJ (2004c) Plant sulphate transporters: co-ordination of uptake, intracellular and long -distance transport. J Exp Bot 55:1765–1773

Buchner P, Parmar S, Kriegel A, Carpentier M, Hawkesford MJ (2010) The sulfate transporter family in wheat: tissue-specific gene expression in relation to nutrition. Mol Plant 3(2):374–389

Burandt P, Papenbrock J, Schmidt A, Bloem E, Haneklaus S, Schnug E (2001) Genotypical differences in total sulfur contents and cysteine desulf-hydrase activities in *Brassica napus* L. Phyton (Horn, Austria) 41:75–86

Davidiana J-C, Koprivab S (2010) Regulation of sulfate uptake and assimilation–the same or not the same? Mol Plant 3(2):314–325

Falk KL, Tokuhisa JG, Gershenzon J (2007) The effect of sulfur nutrition on plant glucosinolate content: physiology and molecular mechanisms. Plant Biol 9:573–581

Gilbert S, Clarkson DT, Cambridge M, Lambers H, Hawkesford MJ (1997) Sulphate-deprivation has an early effect on the content of ribulose 1,5-bisphosphate carboxylase/oxygenase and photosynthesis in young leaves of wheat. Plant Physiol 115:1231–1239

Godwin RM, Rae AL, Carroll BJ, Smith FW (2003) Cloning and characterization of two genes encoding sulfate transporters from rice (*Oryza sativa* L.). Plant and Soil 257(1):113–123

Hawkesford MJ (2000) Plant responses to sulphur deficiency and the genetic manipulation of sulphate transporters to improve S-utilization efficiency. J Exp Bot 51(342):131–138

Hawkesford MJ (2003) Transporter gene families in plants: the sulphate transporter gene family – redundancy or specialization? Physiol Plant 117:155–165

Hawkesford MJ, Davidian J-C, Grignon C (1993) Sulphate/H+ co-transport in plasma membrane vesicles isolated from Brassica napus: increased transport in membranes isolated from sulphur-starved plants. Planta 190:297–304

Hirai MY, Fujiwara T, Awazuhara M, Kimura T, Noji M, Saito K (2003) Global expression profiling of sulfur-starved *Arabidopsis* by DNA macroarray reveals the role of O-acetyl-l-serine as a general regulator of gene expression in response to sulfur nutrition. Plant J 33(4):651–663

Howarth J, Fourcroy P, Davidian J-C, Smith FW, Hawkesford MJ (2003) Cloning of two contrasting high-affinity sulphate transporters from tomato induced by low sulphate and infection by the vascular pathogen *Verticillium dahlia*. Planta 218:58–64

Howarth JR, Parmar S, Barraclough PB, Hawkesford MJ (2009) A sulphur deficiency-induced gene sdi1, involved in the utilization of stored sulphate pools under sulphur-limiting conditions has potential as a diagnostic indicator of sulphur nutritional status. Plant Biotech J 7:200–209

Kataoka T, Hayashi N, Yamaya T, Takahashi H (2004a) Root-to-shoot transport of sulfate in *Arabidopsis*: evidence for the role of SULTR3;5 as a component of low-affinity sulfate transport system in the root vasculature. Plant Physiol 136:4198–4204

Kataoka T, Watanabe-Takahashi A, Hayashi N, Ohnishi M, Mimura T, Buchner P, Hawkesford MJ, Yamaya T, Takahashi H (2004b) Vacuolar sulfate transporters are essential determinants controlling internal distribution of sulfate in *Arabidopsis*. Plant Cell 16:2693–2704

Kawashima CG, Yoshimoto N, Maruyama-Nakashita A, Tsuchiya YN, Saito K, Takahashi H, Dalmay T (2009) Sulphur starvation induces the expression of microRNA-395 and one of its target genes but in different cell types. Plant J 57(2):313–321. doi:10.1111/j.1365-313X.2008.03690.x. Epub 2008 Oct 14

Krusell L, Krause K, Ott T et al (2005) The sulfate transporter SST1 is crucial for symbiotic nitrogen fixation in *Lotus japonicus* root nodules. Plant Cell 17:1625–1636

Kumar S, Asif MH, Chakrabarty D, Tripathi RD, Trivedi PK (2011) Differential expression and alternative splicing of rice sulphate transporter family members regulate sulphur status during plant growth, development and stress conditions. Funct Integr Genomics 11(2):259–273

Kuo H-F, Chiou T-J (2011) The role of microRNAs in phosphorus deficiency signaling. Plant Physiol 156(3):1016–1024

Liang G, Yang F, Yu D (2010) MicroRNA395 mediates regulation of sulfate accumulation and allocation in *Arabidopsis thaliana*. Plant J 62:1046–1057

Logan HM, Cathala N, Grignon N, Davidian J-C (1996) Cloning of a cDNA encoded by a member of the *Arabidopsis thaliana* ATP sulfurylase multigene family: expression studies in yeast and in relation to plant sulfur nutrition. J Biol Chem 271:12227–12233

Marschner H (1995) Mineral nutrition of higher plants, 2nd edn. Academic, London

Maruyama-Nakashita A, Nakamura Y, Watanabe-Takahashi A, Yamaya T, Takahashi H (2004a) Induction of SULTR1;1 sulfate transporter in *Arabidopsis* roots involves protein phosphorylation/dephosphorylation circuit for transcriptional regulation. Plant Cell Physiol 45(3):340–345

Maruyama-Nakashita A, Nakamura Y, Yamaya T, Takahashi H (2004b) Regulation of high-affinity sulphate transporters in plants: towards systematic analysis of sulphur signalling and regulation. J Exp Bot 55:1843–1849

Maruyama-Nakashita A, Nakamura Y, Watanabe-Takahashi A, Inoue E, Yamaya T, Takahashi H (2005) Identification of a novel *cis*-acting element conferring sulfur deficiency response in *Arabidopsis* roots. Plant J 42(3):305–314

Maruyama-Nakashita A, Nakamura Y, Tohge T, Saito K, Takahashi H (2006) *Arabidopsis* SLIM1 is a central transcriptional regulator of plant sulfur response and metabolism. Plant Cell 18:3235–3251

McGrath SP, Zhao FJ, Withers PJA (1996) Development of sulphur deficiency in crops and its treatment. In: Proceedings of the Fertiliser Society, No. 379. The Fertiliser Society, Peterborough

Nocito FF, Lancilli C, Crema B, Fourcroy P, Davidian J-C, Sacchi G-A (2006) Heavy metal stress and sulfate uptake in maize roots. Plant Physiol 141(3):1138–1148

Pasricha NS, Sarkar AK (2002) Secondary nutrients. In: Fundamentals of soil science. Indian Society of Soil Science, New Delhi

Shibagaki N, Rose A, McDermott JP, Fujiwara T, Hayashi H, Yoneyama T, Davies JP (2002) Selenate-resistant mutants of *Arabidopsis thaliana* identify Sultr1;2, a sulfate transporter required for efficient transport of sulfate into roots. Plant J 29:475–486

Shinmachi F, Buchner P, Stroud JL, Parmar S, Zhao F-J, McGrath SP, Hawkesford MJ (2010) Influence of sulfur deficiency on the expression of specific sulfate transporters and the distribution of sulfur, selenium, and molybdenum in wheat. Plant Physiol 153(1):327–336

Sirko A, De KokL J, Haneklaus S, Hawkesford MJ, Rennenberg H, Saito K, Schnug E, Stulen I (eds) (2009) Sulfur metabolism in plants. Backhuys Publishers, Leiden, pp 113–118

Smith FW, Ealing PM, Hawkesford MJ, Clarkson DT (1995) Plant members of a family of sulfate transporters reveal functional subtypes. Proc Natl Acad Sci U S A 92:9373–9377

Smith FW, Hawkesford MJ, Ealing PM, Clarkson DT, van den Berg PJ, Belcher AR, Warrilow AGS (1997) Regulation of expression of a cDNA from barley roots encoding a high affinity sulphate transporter. Plant J 12:875–884

Stoz HU, Thomson JG, Wang Y (2009) Plant defensins, defence, development and application. Plant Signal Behav 4(11):1010–1012

Stuiver CEE, De Kok LJ (2001) Atmospheric H_2S as sulfur source for plant growth: kinetics of H_2S uptake and activity of O-acetylserine(thiol)lyase as affected by sulfur nutrition. Environ Exp Bot 46:29–36

Subrahmanayam K et al (1991) Effects of forms of S on yield and quality of seed, oil and alkaloids of opium poppy (*papaver somniferum* L). Manuscript, cf. Tandon, H. L. S, 1991.

Sunarpi, Anderson JW (1997) Effect of nitrogen on the export of sulphur from leaves in soybean. Plant and Soil 188:177–187

Tabe LM, Venables I, Grootemaat A, Lewis D (2003) Sulfur transport and assimilation in developing embryos of chickpea (Cicer arietinum). In: Davidian J-C, Grill D, de Kok LJ, Stulen I, Hawkesford MJ, Schnug E, Rennenberg H (eds) Sulfur transport and assimilation in plants. Backhuys Publishers, Leiden, pp 335–337

Takahashi H, Sasakura N, Noji M, Saito K (1996) Isolation and characterization of a cDNA encoding a sulfate transporter from *Arabidopsis thaliana*. FEBS Lett 392:95–99

Takahashi H, Yamazaki M, Sasakura N, Watanabe A, Leustek T, de Almeida-Engler J, Engler G, van Montagu M, Saito K (1997) Regulation of sulfur assimilation in higher plants: a sulfate transporter induced in sulfate deprived roots plays a central role in *Arabidopsis thaliana*. Proc Natl Acad Sci U S A 94:11102–11107

Takahashi H, Asanuma W, Saito K (1999) Cloning of an *Arabidopsis* cDNA encoding a chloroplast localizing sulphate transporter isoform. J Exp Bot 50:1713–1714

Takahashi H, Watanabe-Takahasi A, Smith FW, Blake-Kalff M, Hawkesford MJ, Saito K (2000) The roles of three functional sulphate transporters involved in uptake and translocation of sulphate in *Arabidopsis thaliana*. Plant J 23:171–182

Tandon HLS (1991) Sulphur research and agricultural production in India. The Sulphur Institute, Washington, DC

The Sulphur Institute (2008) Sulphur in Indian agriculture. The Sulphur Institute, Washington, DC

Tomatsu H, Takano J, Takahashi H, Watanabe-Takahashi A, Shibagaki N, Fujiwara T (2007) An *Arabidopsis thaliana* high-affinity molybdate transporter required for efficient uptake of molybdate from soil. Proc Natl Acad Sci U S A 104:18807–18812

Williams JS, Cooper RM (2003) Elemental sulphur is produced by diverse plant families as a component of defence against fungal and bacterial pathogens. Physiol Mol Plant Pathol 63:3–16

Yoshimoto N, Inoue E, Saito K, Yamaya T, Takahashi H (2003) Phloem-localizing sulfate transporter, Sultr1;3, mediates re-distribution of sulfur from source to sink organs in *Arabidopsis*. Plant Physiol 131:1511–1517

Yoshimoto N, Takahashi H, Smith FW, Yamaya T, Saito K (2002) Two distinct high-affinity sulfate transporters with different inducibilities mediate uptake of sulfate in *Arabidopsis* roots. Plant J 29:465–473

Zhao F, Bilsborrow PE, Evans EJ, Syers JK (1993) Sulphur turnover in the developing pods of single and double low varieties of oilseed rape (*Brassica napus* L.). J Sci Food Agric 62:111–119

Zhao FJ, Hawkesford MJ, McGrath SP (1999) Sulphur assimilation and effects on yield and quality of wheat. J Cereal Sci 30(1):1–17

Zhao FJ, McGrath SP, Hawkesford MJ (2001) Sulphur nutrition and the sulphur cycle institute of arable crops, Research report, Rothamsted Experimental Station 2000–2001

Zhao FJ, Fortune S, Barbosa VL, McGrath SP, Stobart PE, Billsborrow EJ, Booth A, Robson P (2006) Effects of sulphur on yield and malting quality of barley. J Cereal Sci 43:369–377

8 Definitions of Heavy Metals, Essential and Beneficial Plant Nutrients

Abstract

Heavy metals are defined as metals with a density higher than 5 g cm^{-3}. Of about 90 naturally occurring elements, 53 are considered as heavy metals. A total of 18 elements are considered essential for plant nutrition. A few of the elements are considered beneficial. They are not essential for plant nutrition but provide indirect benefit to the plants and animals who consume the plants.

8.1 Definition of Heavy Metals

Heavy metals are defined as metals with a density higher than 5 g cm^{-3}. Of about 90 naturally occurring elements, 53 are heavy metals (Weast 1984). Based on their solubility under physiological conditions, 17 heavy metals may be available for living cells and of importance for organism and ecosystems (Weast 1984). Among these metals, *Fe, Mn, Zn, Cu, Mo and Co* are considered as micronutrients and are essential for plant metabolism. *Cr, V, W, As, Ag, Hg, Sb, Cd, Pb and U* have no known function in plant metabolism. In the ionic form, Al^{3+}, Au^{+}, Cd^{2+}, Cu^{+}, Cu^{2+}, Co^{2+}, Cr^{3+}, Fe^{2+}, Fe^{3+}, Hg^{2+}, Mn^{2+}, Ni^{2+}, Pb^{2+}, Sn^{2+}, W^{6+} *and* Zn^{2+} become toxic at different threshold concentrations (Godbold and Hüttermann 1985; Breckle 1991; Nies 1999; Schützendübel and Polle. 2002).

The definition of heavy metals based on their density has little significance for plant uptake since they are taken up by plants in the form of their salts (Appenroth 2010). Elements with density of 3.5–7.0 g cm^{3} have been defined as heavy metals by various authors (Duffus 2002). A precise definition of the term 'heavy metal' either based on density or their position in the periodic table (Appenroth 2010) would be at variance with common perception of the term. Heavy metals are perceived to be potentially toxic components of the soil of different origin, which may find their way either through plant route or otherwise into the human or animal food chain and cause toxicity of known or unknown dimensions in the short or long term. The concentration of heavy metals in soil depends on weathering of enriched bed rock and atmospheric inputs. Natural sources are volcanoes and continental dusts. Anthropogenic sources consist of mining, combustion of fossil fuels, metalworking industries, phosphatic fertilisers and uses of industrial by-products as soil amendments and addition of urban wastes as manures (Lantsy and Mackenzie 1979; Galloway et al. 1982; Angelone

G.N. Mitra, *Regulation of Nutrient Uptake by Plants: A Biochemical and Molecular Approach*, DOI 10.1007/978-81-322-2334-4_8, © Springer India 2015

and Bini 1992; Schützendübel and Polle 2002). The anthropogenic emission of Cd has been reported to be in the range of 30,000 t per year (di Toppi et al. 1999). In Great Britain, Cd concentration of soils heavily polluted by use of sewage sludge is 150 mg kg^{-1}, as compared to 0.1–0.5 mg kg^{-1} in unpolluted soils (Jackson and Alloway 1991). However, Cd has been observed to have some stimulating effects on the growth of barley seedlings at a concentration of 5×10^{-8} M. Such effects have also been observed for Pd and Ti at low concentrations on barley leaves (Kovacs et al. 2009; Nyitrai et al. 2007). Heavy metals have been studied more for their toxic effects on plants rather than any stimulating effects. Heavy metals are not toxic to plants per se. Only when their cellular concentrations exceed a certain threshold value they become toxic and they are commonly termed as 'heavy metals' (Appenroth 2010).

8.2 Essential Plant Nutrients

A total of 18 elements are considered essential for plant nutrition (NRCCA 2010).

1. The macronutrients, applied in larger quantities to the plants, consist of:
   (i) Structural elements: *C, H and O*
   (ii) Primary nutrients: *N, P and K*
   (iii) Secondary nutrients: *S, Ca and Mg*
2. The micronutrients, applied in small quantities to plants, consist of *Zn, Fe, Mn, Cu, B, Mo, Cl−, Co and Ni.*

Micronutrients are essential for plant metabolism. Characteristic deficiency symptoms are observed in plants, when their availability in the growth medium decreases below a certain concentration. Heavy metals, which do not have micronutrient function, do not show such deficiency symptoms.

8.3 Beneficial Plant Nutrients

Beneficial plant nutrients are not essentially required for all the plants. Some of them are essential for some of the plants, but others are beneficial to either plants or animals, who consume these plants. Sodium is essential for *halophytes*, which accumulate salt in vacuoles to maintain turgor and growth. A few of the C4 plants (except corn and sorghum) need Na^+ essentially for specific functions, such as in the concentration of CO_2. Silicon strengthens the stem and provides protection to plants from biotic and abiotic stress. Cobalt is involved in nitrogen fixation by root nodule bacteria and other diazotrophs. Consumption of selenium-rich crop plants such as cabbage, mustard, onion and broccoli provides protection to human beings against cancer and heart disease. The importance of V is due to the discovery in 1980 that it can act as an insulin-mimetic agent.

References

Angelone M, Bini C (1992) Trace elements concentrations in soils and plants of western Europe. In: Adriano DC (ed) Biogeochemistry of trace metals. Lewis, Boca Raton

Appenroth K-J (2010) Definition of 'Heavy metals' and their role in biological system.In: Soil heavy metals, soil biology, vol 19, Chapter-2. Springer-Verlag Berlin, Heidelberg, pp 19–60

Breckle CW (1991) Growth under heavy metals. In: Waisel Y, Eshel A, Kafkafi U (eds) Plant roots: the hidden half. Marcel Dekker, New York, pp 351–373

Di Toppi LS, Lambardi M, Pazzagli L, Cappugi G, Durantc M, Gabbrielli R (1999) Response to cadmium in carrot in vitro plants and cell suspension cultures. Plant Sci 137:119–129

Duffus JH (2002) "Heavy metal"– a meaningless term? Pure Appl Chem 74:793–807

Galloway JN, Thornton JD, Norton SA, Volcho HL, McLean RA (1982) Trace metals in atmospheric deposition: a review and assessment. Atmos Environ 16:1677

Godbold DL, Hüttermann A (1985) Effect of zinc, cadmium and mercury on root elongation of *Picea abies* (Karst.) seedlings, and the significance of these metals to forest die-back. Environ Pollut 38:375–381

Jackson AP, Alloway BJ (1991) The transfer of cadmium from sewage sludge amended soils into the edible component of food crops. Water Air Soil Pollut 57:873–881

Kovacs E, Nyitrai P, Czovek P, Ovari M, Keresztes A (2009) Investigation into the mechanism of stimulation by low-concentration stressors in barley seedlings. J Plant Physiol 166:72–79

Lantsy RJ, Mackenzie FT (1979) Atmospheric trace metals: global cycles and assessment of man's impact. Geochim Cosmochim Acta 43:511

Nies DH (1999) Microbial heavy-metal resistance. Appl Microbiol Biotechnol 51:730–750

NRCCA (2010) North East Region Certified Crop Adviser Study Resources (NRCCA), Cornell University, USA

Nyitrai P, Mayer M, Ovari M, Keresztes A (2007) Involvement of the phosphoinositide signalling pathway in the anti-senescence effect of low-concentration stressors on detached barley leaves. Plant Biol 9:420–426

Schützendübel A, Polle A (2002) Plant responses to abiotic stresses: heavy metal-induced oxidative stress and protection by mycorrhization. J Exp Bot 53(372):1351–1365

Weast RC (1984) CRC handbook of chemistry and physics, 64th edn. CRC Press, Boca Raton

Uptake of Heavy Metals

9

Abstract

Heavy metals and metalloids are often taken up in toxic concentrations by plants from contaminated soils rich in these constituents. Higher plants have built-in cellular mechanisms for metal detoxification and tolerance which try to save them from such toxicities. An elaborate membrane transport system regulates movement of metal ions across plasma membrane of root cells.

9.1 Occurrence of Heavy Metals and Soil Reactions

Most of the heavy metals occur naturally in soils, but many of them are getting enriched in soil due to agricultural, industrial and mining activities, the so-called anthropogenic causes. As discussed in Chap. 8, heavy metals are perceived to be potentially toxic components of the soil of different origin, which may find their way either through plant route or otherwise into the human or animal food chain and cause toxicity of known or unknown dimensions in the short or long term. Micronutrients (Fe, Zn, Cu, Mn, B, Mo, Ni) are also heavy metals considered essential for plant metabolism but become toxic to plants when they exceed a threshold concentration in soils.

9.1.1 Aluminium (Al)

Aluminium is one of the third most abundant metals after oxygen and silicon and constitutes 7 % of the earth's crust. It is a component of aluminosilicate minerals such as feldspars of metamorphic and igneous rocks and of clay minerals in weathered soils. It is also present in the soil along with Fe and Mn as its oxides and as a component of organic matter. At a soil pH less than 5, Al partially solubilises to form the highly rhizotoxic Al^{3+} ion.

$$Al(OH)_3 + 3H^+ \rightleftarrows Al^{3+} + 3H_2O$$

With increase in pH, Al^{3+} ion hydrolyses progressively to $Al(OH)^{2+}$. $Al(OH)_2^+$, $Al(OH)_3$ and in alkaline pH, $Al(OH)_4^-$.

$$Al^{3+} + H_2O \rightleftarrows Al(OH)^{2+} + H^+$$
$$Al(OH)^{2+} + H_2O \rightleftarrows Al(OH)_2^+ + H^+$$
$$Al(OH)_2^+ + H_2O \rightleftarrows Al(OH)_3^0 + H^+$$
$$Al(OH)_3^0 + H_2O \rightleftarrows Al(OH)_4^- + H^+$$

G.N. Mitra, *Regulation of Nutrient Uptake by Plants: A Biochemical and Molecular Approach*, DOI 10.1007/978-81-322-2334-4_9, © Springer India 2015

9.1.2 Chromium (Cr)

The stable forms of Cr ions are the trivalent Cr^{3+} and the hexavalent Cr^{6+}. The hexavalent form is highly toxic and occurs as CrO_4^{2+} or $Cr_2O_7^{2+}$. Normally Cr content of soils is in the range of 10–50 mg kg^{-1}. Serpentine soils may contain up to 125 g kg^{-1}. Soils are contaminated with toxic concentrations of Cr from its use in leather industry, electroplating, production of refractory steel and drilling muds.

9.1.3 Cadmium (Cd)

The Cd content of normal soils is less than 0.1 mg kg^{-1}. Agricultural activities through the use of phosphatic fertilisers, urban sewage sludge, mining activities and exhaust gases from automobiles have resulted in increase of Cd content of soils. In Great Britain, Cd concentration of soils heavily polluted by the use of sewage sludge is 150 mg kg^{-1}, as compared to 0.1–0.5 mg kg^{-1} in unpolluted soils (Jackson and Alloway 1991).

9.1.4 Arsenic (As)

Arsenate (H_3AsO_4) and arsenite (H_3AsO_3) are the inorganic phyto-available forms of As present in soil solution. Small amounts of methyl arsenic acid ($CH_3H_2AsO_2$) and dimethyl arsenic acid [$(CH_3)_2HAsO_2$)] are also present in soil solution. The methylated forms are produced by bacteria and fungi.

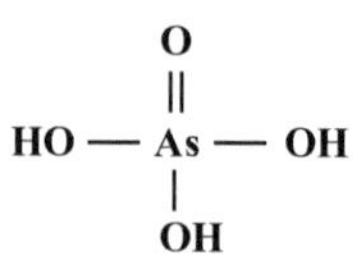

Arsenate

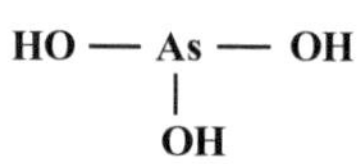

Arsenite

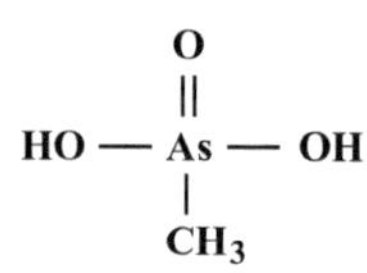

Mono methyl arsenic acid

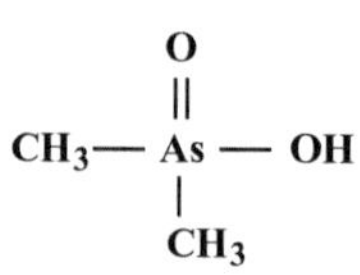

Di methyl arsenic acid

The dominant factors, which control availability of As in soil solution, are redox potential and pH. With increasing pH, the solubility of arsenate increases and solubility of arsenite decreases. The reverse is true with decreasing pH. The safe limit of As in soil is 10 mg kg^{-1} DW. Almost all As in surface water is in the form of arsenate. The sources of As pollution are mining and processing of ores containing Au, Ag, Cu and especially Sn with which As is associated. Earlier use of As containing compounds as pesticides in agriculture also contributed to enrichment of As in soil. However, the largest source of pollution is the recent discovery of high As content in groundwater in Southeast Asian countries, such as Vietnam, Bangladesh and West Bengal, where groundwater is used as a source of drinking water through tube wells and for irrigation in agriculture. The water from these tube wells contains As of about 1,000–3,000 µg L^{-1}. The European Union standard for safe limit of As in drinking water is 10 µg L^{-1} per day.

9.1.5 Lead (Pb)

Lead occurs naturally in all soils, rivers, lakes and sea water and also in air. The Pb content of soils is in the range of 15–40 ppm. Pollution can increase soil lead to several thousand ppm. The major causes of pollution of soils with Pb in populated areas are use of Pb containing paints, gasoline and pesticides. However, uses of most of these materials are phased out in developed countries and are in the process of getting phased out in rest of the countries. If the estimated total Pb level in soil is above 300 ppm, it is considered injurious to young children and pregnant women. Total estimated Pb level in soil above 2,000 ppm is considered hazardous for everybody.

(Soil reactions of heavy metals, which are also considered as micronutrients and of some beneficial elements, are discussed from Chaps. 10, 11, 12, 13, 14, 15, 16, 17, 18, 19, and 20).

9.2 Heavy Metal Content of Plants

9.2.1 Aluminium

The concentration of free Al^{3+} in the symplasm is less than 10^{-10} M due to pH-dependent hydrolysis of Al (as discussed above) and the formation of insoluble $Al(OH)_3$, but it can still be phytotoxic (Ma et al. 2001). Al^{3+} binds ATP 10^7 times more strongly than Mg^{2+}. Even at nanomolar concentration, Al^{3+} outcompetes Mg^{2+} for the binding sites of ATP. Plant species, which accumulate Al^{3+}, have internal mechanisms to detoxify it at higher concentrations. *Hydrangea* can accumulate more than 3,000 ppm Al^{3+} in its leaves. ^{27}Al-NMR spectroscopy showed that the Al in the leaves exists primarily as a 1:1 Al–citrate complex (Ma et al. 1998). Buckwheat accumulates Al as high as 15,000 ppm in leaves when grown on acid soils. In the roots and leaves, most of the Al is complexed with oxalate in a 1:3 Al–oxalate complex (Shen et al. 2004).

9.2.2 Chromium

Cr is toxic to most of the higher plants at concentration of 100 μM kg^{-1} dry weight (Davies et al. 2002). It has been reported by Golovatyj et al. (1999) that Cr distribution in crops does not depend on soil properties or the concentration of this element. Maximum quantity of the element is retained in the root and a minimum in the vegetative and reproductive organs. Huffman and Allaway (1973) reported that in bean 0.1 % of Cr was in seed and 98 % in roots.

9.2.3 Cadmium

Cd is readily taken up by plants from contaminated soil and transported to above-ground parts. There is considerable contribution of atmospheric deposits on Cd content of plants especially in industrialised countries and urban areas. A study in Denmark indicated that atmospheric contribution to Cd content of crop plants could vary between 10 and 60 % depending upon types of crops and soil (Smolders 2001).

Cd concentrations in leaves are higher than storage organs such as fruits or tubers. Zn deficiency and chloride salinity have been reported to increase Cd uptake by plants (Smolders 2001). Some of the crops have been identified as high-Cd crops, such as sunflower kernel, durum wheat and flax as compared to spring wheat, barley, corn and oats (Li et al. 1994).

9.2.4 Arsenic

Inorganic As species are highly toxic to plants. Organic As species are generally less toxic. Plants can accumulate both inorganic and organic forms of As. Roots generally contain more As than shoots. In rice 28–75 times higher As has been found in roots than in shoots (Azizur Rahman et al. 2007). There are varietal differences in As accumulation in different parts of the rice plant. While As has been detected in root, shoot and even the husk, the grains of some of the rice varieties have been found to be free from As (Alam and Rahman 2003).

9.2.5 Lead

Pb is taken up by plants through roots from soil and through leaves from atmospheric deposits.

About 7 % of Pb in soil is taken up by the plants. Finster et al. (2003) reported from a field survey of edible crops grown on Pb-contaminated soils that about 12 % (median) of Pb from the soil was taken up by the roots. On an average, 27 % of Pb from roots is transported to shoots. Detectable Pb concentration in the edible parts was in the range of 11–81 μg g^{-1}. Estimation of Pb in properly washed samples to remove any adhered soil particles show that Pb absorbed by plants does not concentrate in the edible parts of fruit or fruiting vegetables (tomato, peppers, beans, zucchini). Only exception is cucumber with a Pb concentration of 81 ppm. Leafy vegetables, herbs and edible roots (carrot, radish and onion) contain highest levels of Pb.

9.3 Functions of Heavy Metals and Metalloids

Micronutrients, heavy metals and metalloids are taken up by plants from contaminated soils rich in these constituents. Toxicity symptoms are observed when the concentrations of these constituents exceed a certain threshold value. Toxicity is caused by binding of metals to sulphahydryl groups of proteins, which results in inhibition of their activities or/and disruption of their structures. Heavy metals may also displace an existing metal constituent of a complex involved in crucial metabolic pathways. Many enzymes contain metals, which are crucial for their activity. Displacement of these metals by another metal will cause decrease in their activities or complete inhibition of enzyme activity. Divalent cations like Co^{2+}, Ni^{2+} and Zn^{2+} can displace Mg^{2+} from its position in ribulose 1,5-diphosphate carboxylase/oxygenase, which results in loss of its activity (Wildner and Henkel 1979; van Assche and Clijsters 1986). It has been reported that displacement of Ca^{2+} by Cd^{2+} in calmodulin in radish leads to inhibition of the enzyme, calmodulin-dependent phosphodiesterase (Rivetta et al. 1997). Inorganic As species are highly toxic to plants. Arsenate is a phosphate analogue and is transported along with phosphate by phosphate transporters. Inside the cell, it competes with phosphate and replaces it in ATP to form ADP-As. This results in disruption of energy flow in cell (Meharg 1994). Arsenite is also highly toxic and reacts with –SH groups of enzymes, which results in their inactivation. Chromium has been reported to have effect on photosynthesis in terms of CO_2 fixation, electron transport, photo-phosphorylation and enzyme activities (Clijsters and Van Assche 1985). Both chlorophyll a and b have been reported to decrease due Cr toxicity (Vajpayee et al. 1999; Bera et al. 1999).

Heavy metals cause production of reactive oxygen species ($O_2{\cdot}^-$, HO·) due to auto-oxidation and Fenton's reaction, typical for toxicity of micronutrients, Cu and Fe (Polle and Rennenberg 1993; Schützendübel and Polle 2002). ROS produced by As in plants due to conversion of arsenate to arsenite (Meharg and Hartley-Whitaker 2002) results in As toxicity to plants. Reactive oxygen species (ROS) may cause unspecific oxidation of proteins and membrane lipids and may cause DNA injury (Dean et al. 1993; Ames et al. 1993; Schützendübel and Polle 2002).

(Pl. read Chap. 8 for list of heavy metals, metalloids and their ionic forms in which they are found in soil.)

9.4 Mechanism of Heavy Metal Uptake by Higher Plants

9.4.1 Cellular Mechanisms for Metal Detoxification and Tolerance in Higher Plants

Visual symptoms of micronutrient and heavy metal toxicity in crop plants, their effects on plant metabolism and methods for their amelioration have been extensively reported by various authors and institutes (Das et al. 1997; Nable et al. 1997; Asian crops and micronutrient toxicity 2001; Rout et al. 2001; Meharg and Hartley-Whitaker 2002; Reichman 2002; Shanker et al. 2005; Liu et al. 2012).

Plants have developed various detoxification mechanisms to tolerate heavy metal stress and micronutrient toxicity. All of these mechanisms are primarily involved in preventing build-up of toxic concentrations at sensitive sites. Some of these mechanisms may be enumerated as follows (Marschner 1995; Hall 2002):

1. Restriction of metal movement to roots by mycorrhizas
2. Binding to cell wall and root exudates
3. Root border cells
4. Reduced influx across plasma membrane
5. Active efflux into apoplast
6. Chelation in cytosol by various ligands
7. Control of oxidants by antioxidant system
8. Repair and protection of plasma membrane under stress conditions
9. Transport of PC–Cd complex into the vacuole
10. Transport and accumulation of metals in vacuole (Marschner 1995; Hall 2002)

9.4.1.1 Mycorrhiza

Ectomycorrhizas associated with roots of trees and shrubs are reported to have ameliorating effects on heavy metal toxicity on host plants (Marschner 1995; Hüttermann et al. 1999; Jentschke and Godbold 2000; Schützendübel and Polle 2002; Hall 2002). The probable mechanisms suggested include absorption of metals by the hyphal sheath, reduced access to the apoplast due to the hydrophobicity of the fungal sheath, chelation by fungal exudates, adsorption onto the external mycelium and stimulation of the phenolic defence system. Arbuscular mycorrhizas, which can colonise roots of crop plants, have been shown to impart tolerance to heavy metal toxicity by crop plants (Hall 2002). It has been reported that maize grown in the presence of heavy metal with the mycorrhizal strain, *Glomus* isolate Br1, contained considerably lower concentration of heavy metals than plants grown without the mycorrhizal strain. This was due to selective immobilisation of heavy metals within the root tissues containing the mycorrhizal strain (Kaldorf et al. 1999).

9.4.1.2 Cell Wall

While root cell wall is in direct contact with the soil solution, its capacity to absorb metals and provide protection to plasma membrane is limited (Hall 2002). Exposure to Cd has been reported to cause cell wall rigidity of roots. The lignified root tips lose their capacity for nutrient uptake and root growth stops (Punz and Sieghardt 1993; Kahle 1993; Schützendübel and Polle 2002). About 60 % of Cu in the roots of *Lolium multiflorum* (Italian ryegrass) and *T. pratense* are reported to be bound by the cell wall and plasma membrane (Iwasaki et al. 1990). *Minuartia verna ssp. hercynica* grown on heavy metal-contaminated mine dumps have been found to contain high concentrations of Fe, Zn, Cu and Pb associated with Si contained in the cell wall (Neumann et al. 1997).

9.4.1.3 Root Border Cells

The root border cells (formerly considered as sloughed off root cap cells) are living cells programmed to separate from the root cap and from each other by the action of a cell wall-degrading enzyme (Hawes 1991; Hawes et al. 2005). The enzyme solubilises the interconnection among the cells. The cell walls of most of the individual cells remain intact (Hawes and Lin 1990). These cells are released into the environment of root and soil interface. The border cells undergo changes in morphology and gene expression. The cells elongate, form lignified secondary cell walls and excrete proteins into the external medium (Hawes et al. 2003). The number of root cells released as border cells varies among plant species. The cells released in 24 h in *Gossypium hirsutum* (*Malvaceae*) is from 8,000 to 10,000, while no border cells are released by *Brassica rapa* (*Brassicaceae*) (Hawes et al. 2003).

Exposure of border cells to Al induces them to secrete mucilage, which chelates Al and prevents it from entering root tip (Miyasaka and Hawes 2001). An in vitro study on the effect of Fe^{2+} on rice root tip border cells (Xing et al. 2008) indicates that there is an increased cell death of border cells with increased concentration of Fe^{2+}

along with time. This appears to be a protective mechanism to reduce Fe^{2+} toxicity in the rhizosphere and save the root tip from toxic effects. It has been reported that (Cai et al. 2012) iron plaques ubiquitously formed on rice roots along with root border cells, which surround the root caps, have a synergistic effect on protecting rice roots from Al toxicity. Kopittke et al. (2012) reported high concentration of As (V) in cowpea root border cells, when cowpea (*Vigna unguiculata*, 'Red Caloona') seedlings were exposed to 4–20 μg of As (V). They suggested that border cells probably absorbed more As (V) to protect the root tip from toxicity.

9.4.1.4 Plasma Membrane

Heavy metal toxicity rapidly affects function of plasma membrane. Toxicity of Cu increases efflux of ions from plasma membrane of wheat roots (Quartacci et al. 2001). Zn protects membrane integrity and does not cause leakage of ions (Catmack 2000). Damages to plasma membrane by heavy metals are caused by oxidation and cross-linking of protein thiols, inhibition of key membrane proteins such as H^+-ATPase and changes in the composition and fluidity of membrane lipids (Meharg 1993; Hall 2002).

9.4.1.5 Root Exudates

Root exudates do have a role in metal tolerance. Wheat plants secrete oxalic acid in response to stress of light metal Al and accumulate non-toxic aluminium oxalate in the leaves (Ma et al. 1997; 2001). Carboxylic acids such as citric, oxalic and malic acid and amino acid such as histidine are potential ligands of heavy metals. However, their role in detoxification of heavy metals in plants has not been clearly established (Hall 2002).

9.4.1.6 Proteins and Smaller Polypeptides

A number of proteins and smaller peptides are expressed in plants in response to abiotic stress. Three important groups involved in response to heavy metal stress are (A) heat shock proteins, (B) phytochelatins and (C) metallothioneins.

Heat Shock Proteins (HSPs)

Heat shock proteins are expressed in response to a variety of stress conditions including heavy metal stress (Lewis et al. 1999; Hall 2002; Wang et al. 2003). Among the five conserved families of HSPs (HSP100, HSP90, HSP70, HSP60 and sHSP), only small HSPs (sHSPs) are prevalent in plants. sHSPs vary in size from 12 to 40 kDa (Vierling 1991; Lewis et al. 1999). There are several reports on expression of sHSPs in response to heavy metal stress. In rice both heat stress and heavy metal stress increase expression of mRNA of 16–20 kDa, sHSPs (Tseng et al. 1993). Similarly sHSP17 is expressed in roots of *Armeria maritima* plants grown on Cu toxic soils (Neumann et al. 1995). sHSP21 has been reported to be involved in oxidative stress (Härndahl et al. 1999) and may act as antioxidants in protecting complex-I electron transport in mitochondria (Hamilton and Heckathorn 2001). Plant sHSPs show less sequence similarity than HSPs of other organisms. The sequence similarity spans over 100 amino acids proximal to the carboxy-terminal and shows pronounced homology with a-crystallin family (Waters et al. 1996) (*a-crystallin* or α-crystallin is one of the three major structural proteins of eye lens of vertebrates; the other two are β- and γ-crystallin. With ageing the lens proteins undergo various post-translational modifications, which lead to aggregation that predisposes lens to cataract. The chaperone-like activity of α-crystallin prevents aggregation of lens protein and maintains transparency of the lens, Harding 1991). Plant sHSPs form large oligomeric complexes similar to a-crystallins (Suzuki et al. 1998). Plant sHSPs hold and bind denatured substrate in a folding-competent state for subsequent refolding by a chaperon network (Haslbeck et al. 1999; Ding and Candido 2000; Studer and Narberhaus 2000). Some members of sHSPs can reactivate inactive enzymes (Marini et al. 2000; Sun et al. 2001).

Phytochelatins

Phytochelatins (PC) are cysteine-rich peptides synthesised enzymatically and are involved in heavy metal detoxification and accumulation (Pal and Rai 2010). They have general structure (γ Glu-Cys)$_{n-}$Gly, where n=2–11 (Rauser 1995; Zenk 1996; Cobbett 2000; Goldsbrough 2000), but generally (n) is in the range of 2–5 (Cobbett 2000). PCs are synthesised non-translationally from glutathione (GSH) as a substrate, by phytochelatin synthase (PCS), an enzyme that is activated in the presence of metal ions (Cobbett 2000). Synthesis of PCs involves transpeptidation of the γ-Glu-Cyst moiety of GSH to initially include a second molecule of GSH to form PC_2 and in subsequent stages of incubation produces PC oligomers with (n+1) peptides (Cobbett 2000). The *Arabidopsis* PCS (AtPCS1) encodes a polypeptide of 55 kD with 485 amino acids (Tennstedt et al. 2009). Similar PCS activities have been observed in pea (Klapheck et al. 1995), tomato (Chen et al. 1997), and *Arabidopsis* (Howden et al. 1995). It has been reported from a study on peanut plants that exposure to toxic concentration of Cd resulted in expression of different types of PCs (PC2, PC3, PC4), which possibly protects the plant from oxidative damage to macromolecules (Bianucci et al. 2012).

PCs are involved in major detoxification mechanisms of Cd and As ions in various species (Cobbett and Goldsbrough 2002; Tennstedt et al. 2009). PC-metal complexes have been detected in plant cells with Cd, Ag, Cu and As (Maitani et al. 1996; Schmöger et al. 2000). Synthesis of PCs is activated by large number of metal ions both in vivo and in vitro (Vatamaniuk et al. 2000; Oven et al. 2002). The genes of PCS occur in a large number of plants and the enzyme is constitutively expressed. It has been reported recently (Tennstedt et al. 2009) that PC formation contributes significantly to Zn^{2+} detoxification and accumulation in *Arabidopsis*. This explains the occurrence of genes of PCS throughout the plant kingdom and in a wide range of other organisms (Tennstedt et al. 2009).

Metallothioneins

Metallothioneins (MTs) similar to phytochelatins (PCs) are cysteine-rich metal-binding polypeptides, found across most taxonomic groups (Hall 2002; Grennan 2011). MTs are gene-encoded polypeptides, usually classified into two groups. Class 1 MTs contain cysteine residues that align with mammalian (equine) renal MT. Class 2 MTs have similar cysteine clusters, but they do not easily align with Class 1 MTs (Robinson et al. 1993; Prasad 1999). MTs bind metals through formation of mercaptide bonds between several cysteine (Cys) residues present in the protein and the metal. Arrangement of these Cys residues partly determines the metal-binding properties of MTs. MT genes have been identified in a range of higher plants (Prasad 1999; Hall 2002). Plant MTs (including *Arabidopsis*) show large sequence diversity and have been classified into four subfamilies (MT1, MT2, MT3 and MT4) based on the arrangements of Cys residues (Cobbett and Goldsbrough 2002; Blindauer and Leszczyszyn 2010; Grennan 2011; Freisinger 2011). Wheat E-proteins isolated from wheat germ bind Zn^{2+} at a stoichiometry (Zn^{2+}/protein) of approximately 5:1 and are classified as Class 2 metallothioneins (Kagi and Schaffer 1988). E-proteins are encoded by Ec genes located as single copies in the long arms of chromosomes 1A, 1B and 1D of hexaploid wheat, unlike animal MT genes, which are contained in multi-gene clusters (Kawashima et al. 1992). It has been recently reported from the discovery of three-dimensional structure of wheat E(c)-1 MT protein that it has two metal cluster arrangements not observed previously (Freisinger 2011). The C-terminal β(E)-domain consists of four metal ions and a part of the protein consisting of 11 cysteine residues as found in the mammalian counterparts (Peroza and Freisinger 2007). The N-terminal second cluster γ-E (c)-1 consists of two metal ions coordinated by six cysteine residues and is a part of full-length Zn E(c)-1 protein (Loebus et al. 2011). The two domains interact with each other while binding metal ions. Barley MT3 protein is reported to be located in tissues of both maternal and filial origin throughout the period of grain filling, whereas MT4 is confined to embryo and aleurone layer (Hegelund et al. 2012).

The role of MTs in plants is still under investigation (Grennan 2011). Apart from metal binding (Zn, Cd and Cu), MTs have been reported to play a role in other cellular processes such as regulation of cell growth and proliferation, DNA damage repair, scavenging of ROS and a Zn donating role (Cherian and Kang 2006). *Arabidopsis* MTs, 1a, 2a, 2b and 3, are possibly Cu-binding proteins and MTs 4a and 4b Zn binding (Grennan 2011).

MicroRNA (miRNA), Small Interfering RNA (siRNA)

MicroRNA (miRNA) and siRNA (small interfering RNA) have been reported to be involved in response to heavy metal stress in plants. miRNAs regulate various biological processes by negatively controlling the expression of corresponding genes either by (i) post-transcriptional cleavage of target mRNA or inhibition of its translation or (ii) transcriptionally by methylation of target DNA (Gielen et al. 2012).

Functions of siRNA and miRNA and Their Biogenesis

Small non-coding RNAs consisting of 20–24 nucleotides (nt) have been found to be important regulators of protein-coding gene expression. They function either by causing transcriptional gene silencing or post-transcriptional gene silencing (Baulcombe 2004). In plants post-transcriptional gene silencing has been reported to be mediated by RNA slicing (Baumberger and Baulcombe 2006) and translational repression (Lanet et al. 2009). Transcriptional gene silencing is carried out by histone modification and DNA methylation (Schramke and Allshire 2004; Khraiwesh et al. 2010). There are predominately two categories of small RNAs found in plants, such as microRNA (miRNA) and small interfering RNA (siRNA) (Gielen et al. 2012). Both miRNA and siRNA have been found to be highly conserved and function as important regulators of gene expression in plants and animals (Khraiwesh et al. 2012). Several classes of small RNAs identified in plants include miRNAs, repeat-associated small interfering RNAs (*ra-siRNAs*), natural antisense transcript-derived small interfering RNAs (*nat-siRNAs*), transacting small interfering RNAs (*ta-siRNA*), heterochromatic small interfering RNAs (*ha-siRNAs*), secondary transitive siRNAs, primary siRNAs and long small interfering RNAs (*lsiRNAs*) (Chapman and Carrington 2007; Chen 2009; Vazquez et al. 2010). Biogenesis of siRNA is given in Box 9.1.

miRNAs are encoded by endogenous *MIR* genes (see Box 3.1). A number of biological and metabolic processes are regulated by miRNA,

Box 9.1: Biogenesis of siRNA

Biogenesis of siRNA: Perfectly double-stranded RNAs (dsRNAs) originating from different sources such as RNAs transcribed from inverted repeats, natural cis-antisense transcript pairs, dsRNA produced from single-stranded RNA by the action of RNA-dependent RNA polymerase, the replication of RNA viruses and regions of genome rich in retro-elements (Khraiwesh et al. 2012) are processed to produce siRNAs. One of the four DCL (dicer-like homologue) proteins assisted by dsRNA-binding protein HEN1(HUA ENHANCER1) cleaves dsRNA into 21–24 nt siRNAs. Multiple DCLs cleave dsRNA and produce siRNA of different sizes. Similar to miRNA, siRNAs are loaded into AGO (AGRONAUTE) protein-containing RISC (RNA-induced silencing complex) that guide target regulation at the post-transcriptional level or transcription level through a pathway called RNA-directed DNA methylation. In rice (Wu et al. 2010) 24 nt long miRNAs (lmiRNA) are produced from their precursors by DCL3 and loaded into AGO4 clade of proteins according to hierarchical rules, based on their upstream biogenesis machinery and the 5′-terminal nucleotide. lmiRNAs direct DNA methylation at the loci from which they are produced as well as in trans at their target genes and play a role in gene regulation.

such as auxin signalling, meristem boundary formation and organ separation, leaf development and polarity, lateral root formation, transition from juvenile-to-adult vegetative phase and from vegetative-to-flowering phase, floral organ identity and reproduction. They also regulate plant response to biotic and abiotic stress and the miRNA pathway itself (Khraiwesh et al. 2012).

miRNA Expression and Heavy Metal Stress

Effects of heavy metal stress on expression of various miRNAs have been reported for a number of plants. The up- and downregulation of expression of different miRNAs are specific to plants, plant tissues and the stress caused by a particular heavy metal. For rice (Huang et al. 2009), toxic concentrations of Cd cause upregulation of miR601, miR602 and miR603 in leaves and downregulation of miR604 in roots. Expression of miR601 in leaves and miR605 and miR606 in roots is unaffected by Cd toxicity. In *Medicago truncatula* (Zhou et al. 2008), exposure to Cd, Hg and Al upregulates expression of miR171, miR319, miR393 and miR529 in leaves but downregulates expression of miR166 and miR398. *Brassica napus*, when exposed to Cd stress (Huang et al. 2010), results in strong upregulation of expression of miR156a, miR167a and miR167c in roots and miR167a and miR167c in leaves (Ding and Zhu 2009).

miRNA and Heavy Metal-Induced Oxidative Stress

As stated earlier most biotic and abiotic stresses including those due to heavy metals cause production of reactive oxygen species (ROS). Expressions of two closely related Cu/Zn superoxide dismutase (cytosolic CSD1 and chloroplastic CSD2) transcripts (which can detoxify oxidative stress) are induced in response to oxidative stress. Oxidative stress also downregulates transcription of miR398, which otherwise would have cleaved mRNA of CSD1 and CSD2. This results in post-transcriptional accumulation of mRNA of CSD1 and CSD2 (Sunkar et al. 2006). Expression of all the three miR398s (miR398$_a$, miR398$_b$, miR398c) is downregulated in *Arabidopsis*, when exposed to excess of Cu. Expression of miR398s is induced due to Cu deficiency with concurrent downregulation of CSD1 and CSD2. Fe-SOD (FSD) is simultaneously upregulated, which takes over dismutase function (Sunkar et al. 2006; Cuypers et al. 2011; Gielen et al. 2012). Such regulation is carried out by SPL7 (squamosa promoter-binding protein-like 7), which directly binds GTAC motifs of both FSD and miR398b/c promoters and upregulates their expression. This results in positive regulation of FSDs and negative regulation of CSDs (Abdel-Ghany and Pilon 2008; Yamasaki et al. 2009).

Oxidative stress caused by Fe and Zn toxicity also causes downregulation of expression of miR398 and upregulation of CSDs. The genes of miR398a, 398b and 398c are differently expressed in leaves and roots of *Arabidopsis* due to Zn toxicity. Transcription of miR398a decreases in leaves and roots, but transcription of miR398b and miR398c is induced in leaves with no response in roots due to Zn abundance (Remans et al. 2012). A genome-wide study of H_2O_2-regulated miRNA from rice seedlings indicates that miR169, miR397, miR827 and miR1425 are upregulated and miR528 downregulated in response to H_2O_2 treatment as compared to control (Li et al. 2011).

miRNA and Metal Complexation

There appears to be no correlation between miRNA and complexation under metal stress. Exposures to Cd and sulphur deficiency have been reported to upregulate expression of miR395 (Huang et al. 2010). Cd, Hg and other metals have high affinity for thiols, the functional groups of GSH and phytochelatins (PCs). Under metal stress, GSH and PCs form complexes with metal ions and prevent toxic effects of free metal ions to plants (Verbruggen et al. 2009; Carrasco-Gil et al. 2011; Cobbett and Goldsbrough 2002). An increase in PCs has been reported in *Arabidopsis* due to Cd treatment (Semane et al. 2007; Howden et al. 1995).

siRNA

Involvement of siRNAs in abiotic stress response has been reported by Sunkar and Zhu (2004). nat-siRNAs, which are derived from natural cis-antisense transcript pairs of overlapping genes, *SRO5* (an unknown gene in antisense orientation) *and P5CDH* (Δ^1 pyroline-5-carboxylate dehydrogenase), have been reported to provide osmoprotection and manage oxidative stress caused by salt stress in *Arabidopsis* (Borsani et al. 2005). Up- and downregulations of siRNA due to cold, heat, salt and drought stress have been reported (Yao et al. 2010). No report, which directly involves siRNA with metal stress, could be found in the literature scanned.

9.4.2 Membrane Transport Systems Involved in Transport of Micronutrients and Heavy Metals

Apart from various mechanisms adopted by plants to minimise toxic effects of heavy metals including micronutrients to maintain cellular homeostasis within an acceptable physiological range, an elaborate membrane transport system regulates movement of metal ions across plasma membrane (Hall and Williams 2003). These transporters include:

1. *Heavy metal (CPx-type) ATPases (HMAs)*
2. *ABC transporters*
3. *The Nramps (natural resistance-associated macrophage proteins)*
4. *The cation diffusion facilitator (CDF) family*
5. *The ZIP family*
6. *The cation antiporters (CAX family)*

9.4.2.1 P-Type ATPases

The P-type ATPase superfamily shares a common enzymatic mechanism in which ATP hydrolysis supports the transport of ions across plasma membrane (Axelsen and Palmgren 2001). P-type ATPases are found in all types of living organisms and include H^+-ATPases of plants and fungi, Na^+/K^+-ATPases of animals and Ca^{2+}-ATPases found in many organisms (Axelsen and Palmgren 2001; Hall and Williams 2003). Generally P-type ATPases contain 8–12 transmembrane domains with a large cytoplasmic loop, usually between TM-4 and TM-5, which contains a number of conserved motifs including the phosphorylation site (Palmgren and Axelsen 1998; Palmgren and Harper 1999). Heavy metal ATPases (HMAs) have eight transmembrane domains with a large cytoplasmic loop between TM-6 and TM-7 (Mills et al. 2003; Hall and Williams 2003).

P-type ATPase superfamily has been classified into five major families and 10 subfamilies according to their substrate specificity (Palmgren and Axelsen 1998; Axelsen and Palmgren 2001).

Heavy Metal ATPases (HMAs)

HMAs are grouped under P_{1B}-subfamily and have significant sequence similarities among bacteria, plants and humans (Palmgren and Axelsen 1998). This group also has been described as CPx-ATPases, since they contain a conserved intra-membrane sequence of cysteine–proline–cysteine/histidine/serine (Solioz and Vulpe 1996).

The HMA group is subdivided into two clusters, (i) the Cu cluster and (ii) the Zn cluster (Rensing et al. 1999):

(i) The Cu-cluster proteins transport monovalent Cu^+ and Ag+ .

(ii) The Zn-cluster proteins transport divalent Zn^{2+} and other heavy metals such as Co^{2+}, Cd^{2+} and Pb^{2+} (Axelsen and Palmgren 2001).

Arabidopsis has eight P_{1B}-ATPases, four of which belong to Cu^{2+} cluster and the other four to Zn^{2+} cluster (Baxter et al. 2003; Lee et al. 2007). The possible functions of some of the P_{1B}-ATPases in *Arabidopsis* have been reported as follows.

Cu Cluster

AtHMA6/PAA1 is involved in Cu transport system in chloroplast and delivery of cofactor to stomatal Cu/Zn superoxide dismutase (Shikanai et al. 2003). AtHMA8/PAA2 transports Cu into the thylakoid lumen to supply plastocyanin (Abdel-Ghany et al. 2005). AtHMA5 is involved in transmembrane transport and also interacts with Cu-metallochaperones (CCH) (Williams et al. 2000; Andres-Colas et al. 2006). AtHMA7/

RAN1 is possibly associated with the delivery of Cu ions to ethylene receptors (Hirayama et al. 1999). AtHMA1, which phylogenetically falls in Zn cluster, delivers Cu ions to the stroma for chloroplast superoxide dismutase activity (Seigneurin-Berny et al. 2006).

Zn Cluster

AtHMA2 has been reported to drive efflux of Zn^{2+} from the plant cells and also controls level of non-physiological heavy metals, such as Cd^{2+} (Eren and Arguello 2004). AtHMA4 clusters with Zn/Co/Cd/Pb (Mills et al. 2003) and possibly transports Zn (Hussain et al. 2004). AtHMA3 is involved in Cd/Pb transport in yeast. AtHMA3::GUS is localised in vacuole and possibly involved in Cd influx into the vacuole (Gravot et al. 2004).

P_{1B}-Type ATPases in Rice and Barley

A family of nine proteins belonging to P_{1B}-type ATPases has been identified in rice, whereas their numbers in barley is ten (Williams and Mills 2005). Phylogenetic analysis in rice indicates that OsHMA1 to OsHMA3 belong to Zn cluster and OsHMA4 to OsHMA9 belong to Cu cluster. Mills et al. (2012) could identify 9 P_{1B}-ATPases in barley and characterised HvHMA2 with a conserved aspartate phosphorylating site. HvHMA2 functions as a Zn and Cd pump.

9.4.2.2 ABC Transporters (ATP-Binding Cassette)

The ABC (ATP-binding cassette) transporter consists of a large superfamily found in all the three kingdoms. Most of them, but not all, are membrane proteins and are involved in a wide range of transport functions (Davies and Coleman 2000; Theodoulou 2000; Martinoia et al. 2002; Hall and Williams 2003; Rea 2007; Kang et al. 2011). Originally identified as a transporter involved in final detoxification process by deposition in the vacuole (Martinoia et al. 1993), the ABC transporters have been reported to be involved in diverse processes such as response to biotic and abiotic stress, surface lipid deposition, phytate accumulation in seeds and transport multifarious substrates such as ions, sugars, lipids, peptides, pigments, xenobiotics and antibiotic (Hall and Williams 2003; Rea 2007; Kang et al. 2011). In *Arabidopsis* 22 out of 130 ABC transporters have been functionally analysed (Kang et al. 2011).

According to Rea (2007), there are three basic features of plant ABC transport:

(i) Transport is energised by MgATP and not free ATP.
(ii) Transport is insensitive to transmembrane H^+ electrochemical potential difference.
(iii) Transport is extremely sensitive to vanadate.

Structure of ABC Transporters

ABC transporters consist of four core structural domains, two transmembrane domains (TMDs) containing multiple (usually 4–6) membrane-spanning α-helices and two nucleotide-binding folds (NBFs) or nucleotide-binding domains (NBDs) assembled with an internal twofold or pseudo twofold geometry (Rea 2007). The TMDs and NBFs cooperate during ATP hydrolysis to facilitate active transport (Kang et al. 2011). The ABC proteins have a consensus sequence of a signature amino acid motif (alias C motif), [LIVMFY]S[SG]G × 3[RKA][LIVMYA] × [LIVFM][AG] (commonly known as LSSG) with several variations (Rea 2007).

Over 120 ABC transporters have been identified in *Arabidopsis* and 121 ABC transporter open reading frames (ORFs) in rice (Garcia et al. 2004). The *Arabidopsis* ABC proteins, based on their domain structure and phylogenetic relationship, are currently classified into eight subfamilies in analogy with animal ABC proteins as ABCA, ABCB, ABCC, ABCD, ABCE, ABCF, ABCG and ABCI (ABCH is not found in plants) (Verrier et al. 2008; Kretzschmar et al. 2011). The two subunits (TMD and NBD) of ABC transporters are either encoded by individual genes, by two genes each encoding one TMD and one NBD (half-size ABCs) that form heterodimers, by one gene encoding one TMD and one NBD (half size) that form homodimers or by a single gene (full-size ABCs). The subfamilies from ABCA to ABCD have a forward TMD, NBD domain organisation (Verrier et al. 2008). While AtABCA1 (AOH according to previous nomenclature) is the largest full-size ABC protein in *Arabidopsis,* the remaining 11 members of ABCA family (ATH) are half-size proteins.

ABC Transporters and Detoxification of Heavy Metals

ABC transporters have long been reported to be associated with heavy metal and metalloid detoxification (Hanikenne et al. 2005; Kang et al. 2011). The ABC transporter YCF1 (yeast cadmium factor1) in *Saccharomyces cerevisiae* transports bis(glutathione) cadmium complexes (GS_2Cd) and GS_2-As from cytoplasm to the vacuole. The absence of this transporter causes hypersensitivity to Cd, As and Hg (Szczypka et al. 1994; Li et al. 1997; Ghosh et al. 1999; Gueldry et al. 2003). Overexpression of Sc-YCF1 in *Arabidopsis* results in Cd-tolerant plants (Song et al. 2003). The *Arabidopsis* ABC transporters, AtABCC1 and AtABCC2, have been reported to transport phytochelatin–arsenic complexes, As(III)-PC_2 and apoPC, when expressed in yeast (Song et al. 2010). AtABCC1 and AtABCC2 have also been reported to contribute to Cd^{2+} and Hg^{2+} tolerance (Park et al. 2012). Overexpression of AtABCC1 in *Arabidopsis* has been reported to increase Cd accumulation and tolerance (Kang et al. 2011). AtABCC1 in the absence of AtABCC2 (due to its redundant function with AtABCC1) can confer significant tolerance to divalent heavy metals (Kang et al. 2011). AtABCC3 and AtABCC6 are possibly associated with heavy metal tolerance. AtABCB25 (AtATM3) is a mitochondrial ABC transporter involved in biogenesis Fe–S clusters in plants (Kushnir et al. 2001; Bernard et al. 2009). AtABCB25 is strongly upregulated in Cd-treated plants (Bovet et al. 2005).

9.4.2.3 The NRAMPs (Natural Resistance-Associated Macrophage Proteins)

The Nramp gene was first identified in mouse, where it was found in the phagosomes (a membrane-bound cytoplasmic vesicle within the phagocyte that engulfs it) of infected macrophages. It determines sensitivity to bacterial infection by regulating concentrations of essential divalent cations such as Fe and Mn in the macrophage compartment (Hall and Williams 2003; Supek et al. 1996; Nelson 1999). The Nramp genes code for highly conserved family of integrated membrane proteins. These proteins are proton/metal symporters and have broad spectrum of divalent metal cation substrate, such as Fe^{2+}, Mn^{2+}, Cd^{2+}, Co^{2+}, Cu^{2+}, Ni^{2+} and Pb^{2+} (Gunshin et al. 1997; Nevo and Nelson 2006). Nramps are found in all living organisms from bacteria to human beings (Hall and Williams 2003). Three Nramps identified in yeasts regulate uptake of Fe, Mn, Co, Cu and Cd (Supek et al. 1997; Liu et al. 1997; Chen et al. 1999).

Nramp Genes in Plants

The Nramp genes have been identified in several plant species (Williams et al. 2000; Bereczky et al. 2003; Kaiser et al. 2003; Mizuno et al. 2005; Xiao et al. 2008; Oomen et al. 2009; Wei et al. 2009). Similar to other Nramps, the plant Nramps are highly conserved proteins containing 12 predicted transmembrane domains with a characteristic conserved motif between TM-8 and TM-9 (Gunshin et al. 1997; Curie et al. 2000; Williams et al. 2000).

In *Arabidopsis* out of six Nramp genes, five (AtNramp1–4 and AtNramp6) have been characterised at the molecular level (Curie et al. 2000; Thomine et al. 2000; Cailliatte et al. 2009). In *Arabidopsis*, heterologous expressions of AtNramp1, AtNramp3 and AtNramp4 in yeast mutants indicate that these proteins can transport Fe, Mn and Cd (Curie et al. 2000; Thomine et al. 2000). AtNramp3 and AtNramp4 are located on the vacuolar membrane of the embryo and mobilise vacuolar Fe store for early plant development (Lanquar et al. 2005). AtNramp6 contributes to Cd toxicity (Cailliatte et al. 2009). AtNramp1 acts as a Mn transporter for high-affinity Mn uptake by the roots from the soil in conditions of Mn deficiency (Cailliatte et al. 2010).

Rice OsNramps1 to OsNramps3 are the first three Nramps in rice to be reported (Belouchi et al. 1997). OsNramp1 is involved in Cd accumulation in rice, and the level of OsNramp expression is higher in the roots of high-Cd-accumulating *indica* cultivars than low

Cd-accumulating *japonicas* (Takahashi et al. 2011). Recent characterisation of OsNramps5 indicates its involvement in transport and uptake of Mn, Fe and Cd by rice (Ishimaru et al. 2012).

Tomato LeNramp1 is localised in the vascular parenchyma of root hair zone also in the root epidermis and cortex behind the root tip. It possibly plays a role in distribution of Fe in vascular parenchyma under Fe-deficient conditions (Bereczky et al. 2003).

Groundnut (***Arachis hypogaea***) AhNramp1 localised in the epidermis of plasma membrane of groundnut roots is a functional Fe transporter and is involved in uptake of Fe from the soil and distribution within the groundnut plant (Xiong et al. 2012).

MbNramp1 found in a fruit tree *(Malus baccata)* is involved in Fe, Mn and Cd transport (Xiao et al. 2008).

9.4.2.4 The Cation Diffusion Facilitator (CDF) Family

The CDF family of transporters was first identified in bacteria. Subsequently these have been found in yeast, plants and animals. The proteins are involved in efflux of transitional metal cations, Zn^{2+}, Cd^{2+}, Co^{2+}, Ni^{2+} or Mn^{2+}, from cytoplasm to outside of the cell or into subcellular compartments to maintain metal homeostasis and tolerance to their toxic effects (Paulsen and Saier 1997; Eide 1998; van Der Zaal et al. 1999; Hall and Williams 2003; Hanikenne et al. 2005). These proteins have six transmembrane domains with an N-terminal signature sequence and a C-terminal cation-binding domain. They have a Zn-binding histidine-rich domain between TM4 and TM5, a signature sequence between TM1 and TM2 and a cation efflux domain between TM1 and TM6 (Paulsen and Saier 1997; Huang and Gitschier 1997; Williams et al. 2000; Gaither and Eide 2001). The members of this family show a high degree of size variability with 280–740 amino acid residues (van Der Zaal et al. 1999). *Arabidopsis* CDF, ZAT (AtMTP1) contains 398 amino acid residues (van Der Zaal et al. 1999).

Eight genes coding proteins with homology to CDF family have been found in *Arabidopsis* genome. ESTs of CDF family have been found in a number of plants (Mäser et al. 2001).

9.4.2.5 The ZIP (ZRT-, IRT-Like Proteins) Family

The members of ZIP family are involved in influx of Zn, Fe, Mn and Cu from outside the cell or from subcellular compartments into the cytoplasm with variable substrate range and specificity. The ZIP proteins are predicted to have eight transmembrane domains with extra-cytoplasmic C and N termini. There is a variable histidine-rich loop possibly for metal binding, between TM-3 and TM-4. The length and amino acid sequence of the loop is also variable. The transmembrane domains TM-4 and TM-5 are amphipathic and possibly form a polar cavity required for transport of metal cations. The loop between TM-2 and TM-3 possibly is involved in initial binding of the substrate (Guerinot 2000; Mäser et al. 2001; Gaither and Eide 2001; Hall and Williams 2003; Hanikenne et al. 2005; Milner et al. 2013). Members of ZIP family also have been reported to transport heavy metal, Cd, and hence are involved in toxicity of essential and non-essential heavy metals (Guerinot 2000; Pence et al. 2000; Rogers et al. 2000).

ZIP Family Genes in Plants

The ZIP family has been classified into four subfamilies based on sequence conservation (Gaither and Eide 2001). ZIPs from plants and fungi come under subfamily I.

Arabidopsis There are 15 ZIP genes in *Arabidopsis* (Mäser et al. 2001). Some of the ZIP family genes in *Arabidopsis* have been characterised through yeast complementation and expression analysis. AtZIP1, AtZIP2, AtZIP3 and AtZIP4 play a role in cellular Zn uptake. AtZIP1, AtZIP3 and AtZIP4 are induced at transcriptional level under Zn-limiting conditions (Guerinot 2000; Gaither and Eide 2001; Hall and Williams 2003; Hanikenne et al.

2005). AtIRT1 (iron-regulated transporter1) a member of ZIP family (Eide et al. 1996) is possibly the major transporter involved in high-affinity iron uptake by roots (Connolly et al. 2002; Vert et al. 2002). Overexpression of AtIRT1 leads to accumulation of Cd and Zn, indicating a possible role of this transporter in uptake of Cd and Zn (Connolly et al. 2002; Hall and Williams 2003). AtIRT2 also expressed in root epidermal cells under conditions of Fe deficiency can only supplement and not replace AtIRT1 in Zn and Fe uptake (Grotz and Guerinot 2002). AtIRT2 has been recently reported to be involved in Fe homeostasis by transporting Fe into endo-membrane vesicles and is not involved in cellular uptake of Fe (Vert et al. 2009). Recently 11 members of ZIP family of *Arabidopsis* have been studied in detail by Milner et al. (2013). They report from yeast complementation studies that possibly ZIP7 can transport Zn, Mn and Fe; ZIP1 and ZIP2 can transport Zn and Mn; ZIP3, ZIP11 and ZIP12 can transport Zn alone; ZIP5, ZIP6 and ZIP9 can transport Mn alone; and none can transport Cu. According to them (Milner et al. 2013), AtZIP1 does not have a major role in Zn uptake. Due to its localisation in the vacuole, it probably remobilises Zn and Mn from the vacuole to the cytosol. AtZIP2 localised in the plasma membrane is probably not involved in root Zn or Mn uptake from the soil. It translocates Mn (possibly Zn) from root to shoot.

ZIP transporters have been identified from a number of plants mainly dicots (Grotz and Guerinot 2006). These are involved in transport of various metal ions such as Mn^{2+}, Fe^{2+}/Fe^{3+}, Cd^{2+}, Co^{2+}, Cu^{2+}, Ni^{2+} and especially Zn^{2+}.

Barley HvIRT1 of barley is an ortholog of AtIRT1 of *Arabidopsis* and is localised in the plasma membrane. It is involved in uptake of Mn^{2+} and Fe^{2+}/Fe^{3+} (Pedas et al. 2008). HvIRT1 is upregulated under Fe-deficient conditions. It is also upregulated due to Mn deficiency and is correlated with increased Mn uptake in a Mn-efficient variety of barley (Pedas et al. 2008).
Rice Rice under Fe deficiency induces expression of OsIRT1 and OsIRT2 in the root epidermis. Similar to AtIRT1, both transport Fe when expressed in yeast (Ishimaru et al. 2006). As AtIRT1 in *Arabidopsis*, both OsIRT1 and OsIRT2, when expressed in yeast, transport Cd under conditions of Fe deficiency, and there is an increase in Cd uptake and translocation by the rice plant (Nakanishi et al. 2006). OsZIP4 is localised in plasma membrane and is involved in Zn uptake (Ishimaru et al. 2005).

Tomato LeIRT1 and LeIRT2 are expressed predominantly in roots of tomato plants (Eckhardt et al. 2001). Under iron deficiency, LeIRT1, and not LeIRT2, gene is strongly induced along with genes of P and K transporters, which suggest a possible co-regulation of transporter genes. Deficiencies of P and K also upregulate LeIRT1. Both LeIRT1 and LeIRT2 appear to have a broad substrate base (Wang et al. 2002).

Soybean GmZIP1 identified from soybean is highly specific for Zn and expressed in nodules and not in roots, stems or leaves. The protein is localised in the peri-bacteroid membrane, which suggests its possible role in symbiosis (Moreau et al. 2002).

9.4.2.6 The CAX Family (Cation/H^+ Antiporters)

The CAX proteins are divalent cation/H^+ antiporters involved in cation influx into the vacuole. The *Arabidopsis* AtCAX1 is a vacuolar high-affinity Ca^{2+}/H^+ antiporter. AtCAX2 has low affinity for Ca^{2+} and possibly transports Mn^{2+} and Cd^{2+} across the tonoplast (Hirschi et al. 1996; 2000). Both AtCAX1 and CAX2 have 11 putative transmembrane domains and a central hydrophilic region rich in acidic amino acid residues between TM-6 and TM-7 (Hirschi 2001; Maeshima 2001; Hall and Williams 2003). Nine additional CAX genes have been identified in the *Arabidopsis* genome (Mäser et al. 2001). AtMHX1 is an additional member of this family, which is an H^+-coupled antiporter and can transport Mg^{2+} and Zn^{2+} across tonoplast (Shaul et al. 1999). AtCAX4 is expressed in root apex and lateral root primordia. Expression of AtCAX4 is induced by increasing Ni^{2+} and Mn^{2+} levels and decreasing

Ca^{2+} levels. This indicates that AtCAX4 is a cation/H^+ antiporter, which is involved in root growth under heavy metal stress conditions (Mei et al. 2009).

References

Abdel-Ghany SE, Pilon M (2008) Micro RNA-mediated systemic down-regulation of copper protein expression in response to low copper availability in *Arabidopsis*. J Biol Chem 283:15932–15945

Abdel-Ghany SE, Muller-Moule P, Niyogi KK, Pilon M, Shikanai T (2005) Two P-type ATPases are required for copper delivery in *Arabidopsis thaliana* chloroplasts. Plant Cell 17:1233–1251

Alam MZ, Rahman MM (2003) Accumulation of arsenic in rice plant from arsenic contaminated irrigation water and effect on nutrient content. In: Ahmed MF (ed) Proceeding of international symposium on fate of Arsenic in the environment. BUET, Dhaka, pp 131–136

Ames B, Shigenaga M, Hagen T (1993) Oxidants, antioxidants, and the degenerative diseases of ageing. Proc Natl Acad Sci U S A 90:7915–7922

Andres-Colas N, Sancenon V, Rodriguez-Navarro S, Mayo S, Thiele DJ, Ecker JR, Puig S, Penarrubia L (2006) The *Arabidopsis* heavy metal P-type ATPase HMA5 interacts with metallochaperones and functions in copper detoxification of roots. Plant J 45:225–236

Asian crops and micronutrient toxicity (2001) Food and Fertiliser Technology centre for Asian and Pacific Region, Taiepei, Taiwan, 1 Sept 2001

Axelsen KB, Palmgren MG (2001) Inventory of the superfamily of P-type ion pumps in *Arabidopsis*. Plant Physiol 126:696–706

Azizur Rahman M, Hasegawa H, Mahfuzur Rahman M, Nazrul Islam M, Majid Miah MA, Tasmen A (2007) Accumulation of arsenic in tissues of rice plant (*Oryza Sativa* L.) and its distributions in fractions of rice grains. Chemosphere 69:942–948

Baulcombe D (2004) RNA silencing in plants. Nature 431:356–363

Baumberger N, Baulcombe DC (2006) *Arabidopsis* ARGONAUTE1 is an RNA Slicer that selectively recruits micro RNAs and short interfering RNAs. Proc Natl Acad Sci U S A 102:11928–11933

Baxter I, Tchieu J, Sussman MR, Boutry M, Palmgren MG, Gribskov M, Harper JF, Axelsen KB (2003) Genomic comparison of P-type ATPase ion pumps in *Arabidopsis* and rice. Plant Physiol 132:618–628

Belouchi A, Kwan T, Gros P (1997) Cloning and characterization of the OsNramp family from Oryza sativa, a new family of membrane proteins possibly implicated in the transport of metal ions. Plant Mol Biol 33:1085–1092

Bera AK, Kanta-Bokaria AK, Bokaria K (1999) Effect of tannery effluent on seed germination, seedling growth and chloroplast pigment content in mung bean (*Vigna radiata* LWilczek). Environ Ecol 17(4):958–961

Bereczky Z, Wang HY, Schubert V, Ganal M, Bauer P (2003) Differential regulation of *nramp* and *irt* metal transporter genes in wild type and iron uptake mutants of tomato. J Biol Chem 278:24697–24704

Bernard DG, Cheng Y, Zhao Y, Balk J (2009) An allelic mutant series of ATM3 reveals its key role in the biogenesis of cytosolic iron-sulfur proteins in *Arabidopsis*. Plant Physiol 151:590–602

Bianucci E, Sobrino-Plata J, Carpena-Ruiz RO, Tordable M d-C, Fabra A, Hernández LE, Castro S (2012) Contribution of phytochelatins to cadmium tolerance in peanut plants. Metallomics 4:1119–1124

Blindauer CA, Leszczyszyn OI (2010) Metallothioneins: unparalleled diversity in structures and functions for metal ion homeostasis and more. Nat Prod Rep 27:720–741

Borsani O, Zhu J, Verslues PE, Sunkar R, Zhu JK (2005) Endogenous siRNAs derived from a pair of natural *cis*-antisense transcripts regulate salt tolerance in *Arabidopsis*. Cell 123:1279–1291

Bovet L, Feller U, Martinoia E (2005) Possible involvement of plant ABC transporters in cadmium detoxification: A cDNA sub-microarray approach. Environ Int 31:263–267

Cai M-Z, Zhang S-N, Xing C-H, Wang F-M, Zhu L, Wang N, Lin L-Y (2012) Interaction between iron plaque and root border cells ameliorates aluminum toxicity of *Oryza sativa* differing in aluminum tolerance. Plant and Soil 353(1/2):155

Cailliatte R, Lapeyre B, Briat JF, Mari S, Curie C (2009) The NRAMP6 metal transporter contributes to cadmium toxicity. Biochem J 422:217–228

Cailliatte R, Schikora A, Briat J-F, Mari S, Curie C (2010) High-affinity manganese uptake by the metal transporter NRAMP1 is essential for *Arabidopsis* growth in low manganese conditions. Plant Cell 22(3):904–917

Carrasco-Gil L, Álvarez-Fernández A, Sobrino-Plata J, Milán R, Carpena-Ruiz RO, Leduc DL, Andrews JC, Abadía J, Hernández LE (2011) Complexation of Hg with phytochelatins is important for plant Hg tolerance. Plant Cell Environ 34:778–791

Catmack I (2000) Possible roles of zinc in protecting plant cells from damage by reactive oxygen species. New Phytol 146:185–205

Chapman EJ, Carrington JC (2007) Specialization and evolution of endogenous small RNA pathways. Nat Rev Genet 8:884–896

Chen X (2009) Small RNAs and their roles in plant development. Annu Rev Cell Dev Biol 25:21–44

Chen X-Z, Peng J-B, Cohen A, Nelson H, Nelson N, Hediger MA (1999) Yeast SMF1 mediates H^+-coupled iron uptake with concomitant uncoupled cation currents. J Biol Chem 274:35089–35094

Cherian MG, Kang YJ (2006) Metallothionein and liver cell regeneration. Exp Biol Med (Maywood) 231:138–144

Clijsters H, Van Assche F (1985) Inhibition of photosynthesis by heavy metals. Photosynth Res 7:31–40

Cobbett SC (2000) Phytochelatins and their roles in heavy metal detoxification. Plant Physiol 123(3):825–832

Cobbett C, Goldsbrough P (2002) Phytochelatins and metallothioneins: roles in heavy metal detoxification and homeostasis. Annu Rev Plant Biol 53:159–182

Connolly EL, Fett JP, Guerinot ML (2002) Expression of the IRT1 metal transporter is controlled by metals at the levels of transcript and protein accumulation. Plant Cell 14:1347–1357

Curie C, Alonso JM, Le Jean M, Ecker JR, Briat JF (2000) Involvement of NRAMP1 from *Arabidopsis thaliana* in iron transport. Biochem J 347:749–755

Cuypers A, Smeets K, Ruytinx J, Opdenakker K, Keunen E, Remans T, Horemans N, Vanhoudt N, van Sanden S, van Belleghem F et al (2011) The cellular redox state as a modulator in cadmium and copper responses in *Arabidopsis thaliana* seedlings. J Plant Physiol 168:309–316

Das P, Samantaray S, Rout GR (1997) Studies on cadmium toxicity in plants. A review. Environ Pollut 98(1):29–36

Davies TGE, Coleman JOD (2000) The *Arabidopsis thaliana* ATP-binding cassette proteins: an emerging super family. Plant Cell Environ 23:431–443

Davies FT, Puryear JD, Newton RJ, Egilla JN, Grossi JAS (2002) Mycorrhizal fungi increase chromium uptake by sunflower plants: influence on tissue mineral concentration, growth, and gas exchange. J Plant Nutr 25:2389–2407

Dean RT, Gieseg S, Davies M (1993) Reactive species and their accumulation on radical-damaged proteins. Trends Biochem Sci 18:437–441

Ding L, Candido EPM (2000) HSP25, a small heat shock protein associated with dense bodies and M-lines of body wall muscle in *Caenorhabditis elegans*. J Biol Chem 275:9510–9517

Ding Y-F, Zhu C (2009) The role of microRNAs in copper and cadmium homeostasis. Biochem Biophys Res Commun 386:6–10

Eckhardt U, Margues AM, Buckhout TJ (2001) Two iron-regulated cation transporters from tomato complement metal uptake-deficient yeast mutants. Plant Mol Biol 45:437–448

Eide DJ (1998) The molecular biology of metal ion transport in *Saccharomyces cerevisiae*. Annu Rev Nutr 18:441–469

Eide D, Broderius M, Fett J, Guerinot ML (1996) A novel iron-regulated metal transporter from plants identified by functional expression in yeast. Proc Natl Acad Sci U S A 93:5624–5628

Eren E, Arguello JM (2004) Arabidopsis HMA2, a divalent heavy metal-transporting (P_{IB})-type ATPase, is involved in cytoplasmic Zn^{2+} homeostasis. Plant Physiol 136:3712–3723

Finster ME, Gray KA, Binns HJ (2003) Lead levels of edibles grown on lead contaminated residential soils: a field survey. www.sciencedirect.com

Freisinger E (2011) Structural features specific to plant metallothioneins. J Biol Inorg Chem 16(7): 1035–1045

Gaither LA, Eide DJ (2001) Eukaryotic zinc transporters and their regulation. Biometals 14:251–270

Garcia O, Bouige P, Forestier C, Dassa E (2004) Inventory and comparative analysis of rice and *Arabidopsis* ATP-binding cassette (ABC) systems. J Mol Biol 343:249–265

Ghosh M, Shen J, Rosen BP (1999) Pathways of As (III) detoxification in *Saccharomyces cerevisiae*. Proc Natl Acad Sci U S A 96:5001–5006

Gielen H, Remans T, Vangronsveld J, Cuypers A (2012) Micro RNAs in metal stress: specific roles or secondary responses? Int J Mol Sci 13(12):15826–15847

Goldsbrough P (2000) Metal tolerance in plants: the role of phytochelatins and metallothioneins. In: Terry N, Banuelos G (eds) Phytoremediation of contaminated soil and water. Lewis Publishers, Boca Raton, pp 221–233

Golovatyj SE, Bogatyreva EN, Golovatyi SE (1999) Effect of levels of chromium content in a soil on its distribution in organs of corn plants. Soil Res Fert 197–204

Gravot A, Lieutaud A, Verret F, Auroy P, Vavasseur A, Richaud P (2004) AtHMA3, a plant P1B-ATPase, functions as a Cd/Pb transporter in yeast. FEBS Lett 561:22–28

Grennan AK (2011) Metallothioneins, a diverse protein family. Plant Physiol 155(4):1750–1751

Grotz N, Guerinot ML (2002) Limiting nutrients: an old problem with new solutions? Curr Opin Plant Biol 5:158–163

Grotz N, Guerinot ML (2006) Molecular aspects of Cu, Fe and Zn homeostasis in plants. Biochem Biophys Acta 1763:595–608

Gueldry O, Lazard M, Delort F, Dauplais M, Grigoras I, Blanquet S, Plateau P (2003) Ycf1p-dependent Hg (II) detoxification in *Saccharomyces cerevisiae*. Eur J Biochem 270:2486–2496

Guerinot ML (2000) The ZIP family of metal transporters. Biochim Biophys Acta 1465:190–198

Gunshin H, Mackenzie B, Berger UV, Gunshin Y, Romero MF, Boron WF, Nussberger S, Gollan JL, Hediger MA (1997) Cloning and characterization of a mammalian proton-coupled metal-ion transporter. Nature 388:482–488

Hall JL (2002) Cellular mechanisms for heavy metal detoxification and tolerance. J Exp Bot 53:1–11

Hall JL, Williams LE (2003) Transition metal transporters in plants. J Exp Bot 54(393):2601–2613

Hamilton EW III, Heckathorn SA (2001) Mitochondrial adaptations to NaCl. Complex I is protected by antioxidants and small heat shock proteins, whereas complex II is protected by proline and betaine. Plant Physiol 126:1266–1274

Hanikenne M, Motte P, Wu MCS, Wang T, Loppes R, Matagne RF (2005) A mitochondrial half-size ABC transporter is involved in cadmium tolerance in *Chlamydomonas reinhardtii*. Plant Cell Environ 28:863–873

Harding JJ (1991) In cataract; biochemistry, epidemiology and pharmacology. Chapman & Hall, London

Härndahl U, Hall RB, Osteryoung KW, Vierling E, Bornman JF, Sundby C (1999) The chloroplast small heat shock protein undergoes oxidation-dependent conformational changes and may protect plants from oxidative stress. Cell Stress Chaperones 4:129–138

Haslbeck M, Walke S, Stromer T, Ehrnsperger M, White HE, Chen S, Saibil HR, Buchner J (1999) Hsp26: a temperature regulated chaperone. EMBO J 18:6744–6751

Hawes MC (1991) Living plant cells released from the root cap: a regulator of microbial populations in the rhizosphere? In: Keiser DL, Cregan PB (eds) The rhizosphere and plant growth. Kluwer Academic Publishers, Dordrecht, pp 51–59

Hawes MC, Lin H-J (1990) Correlation of pectolytic enzyme activity with the programmed release of cells from root caps of pea (*Pisum sativum*). Plant Physiol 94:1855–1859

Hawes MC, Bengough G, Cassab G, Ponce G (2003) Root caps and rhizosphere. J Plant Growth Regul 21:352–367

Hawes MC, Woo H, Wen F (2005) Root border cells: a delivery system for chemicals controlling plant health. Roots and soil management: interactions between roots and soil, agronomy monograph no. 48, 107–117. American Society of Agronomy, Crop Science Society of America, Soil Science Society of America, Madison

Hegelund JN, Sciller M, Kichey T, Hansen TH, Pedas P, Husted S, Schjoerring JK (2012) Barley metallothioneins: MT3 and MT4 are localised in the grain aleurone layer and show differential Zn binding. Plant Physiol 159:1125–1137

Hirayama T, Kieber JJ, Hirayama N, Kogan M, Guzman P, Nourizadeh S, Alonso JM, Dailey WP, Dancis A, Ecker JR (1999) RESPONSIVE-TO-ANTAGONIST1, a Menkes/Wilson disease-related copper transport, is required for ethylene signaling in Arabidopsis. Cell 97:383–393

Hirschi K (2001) Vacuolar H^+/Ca^{2+} transport: who's directing the traffic? Trends Plant Sci 6:100–104

Hirschi KD, Zhen R-G, Cunningham KW, Rea PA, Fink GR (1996) CAX1, an H^+/Ca^{2+} antiporter from *Arabidopsis*. Proc Natl Acad Sci U S A 93:8782–8786

Hirschi KD, Korenkov VD, Wilganowski NL, Wagner GJ (2000) Expression of *Arabidopsis CAX2* in tobacco. Altered metal accumulation and increased manganese tolerance. Plant Physiol 124:125–134

Howden R, Goldsbrough PB, Andersen CR, Cobbett CS (1995) Cadmium-sensitive, cad1, mutants of *Arabidopsis thaliana* are phytochelatin deficient. Plant Physiol 107:1059–1066

Huang L, Gitschier J (1997) A novel gene involved in zinc transport is deficient in the lethal milk mouse. Nat Genet 17:292

Huang SQ, Peng J, Qiu CX, Yang ZM (2009) Heavy metal-regulated new microRNAs from rice. J Inorg Biochem 103:282–287

Huang SQ, Xiang AL, Che LL, Chen S, Li H, Song JB, Yang ZM (2010) A set of miRNAs from *Brassica napus* in response to sulphate deficiency and cadmium stress. Plant Biotechnol J 8:887–899

Huffman EWD Jr, Allaway HW (1973) Chromium in plants: distribution in tissues, organelles, and extracts and availability of bean leaf Cr to animals. J Agric Food Chem 21:982–986

Hussain D, Haydon MJ, Wang Y, Wong E, Sherson SM, Young J, Camakaris J, Harper JF, Cobbett CS (2004) P-type ATPase heavy metal transporters with roles in essential zinc homeostasis in *Arabidopsis*. Plant Cell 16:1327–1339

Hüttermann A, Arduini I, Godbold DL (1999) Metal pollution and forest decline. In: Prasad NMV, Hagemeyer J (eds) Heavy metal stress in plants: from molecules to ecosystems. Springer, Berlin, pp 253–272

Ishimaru Y, Suzuki M, Kobayashi T, Takahashi M, Nakanishi H, Mori S et al (2005) OsZIP4, a novel zinc-regulated zinc transporter in rice. J Exp Bot 56:3207–3214

Ishimaru Y, Suzuki M, Tsukamoto T, Suzuki K, Nakazono M, Kobayashi T, Wada Y, Watanabe S, Matsuhashi S, Takahashi M, Nakanishi H, Mori S, Nishizawa NK (2006) Rice plants take up iron as an Fe^{3+}-phytosiderophore and as Fe^{2+}. Plant J 45:335

Ishimaru Y, Takahashi R, Bashir K et al (2012) Characterizing the role of rice NRAMP5 in manganese, iron and cadmium transport. Sci Rep 2:286

Iwasaki K, Sakurai K, Takahashi E (1990) Copper binding by the root cell walls of Italian ryegrass and red clover. Soil Sci Plant Nutr 36:431–439

Jackson AP, Alloway BJ (1991) The transfer of cadmium from sewage sludge amended soils into the edible component of food crops. Water Air Soil Pollut 57:873–881

Jentschke G, Godbold DL (2000) Metal toxicity and ectomycorrhizas. Physiol Plant 109:107–116

Kagi JHR, Schaffer A (1988) Biochemistry of metallothionein. Biochemistry 27:8509–8515

Kahle H (1993) Response of roots of trees to heavy metals. Environ Exp Bot 33:99–119

Kaiser BN, Moreau S, Castelli J, Thomson R, Lambert A, Bogliolo S, Puppo A, Day DA (2003) The soybean NRAMP homologue, GmDMT1, is a symbiotic divalent metal transporter capable of ferrous iron transport. Plant J 35:295–304

Kaldorf M, Kuhn AJ, Schröder WH, Hildebrandt U, Bothe H (1999) Selective element deposits in maize colonized by a heavy metal tolerance conferring arbuscular mycorrhizal fungus. J Plant Physiol 154:718–728

Kang J, Park J, Choi H, Burla B, Kretzschmar T, Lee Y, Martinoia E (2011) Plant ABC transporters. Arabidopsis Book 9:e0153. doi:10.1199/tab.0153, Published online 2011 December 6

Kawashima I, Kennedy TD, Chino M, Lane BG (1992) Wheat E_c metallothionein genes. Eur J Biochem 209:971–976

Khraiwesh B, Arif MA, Seumel GI, Ossowski S, Weigel D, Reski R, Frank W (2010) Transcriptional control of gene expression by micro RNAs. Cell 140:111–122

Khraiwesh B, Zhu J-K, Zhu J (2012) Role of miRNAs and siRNAs in biotic and abiotic stress responses of plants. Biochim Biophys Acta 1819(2):137–148

Klapheck S, Schlunz S, Bergmann L (1995) Synthesis of phytochelatins and homo-phytochelatins in *Pisum sativum* L. Plant Physiol 107:515–521

Kopittke PM, de Jong MD, Menzies NW, Wang P, Donner E et al (2012) Examination and distribution of arsenic in hydrated and fresh cowpea roots using two and three dimensional techniques. Plant Physiol 159(3):1149–1158

Kretzschmar T, Burla B, Lee Y, Martinoia E, Nagy R (2011) Functions of ABC transporters in plants. Essays Biochem 50:145–160

Kushnir S, Babiychuk E, Storozhenko S, Davey MW, Papenbrock J, De Rycke R, Engler G, Stephan UW, Lange H, Kispal G (2001) A mutation of the ABC transporter Sta1 leads to dwarfism and chlorosis in the *Arabidopsis* mutant starik. Plant Cell 13:89–100

Lanet E, Delannoy E, Sormani R, Floris M, Brodersen P, Crete P, Voinnet O, Robaglia C (2009) Biochemical evidence for translational repression by *Arabidopsis* micro RNAs. Plant Cell 21:1762–1768

Lanquar V, Lelievre F, Bolte S et al (2005) Mobilization of vacuolar iron by AtNRAMP3 and AtNRAMP4 is essential for seed germination on low iron. EMBO J 24:4041–4051

Lee S, Kim Y-Y, Lee Y, An G (2007) Rice P_{1B}-type heavy-metal ATPase, OsHMA9, is a metal efflux protein. Plant Physiol 145:831–842

Lewis S, Handy RD, Cordi B, Billinghurst Z, Depledge MH (1999) Stress proteins (HSPs): methods of detection and their use as an environmental biomarker. Eco Toxicol 8:351–368

Li YH, Chaney RL, Schneiter AA (1994) Effect of soil chloride level on cadmium concentration in sunflower kernels. Plant Soil 167:275–280

Li ZS, Lu YP, Zhen RG, Szczypka M, Thiele DJ, Rea PA (1997) A new pathway for vacuolar cadmium sequestration in *Saccharomyces cerevisiae*: YCF1-catalyzed transport of bis(glutathionato) cadmium. Proc Natl Acad Sci U S A 94:42–47

Li T, Li H, Zhang Y-X, Liu J-Y (2011) Identification and analysis of seven H_2O_2-responsive miRNAs and 32 new miRNAs in the seedlings of rice (*Oryza sativa* L. ssp. *indica*). Nucleic Acids Res 39(7):2821–2833

Liu XF, Supek F, Nelson N, Culotta VC (1997) Negative control of heavy metal uptake by the *Saccharomyces cerevisiae BSD2* gene. J Biol Chem 272:11763–11769

Liu GE, Simonne H, Li Y (2012) Nickel nutrition in plants, HS1191, one of a series of the horticultural sciences, Florida Cooperative Extension Service, Institute of Food and Agricultural Sciences, University of Florida. Original publication date: June 2011

Loebus J, Peroza EA, Bluthgen N, Meyer-Klauke W, Zerbe O, Freisinger E (2011) Protein and metal cluster structure of the wheat metallothionein domain γ-E (c)-1: the second part of the puzzle. J Biol Inorg Chem 16:683–694

Ma JF, Zheng SJ, Matsumoto H (1997) Detoxifying aluminium with buckwheat. Nature 390:569–570

Ma JF, Hiradate S, Matsumoto H (1998) High aluminum resistance in buckwheat: II. Oxalic acid detoxifies aluminum internally. Plant Physiol 117:753–759

Ma JF, Ryan PR, Delhaize E (2001) Aluminium tolerance in plants and the complexing role of organic acids. Trends Plant Sci 6:273–278

Maeshima M (2001) Tonoplast transporters: organization and function. Annu Rev Plant Physiol Plant Mol Biol 52:469–497

Maitani T, Kubota H, Sato K, Yamada T (1996) The composition of metals bound to class III metallothionein (phytochelatin and its desglycyl peptide) induced by various metals in root cultures of *Rubia tinctorum*. Plant Physiol 110:1145–1150

Marini I, Moschini R, Corso AD, Mura U (2000) Complete protection by a-crystallin of lens sorbitol dehydrogenase undergoing thermal stress. J Biol Chem 275:32559–32565

Marschner H (1995) Mineral nutrition of higher plants. Academic, San Diego

Martinoia E, Grill E, Tommasini R, Kreuz K, Amrhein N (1993) An ATP-dependent glutathione *S*-conjugate 'export' pump in the vacuolar membrane of plants. Nature 364:247–249

Martinoia E, Klein M, Geisler M, Bovet L, Forestier C, Kolukisaoglu U, Müller-Röber B, Schulz B (2002) Multi functionality of plant ABC transporters—more than just detoxifiers. Planta 214:345–355

Mäser P, Thomine S, Schroeder JI et al (2001) Phylogenetic relationships within cation transporter families of *Arabidopsis*. Plant Physiol 126:1646–1667

Meharg AA (1993) The role of the plasmalemma in metal tolerance in angiosperms. Physiol Plant 88: 191–198

Meharg AA (1994) Integrated tolerance mechanisms – constitutive and adaptive plant-responses to elevated metal concentrations in the environment. Plant Cell Environ 17:989–993

Meharg AA, Hartley-Whitaker J (2002) Arsenic uptake and metabolism in arsenic resistant and non-resistant plant species. New Phytol 154:29–43

Mei H, Cheng NH, Zhao J, Park S, Escareno RA, Pittman JK, Hirschi KD (2009) Root development under metal stress in *Arabidopsis thaliana* requires the H+/cation antiporter CAX4. New Phytol 183(1):95–105

Mills RF, Krijger GC, Baccarini PJ, Hall JL, Williams LE (2003) Functional expression of AtHMA4, a P_{1B}-type ATPase in the Zn/Co/Cd/Pb subclass. Plant J 35:164–175

Mills RF, Peaston KA, Runions J, Williams LE (2012) HvHMA2, a P(1B)-ATPase from barley, is highly conserved among cereals and functions in Zn and Cd transport. PLoS One 7(8):e4260

Milner MJ, Seamon J, Craft E, Kochian LV (2013) Transport properties of members of the ZIP family in plants and their role in Zn and Mn homeostasis. J Exp Bot 64(1):369–381

Miyasaka SC, Hawes MC (2001) Possible role of root border cells in detection and avoidance of aluminium toxicity. Plant Physiol 125:1978–1987

Mizuno T, Usui K, Horie K, Nosaka S, Mizuno N, Obata H (2005) Cloning of three ZIP/Nramp transporter genes from a Ni hyperaccumulator plant *Thlaspi japonicum* and their Ni^{2+}-transport abilities. Plant Physiol Biochem 43:793–801

Moreau S, Thomson RM, Kaiser BN, Trevaskis B, Guerinot ML, Udvardi MK, Puppo A, Day DA (2002) GmZIP1 encodes a symbiosis-specific zinc transporter in soybean. J Biol Chem 277:4738–4746

Nable RO, Banuelos GS, Paull JG (1997) Boron toxicity. Plant and Soil 193:181–198

Nakanishi H, Ogawa I, Ishimaura Y, Mori S, Nishizawa NK (2006) Possibility of using other plant species for Cd phytoextraction. Soil Sci Plant Nutr 52:32–42

Nelson N (1999) Metal ion transporters and homeostasis. EMBO J 18:4361–4371

Neumann D, Zur Nieden U, Lichtenberger O, Leopold I (1995) How does *Armeria maritima* tolerate high heavy metal concentrations? J Plant Physiol 146:704–717

Neumann D, zur Nieden U, Schwieger W, Leopold I, Lichtenberger O (1997) Heavy metal tolerance in *Minuartia verna*. J Plant Physiol 151:101–108

Nevo Y, Nelson N (2006) The NRAMP family of metal-ion transporters. Biochim Biophys Acta 1763: 609–620

Oomen RJFJ, Wu J, Lelievre F, Blanchet S, Richaud P, Barbier-Brygoo H, Aarts MGM, Thomine S (2009) Functional characterization of NRAMP3 and NRAMP4 from the metal hyperaccumulator *Thlaspi caerulescens*. New Phytol 181:637–650

Oven M, Page J, Zenk M, Kutchan T (2002) Molecular characterization of the homophytochelatin synthase of soybean *Glycine max*: relation to phytochelatin synthase. J Biol Chem 277:4747–4754

Pal R, Rai JP (2010) Phytochelatins: peptides involved in heavy metal detoxification. Appl Biochem Biotechnol 160(3):945–963

Palmgren MG, Axelsen KB (1998) Evolution of P-type ATPases. Biochim Biophys Acta 1365:37–45

Palmgren MG, Harper JF (1999) Pumping with P-type ATPases. J Exp Bot 50:883–893

Park J, Song WY, Ko D, Eom Y, Hansen TH, Schiller M, Lee TG, Martinoia E, Lee Y (2012) The phytochelatin transporters AtABCC1 and AtABCC2 mediate tolerance to cadmium and mercury. Plant J 69(2): 278–288

Paulsen IT, Saier MH Jr (1997) A novel family of ubiquitous heavy metal ion transport proteins. J Membr Biol 156:99–103

Pedas P, Ytting CK, Fuglsang AT, Jahn TP, Schjoerring JK, Husted S (2008) Manganese efficiency in barley: identification and characterization of the metal ion transporter HvIRT1. Plant Physiol 148:455–466

Pence NS, Larsen PB, Ebbs SD, Lasat MM, Letham DLD, Garvin DF, Eide D, Kochian LV (2000) The molecular basis for heavy metal hyperaccumulation in *Thlaspi caerulescens*. Proc Natl Acad Sci U S A 97:4956–4960

Peroza EA, Freisinger E (2007) Metal ion binding properties of *Triticum* [corrected] *aestivum* Ec-1 metallothionein: evidence supporting two separate metal thiolate clusters. J Biol Inorg Chem 12(3):377–391

Polle A, Rennenberg H (1993) Significance of antioxidants in plant adaptation to environmental stress. In: Mansfield T, Fowden L, Stoddard F (eds) Plant adaptation to environmental stress. Chapman & Hall, London, pp 263–273

Prasad MNV (1999) Metallothioneins and metal binding complexes in plants. In: Prasad MNV, Hagemeyer J (eds) Heavy metal stress in plants: from molecules to ecosystems. Springer, Berlin, pp 51–72

Punz W, Sieghardt H (1993) The response of roots of herbaceous plant species to heavy metals. Environ Exp Bot 33:85–98

Quartacci MF, Cosi E, Navari-Izzo F (2001) Lipids and NADPH-dependent superoxide production in plasma membrane vesicles from roots of wheat grown under copper deficiency or excess. J Exp Bot 52:77–84

Rauser WE (1995) Phytochelatins and related peptides. Structure, biosynthesis, and function. Plant Physiol 109:1141–1149

Rea PA (2007) Plant ATP-binding cassette transporters. Annu Rev Plant Biol 58:347–375

Reichman SM (2002) The responses of plants to metal toxicity: a review focusing on copper, manganese and zinc. Occasional Paper No. 14, Australian Minerals & Energy Environment Foundation, Melbourne

Remans T, Opdenakker K, Guisez Y, Carleer R, Schat H, Vangronsveld J, Cuypers A (2012) Exposure of *Arabidopsis thaliana* to excess Zn reveals a Zn-specific oxidative stress signature. Environ Exp Bot 84:61–71

Rensing C, Ghosh M, Rosen BP (1999) Families of soft-metal-ion-transporting ATPases. J Bacteriol 181:5891–5897

Rivetta A, Negrini N, Cocucci M (1997) Involvement of Ca^{2+}-calmodulin in Cd^{2+} toxicity during the early phases of radish (*Raphanus sativus* L.) seed germination. Plant Cell Environ 20:600–608

Robinson N, Tommey AM, Kuske C, Jackson PJ (1993) Plant metallothioneins. Biochem J 295(1):1–10

Rogers EE, Eide DJ, Guerinot ML (2000) Altered selectivity in an *Arabidopsis* metal transporter. Proc Natl Acad Sci U S A 97:12356–12360

Rout GR, Samantaray S, Das P (2001) Aluminium toxicity in plants: a review. Agronomie 21:3–21

Schmöger M, Oven M, Grill E (2000) Detoxification of arsenic by phytochelatins in plants. Plant Physiol 122:793–801

Schramke V, Allshire R (2004) Those interfering little RNAs! Silencing and eliminating chromatin. Curr Opin Genet Dev 14:174–180

Schützendübel A, Polle A (2002) Plant responses to abiotic stresses: heavy metal-induced oxidative stress and protection by mycorrhization. J Exp Bot 53(372):1351–1365

Seigneurin-Berny D, Gravot A, Auroy P, Mazard C, Kraut A, Finazzi G, Grunwald D, Rappaport F, Vavasseur A, Joyard J et al (2006) HMA1, a new Cu-ATPase of the chloroplast envelope, is essential for growth under adverse light conditions. J Biol Chem 281: 2882–2892

Semane B, Dupae J, Cuypers A, Noben JP, Tuomainen M, Tervahauta A, Kärenlampi S, van Belleghem F, Smeets K, Vangronsveld J (2007) Leaf proteome responses of *Arabidopsis thaliana* exposed to mild cadmium stress. J Plant Physiol 167:247–254

Shanker AK, Cervantes TC, Loza-Tavera H, Avudainayagam S (2005) Chromium toxicity in plants. Environ Int 31:739–753

Shaul O, Hilgemann DW, de-Almeida-Engler J, Van Montagu M, Inzé D, Galili G (1999) Cloning and characterization of a novel Mg^{2+}/H^{+} exchanger. EMBO J 18:3973–3980

Shen R, Iwashita T, Ma JF (2004) Form of Al changes with Al concentration in leaves of buckwheat. J Exp Bot 55(394):131–136

Shikanai T, Muller-Moule P, Munekage Y, Niyogi KK, Pilon M (2003) PAA1, a P-type ATPase of *Arabidopsis*, functions in copper transport in chloroplasts. Plant Cell 15:1333–1346

Smolders E (2001) Cadmium uptake by plants. Int J Occup Med Environ Health 14(2):177–183

Solioz M, Vulpe C (1996) CPx-type ATPases: a class of P-type ATPases that pump heavy metals. Trends Biochem Sci 21:237–241

Song WY, Sohn EJ, Martinoia E, Lee YJ, Yang YY, Jasinski M, Forestier C, Hwang I, Lee Y (2003) Engineering tolerance and accumulation of lead and cadmium in transgenic plants. Nat Biotechnol 21:914–919

Song W-Y, Choic KS, Kimb DY, Geislera M, Park J, Vincenzettia V, Schellenberg M, Kim SH, Limd YP, Nohe EW, Leeb Y, Martinoia E (2010) *Arabidopsis* PCR2 is a zinc exporter involved in both zinc extrusion and long-distance zinc transport. Plant Cell 22(7):2237–2252

Studer S, Narberhaus F (2000) Chaperone activity and homo- and hetero-oligomer formation of bacterial small heat shock proteins. J Biol Chem 275:37212–37218

Sun W, Bernard C, van de Cotte B, Van Montagu M, Verbruggen N (2001) At-HSP17.6A, encoding a small heat-shock protein in *Arabidopsis*, can enhance osmo-tolerance upon over expression. Plant J 27:407–415

Sunkar R, Zhu JK (2004) Novel and stress-regulated micro RNAs and other small RNAs from *Arabidopsis*. Plant Cell 16:2001–2019

Sunkar R, Kapoor A, Zhu JK (2006) Posttranscriptional induction of two Cu/Zn superoxide dismutase genes in *Arabidopsis* is mediated by down regulation of miR398 and important for oxidative stress tolerance. Plant Cell 18:2051–2065

Supek F, Supekova L, Nelson H, Nelson N (1997) Function of metal-ion homeostasis in the cell division cycle, mitochondrial protein processing, sensitivity to mycobacterial infection and brain function. J Exp Biol 200:321–330

Suzuki TC, Denise C, Krawitz DC, Vierling E (1998) The chloroplast small heat-shock protein oligomer is not phosphorylated and does not dissociate during heat stress in vivo. Plant Physiol 116:1151–1161

Szczypka MS, Wemmie JA, Moye-Rowley WS, Thiele DJ (1994) A yeast metal resistance protein similar to human cystic fibrosis trans membrane conductance regulator (CFTR) and multidrug resistance-associated protein. J Biol Chem 269:22853–22857

Takahashi R, Ishimaru Y, Senoura T, Shimo H, Ishikawa S, Arao T, Nakanishi H, Nishizawa NK (2011) The OsNRAMP1 iron transporter is involved in Cd accumulation in rice. J Exp Bot 62:4843–4850

Tennstedt P, Peisker D, Böttcher C, Trampczynska A, Clemens S (2009) Phytochelatin synthesis is essential for the detoxification of excess zinc and contributes significantly to the accumulation of zinc. Plant Physiol 149(2):938–948

Theodoulou FL (2000) Plant ABC transporters. Biochim Biophys Acta 1465:79–103

Thomine S, Wang R, Ward JM, Crawford NM, Schroeder JI (2000) Cadmium and iron transport by members of a plant metal transporter family in *Arabidopsis* with homology to Nramp genes. Proc Natl Acad Sci U S A 97:4991–4996

Tseng TS, Tzeng SS, Yeh KW et al (1993) The heat-shock response in rice seedlings: isolation and expression of cDNAs that include class I low molecular weight heat shock proteins. Plant Cell Physiol 34:165–168

Vajpayee P, Sharma SC, Tripathi RD, Rai UN, Yunus M (1999) Bioaccumulation of chromium and toxicity to photosynthetic pigments, nitrate reductase activity and protein content of *Nelumbo nucifera* Gaertn. Chemosphere 39:2159–2169

Van Assche F, Clijsters H (1986) Inhibition of photosynthesis in *Phaseolus vulgaris* by treatment with toxic concentration of zinc: effect on ribulose-1,5-bisphosphate carboxylase/oxygenase. J Plant Physiol 125:355–360

Van der Zaal BJ, Neuteboom LW, Pinas JE, Chardonnens AN, Schat H, Verkleij JAC, Hooykaas PJJ (1999) Over-expression of a novel *Arabidopsis* gene related to putative zinc-transporter genes from animals can lead to enhanced zinc resistance and accumulation. Plant Physiol 119:1047–1055

Vatamaniuk O, Mari S, Lu Y, Rea P (2000) Mechanism of heavy metal ion activation of phytochelatin (PC) synthase: blocked thiols are sufficient for PC synthase-catalyzed trans peptidation of glutathione and related thiol peptides. J Biol Chem 275:31451–31459

Vazquez F, Legrand S, Windels D (2010) The biosynthetic pathways and biological scopes of plant small RNAs. Trends Plant Sci 15:337–345

Verbruggen N, Hermans C, Schat H (2009) Mechanisms to cope with arsenic or cadmium excess in plants. Curr Opin Plant Boil 12:364–372

Verrier PJ, Bird D, Burla B, Dassa E, Forestier C, Geisler M, Klein M, Kolukisaoglu U, Lee Y, Martinoia E, Murphy A, Rea PA, Samuels L, Schulz B, Spalding EJ, Yazaki K, Theodoulou FL (2008) Plant ABC proteins-a unified nomenclature and updated inventory. Trends Plant Sci 13:151–159

Vert G, Grotz N, Dédaldéchamp F, Gaymard F, Guerinot ML, Briat J-F, Curie C (2002) IRT1, an *Arabidopsis* transporter essential for iron uptake from the soil and for plant growth. Plant Cell 14:1223–1233

Vert G, Barberon M, Zelazny E, Séguéla M, Briat JF, Curie C (2009) *Arabidopsis* IRT2 cooperates with the high-affinity iron uptake system to maintain iron homeostasis in root epidermal cells. Planta 229:1171–1179

Vierling E (1991) The roles of heat shock proteins in plants. Annu Rev Plant Physiol Plant Mol Biol 42:579–620

Wang Y-H, Garvin DF, Kochian LV (2002) Rapid induction of regulatory and transporter genes in response to phosphorus, potassium, and iron deficiencies in tomato roots. Evidence for cross talk and root/rhizosphere-mediated signals. Plant Physiol 130:1361–1370

Wang W, Vinocur B, Altman A et al (2003) Plant responses to drought, salinity and extreme temperatures: towards genetic engineering for stress tolerance. Planta 218:1–14

Waters ER, Lee GJ, Vierling E (1996) Evolution, structure and function of the small heat shock proteins in plants. J Exp Bot 47:325–338

Wei W, Chai T, Zhang Y, Han L, Xu J, Guan Z (2009) The *Thlaspi caerulescens* NRAMP homologue TcNRAMP3 is capable of divalent cation transport. Mol Biotechnol 41:15–21

Wildner GF, Henkel J (1979) The effect of divalent metal ion on the activity of Mg^{2+}-depleted ribulose-1,5-bisphosphate oxygenase. Planta 146:223–228

Williams LE, Mills RF (2005) P_{IB}-ATPases—an ancient family of transition metal pumps with diverse functions in plants. Trends Plant Sci 10:491–502

Williams LE, Pittman JK, Hall JL (2000) Emerging mechanisms for heavy metal transport in plants. Biochim Biophys Acta 1465:104–126

Wu L, Zhou H, Zhang Q, Zhang J, Liu C, Qi Y (2010) DNA methylation mediated by a Micro RNA pathway. Mol Cell 38(3):465–475

Xiao H, Yin L, Xu X, Li T, Han Z (2008) The iron-regulated transporter, MbNRAMP1, isolated from *Malus baccata* is involved in Fe, Mn and Cd trafficking. Ann Bot 102:881–889

Xing C-H, Zhu M-H, Cai M-Z, Liu P, Xu G-D, Wu S-H (2008) Developmental characteristics and response to iron toxicity of root border cells in rice seedlings. J Zhejiang Univ Sci B 9(3):261–264

Xiong H, Kobayashi T, Kakei Y et al (2012) *AhNRAMP1* iron transporter is involved in iron acquisition in peanut. J Exp Bot 63(12):4437–4446

Yamasaki H, Hayashi M, Fukazawa M, Kobayashi Y, Shikanai T (2009) *SQUAMOSA* promoter binding protein-like7 is a central regulator for copper homeostasis in *Arabidopsis*. Plant Cell 21:347–361

Yao Y, Ni Z, Peng H, Sun F, Xin M, Sunkar R, Zhu JK, Sun Q (2010) Non-coding small RNAs responsive to abiotic stress in wheat (*Triticum aestivum* L.). Funct Integr Genomics 10:187–190

Zenk MH (1996) Heavy metal detoxification in higher plants—a review. Gene 179:21–30

Zhou ZS, Huang SQ, Yang ZM (2008) Bioinformatic identification and expression analysis of new micro RNAs from *Medicago truncatula*. Biochem Biophys Res Commun 374:538–542

10 Iron (Fe) Uptake

Abstract

At the normal pH range of arable soils, available Fe is not enough to meet plant requirement. Deficiency of Fe occurs less in acid soils than calcareous soils with higher pH. Iron deficiency is a major health problem for humans around the world.

There are two distinct iron uptake systems based on the response of plants to Fe deficiency, Strategy I and Strategy II. Strategy I plants include all dicots and non-*graminaceous* monocots. Strategy II plants are limited to *graminaceous* monocots. These plants release mugineic acid (MA) family phytosiderophores to the rhizosphere, where they solubilise sparingly soluble iron by chelation. The chelated complex is then absorbed by the roots. The transporters involved in Fe uptake are (i) IRTs of ZIP family, (ii) Nramps, (iii) ABC transporter, (iv) H^+-ATPase and (v) the YSL transporters.

10.1 Occurrence of Iron and Soil Reactions

Iron constitutes about 5 % of the earth's crust. Most of the soils around the world are rich in iron. Iron in soil is present in the form of an amorphous Fe $(OH)_3$ precipitate, which is the immediate source of iron uptake by plants. Availability of Fe to plant roots depends on redox potential and pH of the soil.

$$Fe(OH)_3 + 3H^+ \rightleftarrows Fe^{3+} + 3H_2O$$

In well-drained, oxidised soils, concentration of Fe^{3+} is greater than Fe^{2+}. With increase in each unit of pH, Fe^{3+} concentration decreases 1,000-fold while Fe^{2+} decreases 100-fold similar to other divalent cations such as Mn^{2+}, Cu^{2+} and Zn^{2+}. At the normal pH range of arable soils, available Fe is not enough to meet plant requirement. Deficiency of Fe occurs less in acid soils than calcareous soils with higher pH.

10.2 Iron Content of Plants

Iron (Fe) is an essential micronutrient required for plant metabolism. Plant tissue concentration of 1–5 μM Fe is considered sufficient and a concentration below 1 μM is likely to cause deficiency. A concentration above 10 μM may cause toxicity with reduction of growth parameters (Mitra et al. 2009). However, these limits may

G.N. Mitra, *Regulation of Nutrient Uptake by Plants: A Biochemical and Molecular Approach*,
DOI 10.1007/978-81-322-2334-4_10, © Springer India 2015

vary considerably among different plant species and their genotypes. Rice grown on low land laterite soils often suffer from iron toxicity. In a study conducted on iron toxic soils in several locations with high available soil iron (DTPA) content (105–570 mg kg^{-1}), the rice leaves were found to contain 360–515 mg kg^{-1} of Fe (Mitra et al. 2009).

About 75 % of cell Fe is associated with chloroplast, and up to 90 % of Fe in leaves is associated with lipoproteins of membranes of chloroplast and mitochondria (Havlin et al. 2007).

Table 10.1 Sensitivity of crops to Fe deficiency

Sensitivity to Fe deficiency	Name of the crops
Highly sensitive	Cauliflower, citrus, field beans, sorghum, groundnut, soybean, spinach, vegetables
Medium sensitivity	Cabbage, corn, cotton, flax, grasses, oats, tomato, rice, wheat
Low sensitivity	Barley, corn, cotton, flax, grasses, oats, rice, soybean, sugar beet, wheat

Table 10.2 Distribution of micronutrients between husk and grain of rice (%) (av. of 15 cultivars) micronutrients

	Fe	Zn	Cu	B
Grain	23	54	47	45
Husk	77	46	53	55

From Jena et al. (2008)

10.3 Functions of Iron in Plants

10.3.1 Iron Deficiency

Iron deficiency causes a number of biochemical and morphological changes. These include root hair morphogenesis, differentiation of rhizodermal cells into transfer cells, chlorosis of leaves, ultrastructural disorganisation of chloroplasts and mitochondria, increased synthesis of organic acids and phenolics and activation of root system to increase its iron uptake capacity. Iron is not mobile in plants. Iron deficiency causes interveinal chlorosis in young leaves, which turn completely chlorotic and necrotic with time in severe deficient conditions.

Iron deficiency is a major health problem for humans around the world. According to one estimate, about 1.62 billion people (about 25 % of world population) suffer from iron deficiency (McLean et al. 2009).

Sensitivity of some of the crops to Fe deficiency is given in Table 10.1. Same crop sometimes is repeated in different columns. This happens due to varietal differences and varying growing conditions.

10.3.2 Iron Toxicity

Iron toxicity in rice causes formation of small brown spots on younger leaves, starting at the tips, which merge with time leading to characteristic leaf bronzing (Mitra et al. 2009). In some rice varieties, the leaves may turn yellow, purple or orange. The leaves of some varieties may roll. Lower leaves may turn brown and die. Growth and tillering are depressed. Root growth is stunted and coated with brown oxides of iron. In severe cases, grains are also tinted.

Laboratory and field studies have indicated that application of high doses of K to rice plants can ameliorate iron toxicity substantially on these soils (Mitra et al. 1990, 2009; Sahu and Mitra 1992; Sahu et al. 2001).

In an interesting study, where iron content of husk (outer coating of grain) and kernel were analysed, it was observed that 77 % of iron is retained by the husk and only 23 % filters into the edible kernel (Table 10.2).

10.3.3 Biochemical Functions of Iron

Iron is involved in a number of metabolic processes in plants. It is a constituent of a number of enzymes and proteins. Several flavoproteins contain firmly bound Fe^{3+} ion such as succinic $FADH_2$ dehydrogenases, dihydroorotic acid dehydrogenase, xanthine and aldehyde oxidases, etc. Iron is a structural component of a number of molecules with porphyrin ring such as cytochromes, hemes, ferrichromes and leghemoglobin. Most of the Fe-containing enzymes are involved in oxidation-reduction reactions in respiration and photosynthesis.

10.4 Mechanism of Iron Uptake by Plants

There are two distinct iron uptake systems based on the response of plants to Fe deficiency, Strategy I and Strategy II (Römheld 1987; Römheld and Marschner 1986; Bughio et al. 2002).

10.4.1 Strategy I Plants

These include all dicots and non-*graminaceous* monocots. Fe deficiency causes a decrease in rhizosphere pH of these plants to facilitate release of Fe^{3+} ion from insoluble sources. The sparingly soluble ferric iron is then reduced at the root surface by membrane-resident NADPH-dependent ferric chelate reductase (Chaney et al. 1972). In *Arabidopsis*, this reductase is encoded by the ferric reductase oxidase gene AtFRO2 (Robinson et al. 1999). Reduced ferrous iron is absorbed into root cells by the high-affinity Fe^{2+}- transporter, IRT1, a member of the ZIP-metal transporter family. IRT1-like Fe^{2+} transporters have been isolated from several dicotyledonous species (Eckhardt et al. 2001; Vert et al. 2001). All the three components of Fe uptake by Strategy I, such as release of protons to lower pH, expression of ferric chelate reductase gene to augment enzyme activity and expression of IRT1 transporter for absorption of Fe by root cells, increase substantially when plants are grown under Fe-deficient conditions (Conte and Walker 2011).

10.4.2 Strategy II Plants

These are limited to *graminaceous* monocots. These plants release mugineic acid (MA) family phytosiderophores to the rhizosphere, where they solubilise sparingly soluble iron by chelation. The chelated complex is then absorbed into the roots. Rice plants use MAs to acquire Fe from the rhizosphere. Synthesis of MAs and uptake of MA-chelated iron are strongly induced under iron-deficient conditions (Kobayashi et al. 2005). An induction ratio greater than 2.0 has been observed for 57 genes in roots, many of which are involved in Fe acquisition mechanism and methionine cycle as well as biosynthesis of MAs from methionine. It has been reported that all the Fe-deficiency-induced genes involved in Fe uptake have a higher incidence of homologous sequences of IDE1 and IDE2 (iron-deficiency-responsive cis-acting elements) in their promoter regions (Kobayashi et al. 2005). Genes for the synthesis of mugineic acid phytosiderophores (PS) have been isolated from barley, wheat, rice, maize and number of other monocots (Higuchi et al. 1999; Kobayashi et al. 2001; Okumura et al. 1994; Takahashi et al. 1999; Inoue et al. 2008). Rice produces less phytosiderophores than wheat and maize and hence less suitable to grow in calcareous soils under Fe-deficient conditions. In an experiment, enzymes involved in PS synthesis of barley were overexpressed in *Oryza sativa* var. *japonica* grown on an Fe-deficient soil. This lead to increased excretion of PS and there was a fourfold increase in rice yield (Takahashi et al. 2001; Morrissey and Guerinot 2009). Zn deficiency has also been reported to induce the synthesis and secretion of MAs in barley (Suzuki et al. 2006).

10.4.2.1 Mugineic Acid (MA) Family Phytosiderophores (PS)

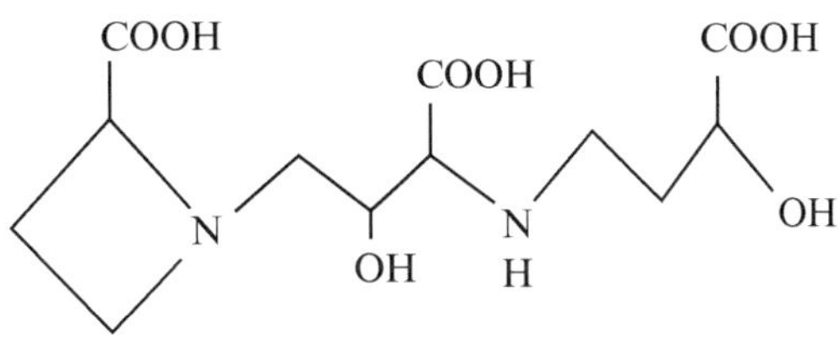

Mugineic acid

Mugineic acid (MA) family phytosiderophores (PS) are secreted by roots of *graminaceous* plants in response to iron deficiency. Although MAs differ in their structures among plant species and even cultivars within the same species, all of

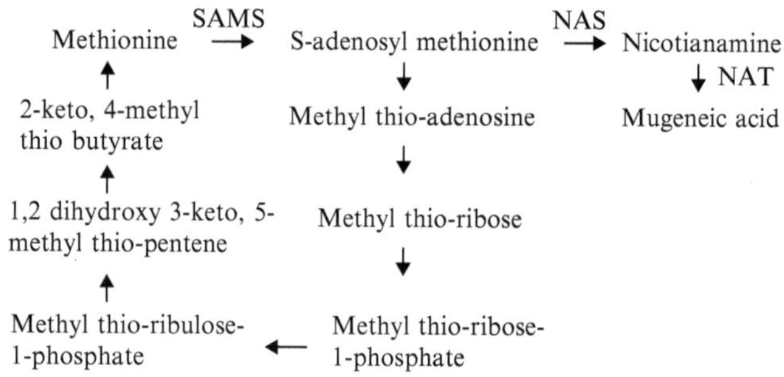

Fig 10.1 Methionine cycle. *SAMS* S-adenosyl methionine synthetase, *NAS* nicotianamine synthase, *NAT* nicotianamine transferase

them contain the same six functional groups which coordinate with the Fe^{3+} ion. All of the MAs share the same pathway from L-methionine to deoxymugineic acid. The subsequent steps differ among plant species and their cultivars. 2′-Deoxymugineic acid is synthesised from three molecules of L-methionine. Wheat roots secrete deoxymugineic acid (Ma et al. 1995). Barley roots secrete 2′-deoxymugineic acid (DMA), mugineic acid (MA) and epi-hydroxy mugineic acid (epi-HMA) (Suzuki et al. 2006). The synthesis of MAs proceeds throughout the day and stored in the roots (as much as 1–2 % of root-dry weight) and secreted to the rhizosphere next morning (Ma et al. 1995). However concentration of L-methionine (the precursor of MAs) is found to be very low in the roots to sustain a high rate of synthesis of MAs. This led to the discovery of methionine cycle, which continuously regenerates methionine to sustain synthesis of MAs (Ma et al. 1995). All enzymes involved in the methionine cycle have been reported for *Bacillus subtilis* (Sekowska et al. 2004; Suzuki et al. 2006). The cycle as described by Suzuki et al. (2006) is given in Fig. 10.1.

10.4.2.2 Nicotianamine (NA)

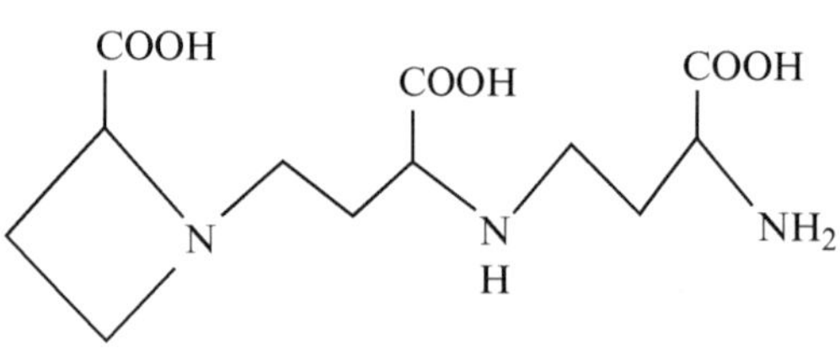

Nioctianamine

Structure and Metal-Binding Properties of NA

Nicotianamine (MW: 303) is an essential metabolite present in most of the plants. NA is a molecule with an azetidine ring produced by enzymatic condensation of three molecules of S-adenosyl methionine by the enzyme nicotianamine synthase (Curie et al. 2009). The molecule has three carboxy and three amino groups, which form an octahedral stable complex with a central metal ion. NA forms stable complexes with Fe^{2+}, Co^{2+}, Zn^{2+}, Ni^{2+} and Cu^{2+} in vitro at an increasing order of affinity (Curie et al. 2009). NA also forms complexes with Fe^{3+} with a higher affinity than Fe^{2+}, but the Fe^{2+} complex is more stable (von Wiren et al. 1999). The chelation properties of NA are highest at neutral or mildly basic pH. It is higher than other ligands such as organic acids in affinity and stability of the complex (Curie et al. 2009).

NAS Genes and Fe Deficiency

Arabidopsis has four nicotianamine synthase (NAS) genes. The genes NAS2 and NAS4 are upregulated due to iron deficiency in roots, which suggests a role of NA in Fe translocation from roots to shoots (Klatte et al. 2009). When the plant transits from vegetative to reproductive stage, NAS3 expression increases fourfold, suggesting a role of NA in translocation of Fe to flowers (Klatte et al. 2009). In barley, NAS1 gene is upregulated due to iron deficiency since NA is a precursor of mugineic acid (MA) (Higuchi et al. 2001). In rice, all the three NAS genes are upregulated due to Fe deficiency in the vascular tissues of roots (Inoue et al. 2003). Barley roots accumulate a higher concentration of MA than rice. Rice roots however accumulate more of NA under Fe deficiency or sufficiency. This indicates that NAS expression is used by rice to produce NA primarily for long-distance translocation of Fe. Rice leaves contain higher levels of NA and MA as compared to barley leaves, where these are present in traces (Higuchi et al. 2001). This is possibly due to the role of both NA and MA in Fe translocation in rice (Morrissey and Guerinot 2009). In tomato root and shoot sections, NA increases within the cells in response to increase in Fe supply (Pich et al. 2001).

Role of NA on Fe Homeostasis

NA levels increase in tobacco and *Arabidopsis*, when NAS is overexpressed in these plants. This results in accumulation of Fe, Zn, Mn and Ni in shoots (Ling et al. 1999; Takahashi et al. 2003; Douchkov et al. 2005) of these plants. Increasing NA levels in Fe-starved roots of rice and *Arabidopsis* by impairing PS synthesis results in seeds with significantly higher Fe content. NA is essential for movement of Fe from vascular to interveinal tissues. Depletion of NA in tomato and *Arabidopsis* due to loss of function of NAS results in interveinal chlorosis, reduced growth and sterility, which are characteristic symptoms of Fe deficiency (Takahashi et al. 2003; Douchkov et al. 2005; Kim et al. 2005).

10.4.3 Iron Transporters

10.4.3.1 The ZIP (ZRT, IRT Like Proteins) Family

AtIRT1 (iron-regulated transporter1), a member of ZIP family (Eide et al. 1996), is possibly the major transporter involved in high-affinity iron uptake by roots (Connolly et al. 2002; Vert et al. 2001) (See Sect. 9.4.2.5). Overexpression of AtIRT1 leads to accumulation of Cd and Zn indicating a possible role of this transporter in uptake of Cd and Zn (Connolly et al. 2002). AtIRT2 also expressed in root epidermal cells under conditions of Fe deficiency can only supplement and not replace AtIRT1 in Zn and Fe uptake (Grotz and Guerinot 2002). AtIRT2 has been recently reported to be involved in Fe homeostasis. AtIRT2 sequesters excess Fe taken up through strongly increased AtIRT1 activity and transports them into root endo-membrane vesicles. It is not involved in cellular uptake of Fe (Vert et al. 2009). It has been mentioned earlier that rice produces less Fe–PS than maize and barley, which makes it less tolerant to Fe deficiency in calcareous soils. This is compensated by the overexpression of OsIRT1 and OsIRT2 in the root epidermis of rice in response to Fe deficiency (Morrissey and Guerinot 2009). OsIRT1 is the first member of ZIP-metal transporter family to be isolated from a *graminaceous* plant. Typical features of ZIP family such as eight transmembrane domains and a variable region with a histidine-rich metal-binding domain are found in OsIRT1 (Bughio et al. 2002). Rice possesses both the Fe^{3+} (Fe–PS)- and Fe^{2+} (OsIRT1)-mediated Fe transport systems. Rice roots have very low ferric reductase activity, since most of the rice varieties are grown under waterlogged anaerobic conditions where Fe (II) ions are readily available.

10.4.3.2 The NRAMPs (Natural Resistance-Associated Macrophage Proteins)

Characteristics of this family of transporters have been discussed in Chapter 9 (see Sect. 9.4.2.3). These proteins are proton/metal symporters and have broad spectrum of divalent metal cation substrate, such as Fe^{2+}, Mn^{2+}, Cd^{2+}, Co^{2+}, Cu^{2+}, Ni^{2+} and Pb^{2+} (Gunshin et al. 1997; Nevo and Nelson 2006). In *Arabidopsis*, heterologous expressions of AtNramp1, AtNramp3 and AtNramp4 in yeast mutants indicate that these proteins can transport Fe, Mn and Cd (Curie et al. 2000; Thomine et al. 2000). AtNramp3 and AtNramp4 are located on the vacuolar membrane of the embryo and mobilise vacuolar Fe store for early plant development (Lanquar et al. 2005).

Recent characterisation of OsNramps5 indicates its involvement in transport and uptake of Mn, Fe and Cd by rice (Ishimaru et al. 2012).

LeNramp1 from tomato is localised in the vascular parenchyma of root hair zone also in the root epidermis and cortex behind the root tip. It possibly plays role in distribution of Fe in vascular parenchyma under Fe-deficient conditions (Bereczky et al. 2003).

AhNramp1 localised in the epidermis of plasma membrane of groundnut roots is a functional Fe transporter and is involved in uptake of Fe from the soil and distribution within the groundnut plant. AhNramp1 is induced in both roots and leaves due to Fe deficiency (Xiong et al. 2012).

10.4.3.3 ABC Transporters (ATP-Binding Cassette)

The characteristics of this group of transporters have already been discussed (See Sect. 9.4.2.2). AtABCB25 (AtATM3) is a mitochondrial ABC transporter involved in biogenesis Fe–S clusters (see Box 10.1) in plants (Kushnir et al. 2001; Bernard et al. 2009). AtABCB25 is strongly upregulated in Cd-treated plants (Bovet et al. 2005).

> **Box 10.1: Fe–S Clusters**
>
> Iron (Fe)–sulphur (S) clusters constitute a group of cofactors of proteins that are involved in various biological processes such as electron transfer, substrate binding/activation, iron–sulphur storage, redox and non-redox catalysis, regulation of gene expression and as sensors for iron and oxygen in all living organisms (Abdel-Ghany et al. 2005; Johnson et al. 2005; Lill 2009). Biogenesis of Fe–S cluster is mediated by multiple gene products encoded by the *isc* (iron–sulphur assembly) *and nif* (nitrogen fixation) operons (Tong et al. 2003). The *isc* operon encodes the translational regulator IscR, the cysteine desulphurase (IscS), scaffold proteins IscU and IscA, chaperone proteins HscA and HscB and ferredoxin (Fdx). The cysteine desulphurase IscS acts on a cysteine residue and forms alanine and elemental sulphur through formation of a persulphide intermediate. The sulphur is directly transferred to the scaffold protein IscU for the subsequent assembly of $[2Fe\text{-}2S]^{2+}$ and $[4Fe\text{-}4S]^{2+}$ clusters (Tong et al. 2003). Three types of Fe–S clusters have been identified, NIF (nitrogen fixation), ISC (iron–sulphur cluster) and SUF. There appears to be two different Fe–S assembly mechanisms in mitochondria and chloroplasts since mitochondria are involved in oxygen consumption, whereas chloroplasts generate oxygen. In mitochondria, IscU is considered as the major scaffold protein for Fe–S clusters. A NifS-like protein (AtCpNifS) with cys-desulphurase activity is localised in chloroplast (Leon et al. 2002; Pilon-Smits et al. 2002; Abdel-Ghany et al. 2005). Two SufE-like proteins have been identified from the chloroplast of *Arabidopsis thaliana* (Narayan Murthy et al. 2007). In *Arabidopsis*, CpSufE activates cysteine desulphurisation by NifS-like proteins and mobilises sulphur for Fe–S biosynthesis. Apart from CpSufE, the *Arabidopsis* genome encodes two other NifS-like proteins, SufE2 and SufE3 with SufE domains with plastid targeting information. The cysteine desulphurase activity of CpSufE can be increased 40-fold by SufE2. The full-length SufE3 protein carries the highly oxygen-sensitive $(4Fe\text{–}4S)^{2+}$ cluster at its NadA domain and is probably involved in a critical step of NAD biosynthesis in *A. thaliana* (Narayan Murthy et al. 2007).

10.4.3.4 H^+-ATPase

H^+-ATPases of plants are part of P-type ATPase superfamily. In response to iron-deficiency H^+-ATPases expressed in the root epidermis release protons to the rhizosphere, which lowers pH and makes iron more soluble. Fe deficiency upregulates the H^+-ATPases, AHA1, AHA2 and AHA7, in the root epidermis. AHA2 is considered as the primary root H^+-ATPase, which releases proton for rhizosphere acidification. Loss of expression of AHA2 has been found to reduce rhizosphere acidification caused by Fe deficiency (Morrissey and Guerinot 2009). CsHA2 gene transcripts have been reported to be unaffected by Fe concentration in *Cucumis sativus* L. (Cesco et al. 2005).

10.4.3.5 The YSL Transporters

Yellow-striped-like (YSL) transporters are so named due to formation of yellow striped leaves in a mutant phenotype of maize, when the gene encoding these transporters is disrupted (von Wirén et al. 1994; Curie et al. 2001). YSL transporters are distantly related to the OPT family of transporters. They transport tri-, tetra-, penta- and hexapeptides as well as amino acid derivatives (Yen et al. 2001; Curie et al. 2009; Conte and Walker 2011).

YS1 found in Poaceae roots is a proton-coupled symporter of Fe (III)–PS complexes (Schaaf et al. 2004). The transport activity and specificity of YS1 transporters have been examined from maize (ZmYS1), barley (HvYS1) and rice (OsYSL15). Expression of ZmYS1, HvYS1 and OsYSL15 is strongly upregulated under Fe-deficient conditions (Curie et al. 2009).

Out of 18 YSL genes in rice, only OsYSL15 is involved in Fe–PS (phytosiderophores of mugineic acid family) uptake from the rhizosphere. Expression of OsYSL15 is upregulated in response to Fe deficiency in the plasma membrane of epidermis of roots and also in the stele, flowers and developing seeds (Inoue et al. 2009; Lee et al. 2009; Morrissey and Guerinot 2009).

HvYS1 specifically transports Fe (III)–PS complexes. This specificity is possibly related to the formation of a highly variable loop between sixth and seventh transmembrane domain of the protein, to form a coiled domain (Harada et al. 2007). ZmYS1 and OsYSL15 have broader transporter capacities with possibilities of transporting Ni (II), Zn (II) and Cu (II)–PS complexes. ZmYS1 and HvYS1 do not transport Fe–PS from soil. These are probably involved in transport of Fe–PS complexes through phloem and play significant role in Fe homeostasis.

Arabidopsis has eight YSL transporters and three of them AtYSL1, AtYSL2 and AtYSL3 have been characterised. AtYSL3 is involved in transport of Fe–PS precursor Fe–NA complex in and out of phloem (Curie et al. 2009). In rice, OsYSL2 is upregulated due to Fe deficiency along with OsYSL15, but is present in phloem companion cells in the shoot. OsYSL2 is involved in Fe transport along with OsYSL15 in long-distance transport of Fe from root to shoot and then to seeds (Inoue et al. 2009; Koike et al. 2004). IDEF2, a NAC transcription factor, has been reported to regulate expression of YSL2 (Ogo et al. 2008). YSLs are present in a large number of tissues and are probably involved in transport of Fe–PS from xylem to phloem for further transport to the growing tissues (Curie et al. 2001, 2009; Di Donato et al. 2004; Schaaf et al. 2005; Waters et al. 2006).

10.4.4 Reutilisation of Apoplastic Fe

10.4.4.1 Phenolics, Organic Acids and Mobilisation of Apoplastic Fe

About 75 % of Fe in the roots is attached to apoplast. The negatively charged carboxyl groups in the cell wall act as a cation sink (Marshner 1995; Bienfait et al. 1985; Morrissey and Guerinot 2009). There is a decrease in this pool due to Fe deficiency. It has been reported that phenolics exuded by roots in response to Fe deficiency in red clover facilitate mobilisation of apoplastic Fe and help in recovery from Fe deficiency. The excreted phenolics have been found to significantly desorb Fe from the cell wall. This reutilisation is not mediated by proton extrusion or root ferric chelate reductase activity (Jin et al. 2007). However, an Fe-phenolic transporter is yet to be identified (Morrissey and Guerinot 2009).

There is an increase in organic acids (citrate, malate and succinate) in the xylem due to Fe deficiency (Lopez-Millan et al. 2000). Citrate is considered as the primary chelator of Fe in preference to others, organic acids, amino acids and nicotianamine at a pH of 5.5 inside the xylem for long-distance transport (von Wiren et al. 1999; Rellan-Alvarez et al. 2008). The naturally occurring Fe–citrate complex in xylem sap of tomato is an oxo-bridged tri-ferric tri-citrate (Fe_3Cit_3). A second Fe–citrate complex, the binuclear ferric-citrate complex (Fe_2Cit_2), may also exist depending upon Fe–Citrate ratio (Rellan-Alvarez et al. 2010).

10.4.4.2 FRD3 MATE Transporter

Ferric reductase defective3 (FRD3) genes encode multidrug and toxin efflux (MATE) transporters, which are found in all living organisms (Omote et al. 2006). FRD3 is known to be involved in Fe homeostasis in plants. FRD3 gene has been cloned and characterised in *Arabidopsis* (Rogers and Guerinot 2002). FRD3 has been reported to have an iron-responsive 27 bp *cis*-regulatory element in its promoter region (Pineau et al. 2012). FRD3 releases citrate into the xylem, which solubilises apoplastic Fe to be transported to shoots and deposited in leaf cells (Durrett et al. 2007, Pineau et al. 2012). The mRNA of FRD3 is detected even under adequate presence of Fe, but it increases twofold under Fe deficiency. Loss of FRD3 results in severe chlorosis and constitutive Fe-deficiency symptoms (Rogers and Guerinot 2002). FRD3 is also involved in regulation of negative interaction between Fe and Zn (Pineau et al. 2012).

10.4.5 FIT1 (Fe-Deficiency-Induced Transcription Factor1)

FIT1 is a bHLH (basic helix–loop–helix) transcription factor (see Sect. 3.4.7.2), which regulates response to iron deficiency in *Arabidopsis*. FIT1 is required for proper regulation of ferric chelate reductase (FRO2) activity and iron transport to the roots of plants. FIT1 regulates FRO2 at the level of steady-state accumulation of mRNA and by controlling expression of ferrous transporter IRT1. Regulation of FIT1 is higher under Fe deficiency than Fe sufficiency. *Arabidopsis* FIT1 is localised in the root hairs at the differentiation zone (Colangelo and Guerinot 2004). FIT1 is the closest homologue of *fer* gene of tomato and has similar but not identical functions. FIT1 bHLHs recognise the E-box 5′-CANNTG-3′ in the promoters of target genes (Colangelo and Guerinot 2004).

10.4.6 FPN Genes

The FPN (ferroportin) family of genes in *Arabidopsis* are involved in efflux of Fe into the xylem for long-distance transport (Morrissey et al. 2009). The three FPN genes in *Arabidopsis* have distinct subcellular localisation (Conte and Walker 2011). The FPN/IREG1 (iron-regulated protein1) is located in the plasma membrane (Morrissey et al. 2009), FPN2/IREG2 is located in the tonoplast (Morrissey et al. 2009) and FPN3/MAR1/RTS3/IREG3 is located on the chloroplast envelops (Conte et al. 2009).

10.4.7 Iron Homeostasis in Subcellular Organelles

10.4.7.1 Mitochondria

Plant mitochondria need iron for heme biosynthesis, respiration and synthesis of Fe–S clusters. Since free Fe is toxic, it needs to be sequestered with proteins to perform its essential functions (Morrissey and Guerinot 2009). A number of proteins are involved in chelating iron in mitochondria to maintain Fe homeostasis.

Frataxin

Frataxin is a highly conserved protein found in all living organisms. It is reported to act as an Fe-chaperon protein donating Fe to the protein involved in Fe–S cluster formation, heme synthesis and mitochondrial iron homeostasis. It is possibly associated with Fe detoxification and storage (Vazzola et al. 2007). Plant frataxin has five segments of beta regions and two alpha helices as well as a potential N-terminal targeting peptide for localisation in mitochondria, which are characteristics of human frataxin. Frataxin is nuclear coded in higher organisms and is required for maintenance of normal Fe level in mitochondria for respiration. The *Arabidopsis* frataxin gene AtFH is a single nuclear-coded gene targeted at mitochondria and has 65 % homology with animal frataxin family. Plant frataxin plays a significant role during embryogenesis (Vazzola et al. 2007). AtFH has been reported to be involved in mitochondrial respiration and survival of plants under oxidative stress (Busi et al. 2004).

AtABCB25 (AtATM3)

There are three ABC transporters in *Arabidopsis*, AtATM1, AtATM2 and AtATM3, which are homologous to yeast mitochondrial ABC transporter (ScATM1) involved in Fe–S cluster assembly (Chen et al. 2007). Only AtATM3 (AtABCB25) is involved in biogenesis Fe–S clusters in plants (Kushnir et al. 2001; Bernard et al. 2009). AtABCB25 is strongly upregulated in Cd-treated plants (Bovet et al. 2005).

10.4.7.2 Chloroplast

Up to 90 % of Fe in leaves is associated with lipoproteins of membranes of chloroplast and mitochondria (Havlin et al. 2007). About 50 % of Fe in chloroplast is located in the stroma and 50 % in the thylakoid membranes. Iron is required as a cofactor in photosynthetic electron transport chain, biosynthesis of heme and Fe–S cluster formation in the chloroplast. In most of the plants, chloroplasts constitute the largest sink for Fe. However, Fe may cause oxidative damage to the chloroplast through Fenton's reaction.

YSL4 and YSL6 Transporters

Apart from ferritin, chloroplasts contain specific transporter proteins which remove Fe and do not allow Fe to accumulate in toxic concentrations. It has been reported that in *A. thaliana*, two Fe transporters YSL4 and YSL6 remove iron from chloroplast and prevent it from iron toxicity. YSL6 is localised in the chloroplast envelop (Divol et al. 2013).

PIC1 (Permease Chloroplast1)

PIC1 is a metal transporter identified in plastids of *A. thaliana* and contains four predicted alpha helices. PIC1 is reported to be involved in transport of Fe across inner envelop of chloroplast and in cellular Fe homeostasis (Duy et al. 2007).

Ferritin

Many cereals and legumes contain ferritin such as corn, beans, peas, lentils and soybean (Lonnerdal 2009). The plant ferritins are similar to animal ferritins and are large proteins with molecular weights of around 480 kDa. They consist of multiple smaller subunits with M.W $\approx$ 24 kDa. The subunits are complexed with oxygen and phosphate, and ferric iron is held inside a shell formed by these subunits forming an iron core. The Fe core appears to be formed inside the plastids after transport of proteins into the plastids (Liu and Theil 2005).

Plant ferritins are nuclear coded and located primarily in the plastids (Waldo et al. 1995). It is reported that ferritin accumulates in the amyloplast of embryonic cells and ferritin-bound iron is the principal source of iron during early germination. The primary role of ferritin in *A. thaliana* is protection from reactive oxygen species (ROS) (Waldo et al. 1995). *A. thaliana* has been reported to have four ferritin genes, AtFER1, AtFER2, AtFER3 and AtFER4 (Petit et al. 2001). The polypeptide sequences deduced from these genes are consistent with the reported ferritin subunits found in different plants. The targets of all the genes are the plastids, and the coded polypeptides have putative transit peptides at the N-terminal extremity. The *A. thaliana* ferritin polypeptides have conserved specific residues for ferroxidase activity and for Fe nucleation as found in animals. AtFER1 and AtFER3 are expressed in response to treatment with excess iron. AtFER2 is not responsive to Fe treatment and expressed in seeds. AtFER1, AtFER3 and AtFER4 are expressed in various vegetative organs but not in seeds (Petit et al. 2001).

Ferric Chelate Reductase

Ferric chelate reductase activities have been observed in leaf discs and leaf protoplasts (Larbi et al. 2001; Brüggemann et al. 1993). Out of the eight members of FRO (ferric reductase oxidase) family in *Arabidopsis*, FRO7 has been reported to be involved in Fe homeostasis in the chloroplasts of young seedlings and is essentially required for survival under iron-limiting conditions. FRO7 is highly expressed in photosynthetic tissues of younger plants. No other FRO proteins are located in the chloroplast (Jeong et al. 2008). FRO7 is not regulated by Fe, and hence, its main role is probably not to supply Fe under iron-limiting conditions, which induce Strategy I response. It has been proposed that FRO7 is possibly involved in Fe supply to young growing plants in response to developmental cues (Jeong et al. 2008).

10.4.7.3 Vacuole

Vacuole is an important storage organelle for Fe. It acts as an initial source of Fe for germinating seeds before the seedlings can acquire Fe from external source.

VIT1 (Vacuolar Iron Transporter)

VIT1 is an Fe–Mn transporter located in the vacuole and transports these metals into the vacuole (Kim et al. 2006). VIT1 shows 62 % amino acid similarity with the Fe–Mn transporter of yeast, CCC1 (cross-complement Ca (II) phenotype of csg1), which transports these metals into the vacuole. VIT1 of *Arabidopsis* and LeVIT1 of tomato are sensitive to high concentration of Fe due to their inability to sequester Fe (Li et al. 2001). Loading of Fe through VIT1 and its proper distribution in the embryo are essential for seedling viability under low Fe conditions (Morrissey et al. 2009). VIT1 expression is not affected by iron availability unlike other proteins involved in Fe metabolism such as IRT1, FRO2, FIT1 and FRD3 (Eide et al. 1996; Rogers and Guerinot 2002; Colangelo and Guerinot 2004).

AtNramp3 and AtNramp4

AtNramp3 and AtNramp4 located on the vacuolar membrane of the embryo are upregulated due to Fe deficiency and remobilise vacuolar Fe store during germination for early plant development (Lanquar et al. 2005).

References

Abdel-Ghany SE, Gulnara HY, Garifullina F, Zhang L, Elizabeth AH, Pilon-Smits EA, Pilon M (2005) Iron-sulfur cluster biogenesis in chloroplasts, involvement of the scaffold protein CpIscA. Plant Physiol 138(1):161–172

Bereczky Z, Wang HY, Schubert V, Ganal M, Bauer P (2003) Differential regulation of Nramp and IRT metal transporter genes in wild type and iron uptake mutants of tomato. J Biol Chem 278:24697–24704

Bernard DG, Cheng Y, Zhao Y, Balk J (2009) An allelic mutant series of ATM3 reveals its key role in the biogenesis of cytosolic iron-sulfur proteins in Arabidopsis. Plant Physiol 151:590–602

Bienfait HF, van den Briel W, Mesland-Mul NT (1985) Free space iron pools in roots. Generation and mobilisation. Plant Physiol 78:596

Bovet L, Feller U, Martinoia E (2005) Possible involvement of plant ABC transporters in cadmium detoxification: a cDNA sub-microarray approach. Environ Int 31:263–267

Brüggemann W, Maas-Kantel K, Moog PR (1993) Iron uptake by leaf mesophyll cells: the role of the plasma membrane-bound ferric-chelate reductase. Planta 190:151–155

Bughio N, Yamaguchi H, Nishizawa NK, Nakanishi H, Mori S (2002) Cloning an iron-regulated metal transporter from rice. J Exp Bot 53(374):1677–1682

Busi MV, Zabaleta EJ, Araya A, Gomez-Casati DF (2004) Functional and molecular characterization of the frataxin homolog from *Arabidopsis thaliana*. FEBS Lett 576(1–2):141–144

Cesco SS, Varanini Z, Pinton R (2005) Two plasma membrane H(+)-ATPase genes are differentially expressed in iron-deficient cucumber plants. Plant Physiol Biochem 43(3):287–292

Chaney R, Brown JC, Tiffin LO (1972) Obligatory reduction of ferric chelates in iron uptake by soybeans. Plant Physiol 50:208–213

Chen S, Sanchez-Fernandez R, Lyver E, Dancis A, Rea P (2007) Functional characterization of AtATM1, AtATM2, and AtATM3, a subfamily of *Arabidopsis* half-molecule ATP-binding cassette transporters implicated in iron homeostasis. J Biol Chem 282:21561–21571

Colangelo E, Guerinot ML (2004) The essential basic helix-loop-helix protein FIT1 is required for the iron deficiency response. Plant Cell 16(12):3400–3412

Connolly EL, Fett JP, Guerinot ML (2002) Expression of the IRT1 metal transporter is controlled by metals at the levels of transcript and protein accumulation. Plant Cell 14:1347–1357

Conte SS, Walker EL (2011) Transporters contributing to iron trafficking in plants. Mol Plant 4(3):464–476

Conte S, Stevenson D, Furner I, Lloyd A (2009) Multiple antibiotic resistance in Arabidopsis thaliana is conferred by mutations in a chloroplast-localized transport protein. Plant Physiol 151:559–573

Curie C, Alonso JM, Le Jean M, Ecker JR, Briat J-F (2000) Involvement of NRAMP1 from *Arabidopsis thaliana* in iron transport. Biochem J 347:749–755

Curie C, Panaviene Z, Loulergue C, Dellaporta SL, Briat JF, Walker EL (2001) Maize yellow stripe1 encodes a membrane protein directly involved in Fe(III) uptake. Nature 409:346

Curie C, Cassin G, Couch D, Divol F, Higuchi K, Jean ML, Misson J, Schikora A, Czernic P, Mari S (2009) Metal movement within the plant: contribution of nicotianamine and yellow stripe 1-like transporters. Ann Bot 103(1):1–11

Di Donato, Jr RJ, Roberts LA, Sanderson T, Eisley RB, Walker EL (2004) *Arabidopsis* Yellow Stripe-Like2 (YSL2): a metal-regulated gene encoding a plasma membrane transporter of nicotianamine-metal complexes. Plant J 39:403

Divol F, Couch D, Conejero G, Roschzttardtz H, Mari S, Curie C (2013) The *Arabidopsis* YELLOW STRIPE LIKE 4 and 6 Transporters control iron release from the chloroplast. Plant Cell 25(3):1040–1055

Douchkov D, Gryczka C, Stephan UW, Hell R, Baumlein H (2005) Ectopic expression of nicotianamine synthase genes results in improved iron accumulation and increased nickel tolerance in transgenic tobacco. Plant Cell Environ 28:365

Durrett TP, Gassmann W, Rogers EE (2007) The FRD3-mediated efflux of citrate into the root vasculature is necessary for efficient iron translocation. Plant Physiol 144:197–205

Duy D, Wanner G, Meda AR, von Wiren N, Soll J, Philippar K (2007) PIC1, an ancient permease in *Arabidopsis* chloroplasts, mediates iron transport. Plant Cell 19(3):986–1006

Eckhardt U, Marques AM, Buckhout TJ (2001) Two iron-regulated cation transporters from tomato complement metal uptake-deficient yeast mutants. Plant Mol Biol 45:437–448

Eide D, Broderius M, Fett J, Guerinot ML (1996) A novel iron-regulated metal transporter from plants identified by functional expression in yeast. Proc Natl Acad Sci U S A 93:5624–5628

Grotz N, Guerinot ML (2002) Limiting nutrients: an old problem with new solutions? Curr Opin Plant Biol 5:158–163

Gunshin H, Mackenzie B, Berger UV, Gunshin Y, Romero MF, Boron WF, Nussberger S, Gollan JL, Hediger MA (1997) Cloning and characterization of a mammalian proton-coupled metal-ion transporter. Nature 388:482–488

Harada E, Sugase K, Namba K, Iwashita T, Murata Y (2007) Structural element responsible for the Fe (III)-phytosiderophore specific transport by HvYS1 transporter in barley. FEBS Lett 581:4298

Havlin JL, Tisdale SL, Beaton JD, Nelson WL (2007) In soil fertility and fertilizers. Prentice Hall (India), New Delhi

Higuchi K, Suzuki K, Nakanishi H, Yamaguchi H, Nishizawa NK, Mori S (1999) Cloning of nicotianamine synthase genes, novel genes involved in the biosynthesis of phytosiderophores. Plant Physiol 119:471–479

Higuchi K, Watanabe Y, Takahashi M, Kawasaki S, Nakanishi H, Nishizawa NK, Mori S (2001) Nicotianamine synthase gene expression differs in barley and rice under Fe-deficient conditions. Plant J 25(2):159–167

Inoue H, Higuchi K, Takahashi M, Nakanishi H, Mori S, Nishizawa NK (2003) Three rice nicotianamine synthase genes, OsNAS1, OsNAS2, and OsNAS3 are expressed in cells involved in long-distance transport of iron and differentially regulated by iron. Plant J 36:366

Inoue H, Takahashi M, Kobayashi T, Suzuki M, Nakanishi H, Mori S, Nishizawa NK (2008) Identification and localisation of the rice nicotianamine aminotransferase gene OsNAAT1 expression suggests the site of phytosiderophore synthesis in rice. Plant Mol Biol 66(1–2):193–203

Inoue H, Kobayashi T, Nozoye T, Takahashi M, Kakei Y, Suzuki K, Nakazono M, Nakanishi H, Mori S, Nishizawa NK (2009) Rice OsYSL15 is an iron-regulated iron(III)-deoxymugineic acid transporter expressed in the roots and is essential for iron uptake in early growth of the seedlings. J Biol Chem 284:3470

Ishimaru Y, Takahashi R, Bashir K et al (2012) Characterising the role of rice NRAMP5 in manganese, iron and cadmium transport. Sci Rep 2:286

Jena D, Singh MV, Pattanayak MR (2008) Scenario of micro and secondary nutrient deficiencies in soils of Orissa and their management,Tech. Bulletin 1, Department of Soil Science and Agricultural Chemistry, Orissa University of Agriculture & Tech Bhubaneswar, India

Jeong J, Cohu C, Kerkeb L, Pilon M, Connolly EL, Guerinot ML (2008) Chloroplast Fe(III) chelate reductase activity is essential for seedling viability under iron limiting conditions. Proc Natl Acad Sci U S A 105(30):10619–10624

Jin CW, You GY, He YF, Tang C, Wu P, Zheng SJ (2007) Iron deficiency induced secretion of phenolics facilitates the reutilisation of root apoplastic iron in red clover. Plant Physiol 144:278

Johnson DC, Dean DR, Smith AD (2005) Structure, function, and formation of biological iron-sulfur clusters. Ann Rev Biochem 74:247–281

Kim S, Takahashi M, Higuchi K, Tsunoda K, Nakanishi H, Yoshimura E, Mori S, Nishizawa NK (2005) Increased nicotianamine biosynthesis confers enhanced tolerance of high levels of metals, in particular nickel, to plants. Plant Cell Physiol 46:1809

Kim SA, Punshon T, Lanzirotti A et al (2006) Localization of iron in *Arabidopsis* seed requires the vacuolar membrane transporter VIT1. Science 314:1295–1298

Klatte M, Schuler M, Wirtz M, Fink-Straube C, Hell R, Bauer P (2009) The analysis of *Arabidopsis* nicotianamine synthase mutants reveals functions for nicotianamine in seed iron loading and iron deficiency responses. Plant Physiol 150:257

Kobayashi T, Nakanishi H, Takahashi M, Kawasaki S, Nishizawa NK, Mori S (2001) *In vivo* evidence that *Ids3* from *Hordeum vulgare* encodes a dioxygenase that converts 2′-deoxymugineic acid to mugineic acid in transgenic rice. Planta 212:864–871

Kobayashi T, Suzuki M, Inoue H, Nakanishi I, Takahash M, Nakanishi H, Mori S, Nishizawa NK (2005) Expression of iron-acquisition-related genes in iron-deficient rice is co-ordinately induced by partially conserved iron-deficiency-responsive elements. J Exp Bot 56(415):1305–1316

Koike S, Inoue H, Mizuno D, Takahashi M, Nakanishi H, Mori S, Nishizawa NK (2004) OsYSL2 is a rice metal-nicotianamine transporter that is regulated by iron and expressed in the phloem. Plant J 39:415

Kushnir S, Babiychuk E, Storozhenko S, Davey MW, Papenbrock J, De Rycke R, Engler G, Stephan UW, Lange H, Kispal G (2001) A mutation of the ABC transporter Sta1 leads to dwarfism and chlorosis in the *Arabidopsis* mutant starik. Plant Cell 13:89–100

Lanquar V, Lelievre F, Bolte S et al (2005) Mobilization of vacuolar iron by AtNRAMP3 and AtNRAMP4 is essential for seed germination on low iron. EMBO J 24:4041–4051

Larbi A, Morales F, Lopez-Milan A, Gogorcenal Y, Abadia A, Moog P, Abadia J (2001) Technical advance: Reduction of Fe (III)-chelates by mesophyll leaf disks of sugar beet- Multi-component origin and effects of Fe deficiency. Plant Cell Physiol 42:94–105

Lee S, Chiecko JC, Kim SA, Walker EL, Lee Y, Guerinot ML, An G (2009) Disruption of OsYSL15 leads to iron inefficiency in rice plants. Plant Physiol 150(2):786–800

Leon S, Touraine B, Briat JF, Lobreaux S (2002) The AtNFS2 gene from *Arabidopsis thaliana* encodes a NifS-like plastidial cysteine desulphurase. Biochem J 366:557–564

Li L, Chen OS, McVey WD, Kaplan J (2001) CCC1 is a transporter that mediates vacuolar iron storage in yeast. J Biol Chem 276:29515–29519

Lill R (2009) Function and biogenesis of iron-sulphur proteins. Nature 460:831–838

Ling HQ, Koch G, Baumlein H, Ganal MW (1999) Map-based cloning of *chloronerva*, a gene involved in iron uptake of higher plants encoding nicotianamine synthase. Proc Natl Acad Sci U S A 96:7098–7103

Liu X, Theil EC (2005) Ferritins: dynamic management of biological iron and oxygen chemistry. Acc Chem Res 38(3):167–175

Lonnerdal B (2009) Soybean ferritin: implications for iron status of vegetarians. Am J Clin Nutr 89(5):1680S–1685S

Lopez-Millan AF, Morales F, Abadia A, Abadia J (2000) Effects of iron deficiency on the composition of the leaf apoplastic fluid and xylem sap in sugar beet: implications for iron and carbon transport. Plant Physiol 124:873–884

Ma JF, Shinada T, Matsuda C, Nomoto K (1995) Biosynthesis of phytosiderophores, mugineic acids, associated with methionine cycling. J Biol Chem 270:16549–16554

Marshner H (1995) Mineral nutrition of higher plants. Academic, London, pp 313–323

McLean E, Cogswell M, Egli I, Woidyla D, de Benoist B (2009) Worldwide prevalence of anaemia. WHO Vitamin and Mineral Nutrition Information System, 1993–2005. Public Health Nutr 12:444–454

Mitra GN, Sahu SK, Dev G (1990) Potassium chloride increases rice yield and reduces symptoms of iron toxicity. Better Crop Int 6(2):14–15

Mitra GN, Sahu SK, Nayak RK (2009) Characterization of iron toxic soils of Orissa and ameliorating effects of potassium on iron toxicity. Proceedings of the IPI-OUAT-IPNI international symposium, Bhubaneswar. vol. I: Invited papers. IPI/IPNI, Horgen/Norcross, p 215

Morrissey J, Guerinot ML (2009) Iron uptake and transport in plants: the good, the bad, and the ionome. Chem Rev 109(10):4553–4567

Morrissey J, Baxter IR, Lee J, Li L, Lahner B, Grotz N, Kaplan J, Salt DE, Guerinot ML (2009) The ferroportin metal efflux proteins function in iron and cobalt homeostasis in *Arabidopsis*. Plant Cell 21(10):3326–3338

Narayan Murthy UM, Ollagnier-de-Choudens S, Sanakis Y, Abdel-Ghany SE, Rousset C, Ye H, Fontecave M, Elizabeth AH, Pilon S, Pilon M (2007) Characterization of *Arabidopsis thaliana* SufE2 and SufE3. Functions in chloroplast iron-sulfur cluster assembly and NAD synthesis. J Biol Chem 282:18254–18264

Nevo Y, Nelson N (2006) The NRAMP family of metal-ion transporters. Biochim Biophys Acta 1763:609–620

Ogo Y, Kobayashi T, Nakanishi RI, Nakanishi H, Kakei Y, Takahashi M, Toki S, Mori S, Nishizawa NK (2008) A novel NAC transcription factor, IDEF2, that recognizes the iron deficiency-responsive element 2 regulates the genes involved in iron homeostasis in plants. J Biol Chem 283(19):13407–13417

Okumura N, Nishizawa NK, Umehara Y, Ohata T, Nakanishi H, Yamaguchi H, Chino M, Mori S (1994) A dioxygenase gene (*Ids2*) expressed under iron deficiency condition in the roots of *Hordeum vulgare*. Plant Mol Biol 25:705–719

Omote H, Hiasa M, Matsumoto T, Otsuka M, Moriyama Y (2006) The MATE proteins as fundamental transporters of metabolic and xenobiotic organic cations. Trends Pharmacol Sci 27:587–593

Petit JM, Briat JF, Lobreaux S (2001) Structure and differential expression of the four members of the *Arabidopsis thaliana* ferritin gene family. Biochem J 359(3):578–582

Pich A, Manteuffel R, Hillmer S, Scholz G, Schmidt W (2001) Fe homeostasis in plant cells: does nicotianamine play multiple roles in the regulation of cytoplasmic Fe concentration? Planta 213:967–976

Pilon-Smits EAH, Garifullina GF, Abdel-Ghany S, Kato S, Mihara H, Hale KL, Burkhead JL, Esaki N, Kurihara T, Pilon M (2002) Characterization of a NifS-like chloroplast protein from *Arabidopsis*. Implications for its role in sulfur and selenium metabolism. Plant Physiol 130:1309–1318

Pineau C, Loubet S, Lefoulon C, Chalies C, Fizames C, Lacombe B, Ferrand M, Loudet O, Berthomieu P, Richard O (2012) Natural variation at the *FRD3* MATE transporter locus reveals cross-talk between Fe homeostasis and Zn tolerance in *Arabidopsis thaliana*. PLoS Genet 8(12):e10003120

Rellan-Alvarez R, Abadia J, Alvarez-Fernandez A (2008) Formation of metal–nicotianamine complexes as affected by pH, ligand exchange with citrate and metal exchange: a study by electrospray ionization time-of-flight mass spectrometry. Rapid Commun Mass Spectrom 22:1553–1562

Rellan-Alvarez R, Giner-Martinez-Sierra J, Orduna J, Rodriguez-Castrillon JA, Garcia-Alonso JL, Abadia J, Alvarez-Fernandez A (2010) Citrate complex in the xylem Sap of iron-deficient tomato resupplied with iron: new insights into plant iron long-distance transport. Plant Cell Physiol 51(1):91–102

Robinson NJ, Procter CM, Connolly EL, Guerinot ML (1999) A ferric-chelate reductase for iron uptake from soil. Nature 397:694–697

Rogers EE, Guerinot ML (2002) FRD3, a member of the multidrug and toxin efflux family, controls iron deficiency responses in *Arabidopsis*. Plant Cell 14:1787–1799

Römheld V (1987) Different strategies for iron acquisition in higher plants. Physiol Plant 70:231–234

Römheld V, Marschner H (1986) Evidence for a specific system for iron phytosiderophores in roots of grasses. Plant Physiol 80:175–180

Sahu SK, Mitra GN (1992) Iron potassium interaction in rice cv. Daya. J Pot Res 8(4):311–319

Sahu SK, Dev G, Mitra GN (2001) Iron toxicity in rice as affected by applied potassium in lateritic soils. J Res Orissa University of Agriculture & Tech 19:62–67

Schaaf G, Ludewig U, Erenoglu BE, Mori S, Kitahara T, von Wiren N (2004) ZmYS1 functions as a proton-coupled symporter for phytosiderophore- and nicotianamine-chelated metals. J Biol Chem 279:9091

Schaaf G, Schikora A, Haberle J, Vert G, Ludewig U, Briat JF, Curie C, von Wiren N (2005) A putative function for the *Arabidopsis* Fe-phytosiderophore transporter homolog AtYSL2 in Fe and Zn homeostasis. Plant Cell Physiol 46:762

Sekowska A, Denervaud V, Ashida H, Michoud K, Haas D, Yokota A, Danchin A (2004) Bacterial variations on the methionine salvage pathway. BMC Microbiol 4:9–25

Suzuki M, Takahashi M, Tsukamoto T, Watanabe S, Matsuhashi S, Yazaki J, Kishimoto N, Kikuchi S, Nakanishi H, Mori S, Nishizawa NK (2006) Biosynthesis and secretion of mugineic acid family phytosiderophores in zinc-deficient barley. Plant J 48:85–97

Takahashi M, Yamaguchi H, Nakanishi H, Nishizawa NK, Mori S (1999) Cloning two genes for nicotianamine aminotransferase, a critical enzyme in iron acquisition (Strategy II) in *graminaceous* plants. Plant Physiol 121:947–956

Takahashi MT, Nakanishi H, Kawasaki S, Nishizawa NK, Mori S (2001) Enhanced tolerance of rice to low iron availability in alkaline soils using barley nicotianamine aminotransferase genes. Nat Biotechnol 19:466

Takahashi M, Terada Y, Nakai I, Nakanishi H, Yoshimura E, Mori S, Nishizawa N (2003) Role of nicotianamine in the intracellular delivery of metals and plant reproductive development. Plant Cell 15:1263

Thomine S, Wang R, Ward JM, Crawford NM, Schroeder JI (2000) Cadmium and iron transport by members of a plant metal transporter family in *Arabidopsis* with homology to Nramp genes. Proc Natl Acad Sci U S A 97:4991–4996

Tong WH, Jameson GN, Huynh BH, Rouault TA (2003) Sub-cellular compartmentalization of human Nfu, an iron-sulfur cluster scaffold protein, and its ability to assemble a [4Fe-4S] cluster. Proc Natl Acad Sci U S A 100:9762–9767

Vazzola V, Losa A, Soave C, Murgia I (2007) Knockout of frataxin gene causes embryo lethality in *Arabidopsis*. FEBS Lett 581(4):667–672

Vert G, Briat JF, Curie C (2001) *Arabidopsis IRT2* gene encodes a root-periphery iron transporter. Plant J 26:181–189

Vert G, Barberon M, Zelazny E, Seguela M, Briat JF, Curie C (2009) Arabidopsis IRT2 cooperates with the high-affinity iron uptake system to maintain iron homeostasis in root epidermal cells. Planta 229(6): 1171–1179

von Wirén N, Mori S, Marschner H, Römheld V (1994) Iron inefficiency in maize mutant ys1 (*Zea mays* L. cv yellow-stripe) is caused by a defect in uptake of iron phyto siderophores. Plant Physiol 106(1):71–77

von Wiren N, Klair S, Bansal S, Briat J-F, Khodr H, Shioiri T, Leigh RA, Hider RC (1999) Nicotianamine chelates both FeIII and FeII. Implications for metal transport in plants. Plant Physiol 119:1107

Waldo GS, Wright E, Whang ZH, Briat JF, Theil EC, Sayers DE (1995) Formation of ferritin iron mineral occurs in plastids (an X-ray absorption spectroscopy study). Plant Physiol 109(3):797–802

Waters BM, Chu HH, Didonato RJ, Roberts LA, Eisley RB, Walker EL (2006) Mutations in *Arabidopsis* yellow stripe-like1 and yellow stripe-like3 reveal their roles in metal ion homeostasis and loading of metal ions in seeds. Plant Physiol 141(4):1446–1458

Xiong H, Kobayashi T, Kakei Y, Senoura T, Nakazono M, Takahashi H, Nakanishi H, Shen H, Duan P, Guo X, Nishizawa NK, Zuo Y (2012) *AhNRAMP1* iron transporter is involved in iron acquisition in peanut. J Exp Bot 63(12):4437–4446

Yen MR, Tseng YH, Saier MH Jr (2001) Maize Yellow Stripe1, an iron-phytosiderophore uptake transporter, is a member of the oligopeptide transporter (OPT) family. Microbiology 147:2881–2883

Zinc (Zn) Uptake

11

Abstract

Zinc deficiency is generally observed in lowland soils. In aerobic soils, Zn^{2+} is readily available to plants. About one third of the world's population suffers from mild Zn deficiency. Zinc is an essential catalytic component of over 300 enzymes. Zinc is an essential plant nutrient but is toxic beyond a threshold concentration. Several regulatory mechanisms, such as control of Zn uptake, intracellular binding by metal chelators (mugineic acid, phytochelatins, metallothioneins), efflux from the cell and sequestration into vacuoles, are adopted to maintain Zn homeostasis by plants. There is a coordinated expression of Zn^{2+} transporters, which are involved in Zn^{2+} uptake from the soil, translocation of Zn^{2+} to various organs and tissues and in intracellular sequestration and transport to vacuole.

11.1 Occurrence of Zinc and Soil Reactions

Lithosphere contains about 80 ppm of Zn. The Zn content of soil is within a range of 10–300 ppm. Zinc deficiency is generally observed in lowland soils, since low redox potential causes precipitation of Zn^{2+} as zinc sulphide (ZnS), zinc carbonate ($ZnCO_3$) and zinc oxy-hydroxides. In aerobic soils, Zn^{2+} is readily available to plants on cation exchange sites. However, calcareous soils tend to become Zn deficient even under aerobic conditions. Adsorption of Zn^{2+} by $CaCO_3$ reduces Zn^{2+} concentration in soil. Magnesite ($MgCO_3$) strongly adsorbs Zn^{2+}. Dolomite [Ca, Mg $(CO_3)_2$] adsorbs Zn^{2+} to a lesser extent. Short-term flooding of these soils may make Zn^{2+} readily available at an intermediate redox potential (Rose et al. 2013). Zinc deficiency was first observed in rice grown on calcareous soils of Northern India (Nene 1966; Yoshida and Tanaka 1969). Lowland rice cultivated in Asia suffers extensively from Zn deficiency. Next to N, P and K, Zn is considered as the most deficient nutrient for the lowland rice-growing soils. Concentration of Zn in soil solution is in the range of 2–70 ppb. More than 50 % of Zn is complexed with organic matter. Solubility of Zn depends on soil pH and decreases with increase in pH. While Zn^{2+} is the dominant ion at acidic pH, Zn $(OH)^+$ ion dominates above pH 7.7 (Lindsay 1979).

$$Zn^{2+} + H_2O \rightleftarrows Zn(OH)^+ + H^+$$

G.N. Mitra, *Regulation of Nutrient Uptake by Plants: A Biochemical and Molecular Approach*, DOI 10.1007/978-81-322-2334-4_11, © Springer India 2015

11.2 Zinc Content of Plants

Zinc (Zn) is a micronutrient essential for plant growth. Zinc concentration less than 15–20 μg in leaves per gram of dry leaf tissues leads to Zn deficiency. Zn concentration in plants is within a range of 25–150 $\mu g\ g^{-1}$.

11.3 Functions of Zn in Plants

11.3.1 Zn Deficiency

Zn-deficiency syndromes include chlorotic leaves, early senescence and stunted growth. Young dividing organs need higher zinc concentration for optimal growth (Marschner 1995). The symptoms of Zn deficiency in rice appear 2–3 weeks after transplanting of rice seedlings as brown blotches and streaks in younger leaves, which may cover the entire older leaves; there is substantial delay in maturity and yield reduction (Yoshida and Tanaka 1969; Van Breemen and Castro 1980; Neue and Lantin 1994). Sensitivity of some of the crops to Zn deficiency has been given Table 11.1. There are varietal differences within the same crop and environmental effects.

About one third of the world's population suffers from mild Zn deficiency. Zinc deficiency causes impaired brain development, dysfunction of the reproductive system, anorexia, immune disorders, hair loss, skin lesions and loss of taste and smell (Song et al. 2010).

Table 11.1 Sensitivity of crops to Zn deficiency

Sensitivity to Zn deficiency	Name of the crops
Highly sensitive	Citrus, corn, apples, beans, onion, rice, soybean, sweet corn, grapes. Fruit trees (deciduous), flax, castor bean
Medium sensitivity	Barley, cotton, potato, sorghum, sugar beet, tomato, wheat
Low sensitivity	Carrot, mustard, oats, peas, rye, safflower

11.3.2 Zn Toxicity

Zinc is taken up from the soil as a divalent cation (Zn^{2+}). It is neither oxidised nor reduced inside the plant cell, but has a strong tendency to form tetrahedral complexes (Berg and Shi 1996; Schützendübel and Polle 2002). Zn becomes toxic at higher concentrations, which vary for different plants and the parts of plant such as leaves, shoots and roots. Toxic symptoms generally appear in younger leaves as chlorotic spots, which progress to reddening of leaves due to increased anthocyanin synthesis. Zn toxicity also results in smaller leaves and reduced root growth (Fontes and Cox 1995; Reichman 2002). A precise control of Zn uptake, homeostasis and allocation to different organs of the plants and cellular organelles is essential for optimal plant growth.

11.3.3 Biochemical Functions of Zinc

Zinc is an essential catalytic component of over 300 enzymes such as alkaline phosphatase, alcohol dehydrogenase, carbonic anhydrase, Cu–Zn superoxide dismutase, etc. Zn plays an important role in transcriptional and posttranscriptional processes, in protein degradation and protein–protein interactions (Broadley et al. 2007; Song et al. 2010). Several transcriptional regulatory proteins have motifs stabilised by Zn, such as Zn finger, Zn cluster and RING finger domains.

11.4 Mechanism of Zn Uptake by Plants

Several regulatory mechanisms, such as control of Zn uptake, intracellular binding by metal chelators, efflux from the cell and sequestration into vacuoles, are adopted to maintain Zn homeostasis by plants.

11.4.1 Low Molecular Weight Organic Acids and Ligands

Rice varieties with higher efficiency for Zn uptake secrete citrate (Hajiboland et al. 2005; Hoffland et al. 2006) and malate (Hajiboland et al. 2005; Gao et al. 2009; Rose et al. 2011) at an increased rate in response to Zn deficiency in soil. However, according to Rose et al. (2013), the amount of citrate and malate excreted is not enough to mobilise Zn from Zn-deficient soil. Further, organic acid secretion in response to Zn deficiency might be caused by root membrane damage by free radical of oxygen (Chen et al. 2009; Rose et al. 2011, 2013). A result different from the observation cited above has been reported from a study of two rice varieties, an efficient (RIL46) and an inefficient (IR74) in Zn uptake. There is almost no difference between them with respect to transcript abundance of Zn-responsive root Zn^{2+} transporters. However, under Zn-deficient conditions, the Zn-efficient variety (RIL46) exuded more of low molecular weight organic acids and deoxymugineic acid (DMA) due to increased expression of putative ligand-expression genes in the roots (Widodo et al. 2010).

11.4.2 Mugineic Acid

Mugineic acid (MA) family phytosiderophores (see Sect. 10.4.2.1) are metal chelators produced in *graminaceous* plants in response to iron deficiency. Zn deficiency has been reported to induce the synthesis and secretion of MAs in barley plants. The levels of HvNAS1, HvNAAT-A, HvNAAT-B, HvIDS2 and HvIDS3 transcripts, which encode the enzymes involved in MA synthesis (see Sect. 10.4.2.1.1), increase in Zn-deficient roots of barley. The MAs secreted from barley roots are effective in Zn uptake from the soil (Suzuki et al. 2006). In rice plant, deoxymugineic acid is involved in efficient Zn^{2+} translocation and distribution within the rice shoot under Zn-deficient conditions but not in Zn^{2+} uptake (Suzuki et al. 2008).

11.4.3 Zn-Requiring Enzymes

It has been reported from a solution culture experiment with a Zn-efficient wheat genotype as compared to a Zn-inefficient one that the expression and activity of Zn-requiring enzymes, Cu/Zn-superoxide dismutase and carbonic anhydrase correlates with differences of their Zinc efficiency (ability to maintain significant yield under Zn-deficient conditions).

11.4.4 Phytochelatins (PCs)

PCs act as important chelators (Sect. 9.4.1.6.2) of excess Zn, but their role under Zn-deficient conditions is not known (Tennstedt et al. 2009).

11.4.5 Metallothioneins (MTs)

MTs are low molecular weight cysteine-rich proteins, which play a role in cellular Cu and Zn homeostasis (Sect. 9.4.1.6.3). It has been reported that MT4 confined to the embryo and aleurone layer in barley grains is involved in preferential Zn binding and functions in Zn storage in developing and mature grains (Hegelund et al. 2012).

11.4.6 Root Traits

A few root traits have been found to correlate with efficiency of Zn^{2+} uptake by plants. Some of these reports may be summarised as follows:

1. Studies with Zn-efficient rice show that forming hills with larger number of plants increases Zn nutrition of the crop (Hoffland et al. 2006).
2. The parameters, root length density of rice plants, influence Zn uptake, when Zn^{2+} is taken up as a DMA–Zn complex (Ptashnyk et al. 2011).
3. In dry land, cereal crops, wheat and barley, formation of root hairs and ability to form longer and finer roots correlate with enhanced Zn^{2+} uptake (Dong et al. 1995; Genc et al. 2007).

4. Ability to maintain crown root development under Zn-deficient conditions is a general characteristic of cultivars, which tolerate Zn^{2+} deficiency (Rose et al. 2013).
5. Root inoculation with AM (arbuscular mycorrhiza) improves Zn^{2+} uptake of aerobic rice genotypes with inherently low Zn^{2+} uptake capacity (Gao 2007).
6. Zn-efficient genotypes of rice have lower levels of ROS (reactive oxygen species) and ROS-related damages than inefficient genotypes. Such tolerance is often correlated with increased levels of antioxidant enzymes or metabolites present in Zn-efficient genotypes (Rose et al. 2013) (Sects. 3.4.1.1.1–3.4.1.1.5).

11.4.7 Zn Transporters

Zinc is an essential plant nutrient but is toxic beyond a threshold concentration. It is essential to maintain Zn^{2+} homeostasis within plants and their various organs at an acceptable physiological limit. This is carried out by a coordinated expression of Zn^{2+} transporters, which are involved in Zn^{2+} uptake from the soil, translocation of Zn^{2+} to various organs and tissues, intracellular sequestration and transport to vacuole.

11.4.7.1 P-Type ATPase Superfamily: The Heavy Metal ATPases (HMAs)

The heavy metal ATPases (HMAs) belong to P_{1B} subfamily of P-type ATPase superfamily (Sect. 9.4.2.1.1). P_{1B} subfamily consists of two clusters, the Cu cluster and Zn cluster. While transporters of Cu cluster transport monovalent cations, the Zn-cluster transporters transport divalent cations such as Zn^{2+}, Cd^{2+}, Co^{2+} and Pb^{2+} (Axelsen and Palmgren 2001). AtHMA2 of *Arabidopsis* drives efflux of Zn^{2+} from the plant cell and controls concentration of non-physiological heavy metals such as Cd^{2+} (Eren and Argüello 2004). HMA2 and HMA4 play a key role in transport of Zn^{2+} from cell to cell and in transport of Zn^{2+} from root to shoot. HMA4 is the main Zn^{2+} transporter in *A. thaliana and A. halleri* (Eren and Argüello 2004; Hussain et al. 2004; Hanikenne et al. 2008; Song et al. 2010). In rice, OsHMA1 to OsHMA3 belong to Zn cluster (Williams and Mills 2005). In barley, HvHMA2 with a conserved aspartate phosphorylating site functions as a Zn^{2+}/Cd^{2+} pump (Mills et al. 2012).

11.4.7.2 Plant Cadmium Resistance (PCR) Transporters

Arabidopsis: PCRs, which provide Cd^{2+} resistance to plants, constitute a small gene family with 12 members and code proteins that differ in their N-terminal domains. They are subdivided into three clades: the first clade includes only PCR10, the second clade consists of seven members (not characterised) and the third clade consists of PCR1, PCR2, PCR3 and PCR11. Among these, PCR1 is strongly expressed in leaves and PCR2 in roots and leaves. According to Song et al. (2010), PCR genes encode functional transporters, which probably act as secondary active transporters. PCR2, which is expressed in both roots and shoots, is possibly involved both in plant survival under toxic and in deficient conditions of Zn^{2+} (Song et al. 2010). It performs two independent functions:

1. It is involved in loading Zn^{2+} into the xylem.
2. Detoxification of excess Zn^{2+} at the root epidermal cells.

PCR2 is a membrane protein and its involvement in the dual processes is probably due to its dual expression pattern. It is expressed in the xylem parenchyma of the young parts of the roots. It is also expressed in the epidermal cells of roots at the differentiation zone, where root hairs develop. While a PCR2 contains two putative membrane spanning α-helices, it can form homoligomers (Song et al. 2010).

11.4.7.3 CDF Family: The MTPs (Metal Transporter Proteins)

The MTPs (metal transporter proteins), which belong to the CDF family (see Sect. 9.4.2.4), are highly specific for Zn^{2+} (Kramer 2005). In *Arabidopsis*, AtMTP3, which is localised in the vacuole of the epidermal cells of roots, controls Zn^{2+} partitioning and provides basic cellular Zn tolerance under conditions of high rates of influx of Zn^{2+} into the root symplasm (Arrivault et al. 2006).

11.4.7.4 The ZIP (ZRT- and IRT-Like Proteins) Family

There are 15 ZIP genes in *Arabidopsis* (Mäser et al. 2001). AtZIP1 to AtZIP4 play a role in cellular Zn^{2+} uptake. AtZIP1, AtZIP3 and AtZIP4 are induced at transcriptional level under Zn-limiting conditions (Guerinot 2000; Gaither and Eide 2001; Hall and Williams 2003). Recent report from yeast complementation studies (Milner et al. 2013) suggests that possibly, ZIP7 can transport Zn, Mn and Fe; ZIP1 and ZIP2 transport Zn and Mn; ZIP3, ZIP11 and ZIP12 transport Zn alone; ZIP5, ZIP6 and ZIP9 transport Mn alone; and none can transport Cu. According to them (Milner et al. 2013), AtZIP1 does not have a major role in Zn uptake. Due to its localisation in the vacuole, it probably remobilises Zn and Mn from the vacuole to the cytosol. AtZIP2, localised in the plasma membrane, is probably not involved in root Zn or Mn uptake from the soil. It translocates Mn (possibly Zn) from root to shoot.

OsZIP4 in rice is localised in apical cells and is involved in Zn uptake. Its constitutive expression changes the Zn distribution within the rice plant (Ishimaru et al. 2005).

GmZIP1 identified from soybean is highly specific for Zn and expressed in nodules and not in roots, stems or leaves. The protein is localised in the peri-bacteroid membrane, which suggests its possible role in symbiosis (Moreau et al. 2002).

11.4.7.5 The CAX (Cation/H^+ Antiporter) Family

AtMHX1 is an additional member of this family, which is an H^+-coupled antiporter, and can transport Mg^{2+} and Zn^{2+} across tonoplast in *Arabidopsis* (Shaul et al. 1999).

11.4.8 Transcription Factors (TFs)

TFs have been reported to be involved in molecular control of Zn^{2+} homeostasis in plants under Zn^{2+} deficiency.

11.4.8.1 bZIPs (Basic-Region Leucine Zipper)

Two closely related members of bZIPs transcription factor gene families, *b*ZIP19 and bZIP23 isolated from *Arabidopsis*, have been reported to be involved in transcriptional regulation for adaptation to Zn deficiency (Assunção et al. 2010). The bZIP19 and bZIP23 proteins bind to a palindromic 10 bp ZDRE (Zinc deficiency response element, RTG TCG ACA Y) in the upstream region of 8 out of a group of 15 ZIP genes. The functions of bZIP19 and bZIP23 are essential for a proper Zn^{2+}-deficiency response and allow *Arabidopsis* to grow under Zn deficiency (Assunção et al. 2010). The transcription factors, bZIP19 and bZIP23, and their target genes with the characteristic *cis* elements are conserved in higher plants including monocot and dicot crops. The Zn homeostasis mechanism as described above possibly operates in all plants under Zn^{2+}-limiting conditions (Assunção et al. 2010).

11.4.8.2 NAM (No Apical Meristem)

NAM (no apical meristem) is a NAC (no apical meristem, ATAF, CUC-shaped cotyledon) transcription factor, one of the largest TF families in plants. NAMB1 gene has been reported to mobilise N, Fe and Zn from vegetative tissues and cause increased partitioning of these nutrients to developing grains of wheat (Waters et al. 2009).

References

Arrivault S, Senger T, Kramer U (2006) The *Arabidopsis* metal tolerance protein AtMTP3 maintains metal homeostasis by mediating Zn exclusion from the shoot under Fe deficiency and Zn oversupply. Plant J 46:861–879

Assunção AGL, Herrero E, Lin YF, Huettel B, Talukdar S, Smaczniak C et al (2010) The *Arabidopsis thaliana* transcription factors bZIP19 and bZIP23 regulate the adaptation to zinc deficiency. Proc Natl Acad Sci U S A 107:10296–10301

Axelsen KB, Palmgren MG (2001) Inventory of the superfamily of P-type ion pumps in *Arabidopsis*. Plant Physiol 126:696–706

Berg JM, Shi Y (1996) The galvanization of biology: a growing appreciation for the roles of zinc. Science 271:1081–1085

Broadley MR, White PJ, Hammond JP, Zelko I, Lux A (2007) Zinc in plants. New Phytol 173:677–702

Chen WR, He ZL, Yang XE, Feng Y (2009) Zinc efficiency is correlated with root morphology, ultra structure, and anti-oxidative enzymes in rice. J Plant Nutr 32:287–305

Dong B, Rengel Z, Graham RD (1995) Root morphology of wheat genotypes differing in zinc efficiency. J Plant Nutr 18:2761–2773

Eren E, Argüello JM (2004) *Arabidopsis* HMA2, a divalent heavy metal-transporting PIB-type ATPase, is involved in cytoplasmic Zn homeostasis. Plant Physiol 136:3712–3723

Fontes RLF, Cox FR (1995) Effects of sulfur supply on soybean plants exposed to zinc toxicity. J Plant Nutr 18:1893–1906

Gaither LA, Eide DJ (2001) Eukaryotic zinc transporters and their regulation. Biometals 14:251–270

Gao X (2007) Bioavailabilty of zinc to aerobic rice. PhD thesis, Wageningen University, Wageningen, The Netherlands

Gao X, Zhang F, Hoffland E (2009) Malate exudation by six aerobic rice genotypes varying in zinc uptake efficiency. J Environ Qual 38:2315–2321

Genc Y, Huang CY, Langridge P (2007) A study of the role of root morphological traits in growth of barley in zinc-deficient soil. J Exp Bot 58:2775–2784

Guerinot ML (2000) The ZIP family of metal transporters. Biochim Biophys Acta 1465:190–198

Hajiboland R, Yang XE, Römheld V, Neumann G (2005) Effect of bicarbonate on elongation and distribution of organic acids in root and root zone of Zn-efficient and Zn-inefficient rice (*Oryza sativa* L.) genotypes. Environ Exp Bot 54:163–173

Hall JL, Williams LE (2003) Transition metal transporters in plants. J Exp Bot 54(393):2601–2613

Hanikenne M, Talke IN, Haydon MJ, Lanz C, Nolte A, Motte P, Kroymann J, Weigel D, Kramer U (2008) Evolution of metal hyper accumulation required *cis-regulatory* changes and triplication of HMA4. Nature 453:391–395

Hegelund JN, Sciller M, Kichey T, Hansen TH, Pedas P, Husted S, Schjoerring JK (2012) Barley metallothioneins: MT3 and MT4 are localised in the grain aleurone layer and show differential Zn binding. Plant Physiol 159:1125–1137

Hoffland E, Wei CZ, Wissuwa M (2006) Organic anion exudation by lowland rice (*Oryza sativa* L.) at zinc and phosphorus deficiency. Plant Soil 283: 155–162

Hussain D, Haydon MJ, Wang Y, Wong E, Sherson SM, Young J, Camakaris J, Harper JF, Cobbett CS (2004) P-type ATPase heavy metal transporters with roles in essential zinc homeostasis in *Arabidopsis*. Plant Cell 16:1327–1339

Ishimaru Y, Suzuki M, Kobayashi T, Takahashi M, Nakanishi H, Mori S et al (2005) OsZIP4, a novel zinc-regulated zinc transporter in rice. J Exp Bot 56:3207–3214

Kramer U (2005) MTP1 mops up excess zinc in *Arabidopsis* cells. Trends Plant Sci 10:313–315

Lindsay WL (1979) Chemical equilibria in soils. John Wiley & Sons, New York

Marschner H (1995) Mineral nutrition of higher plants. Academic, San Diego

Mäser P, Thomine S, Schroeder JI et al (2001) Phylogenetic relationships within cation transporter families of *Arabidopsis*. Plant Physiol 126:1646–1667

Mills RF, Peaston KA, Runions J, Williams LE (2012) HvHMA2, a P(1B)-ATPase from barley, is highly conserved among cereals and functions in Zn and Cd transport. PLoS One 7(8):e4260

Milner MJ, Seamon J, Craft F, Kochian LV (2013) Transport properties of members of the ZIP family in plants and their role in Zn and Mn homeostasis. J Exp Bot 64(1):369–381

Moreau S, Thomson RM, Kaiser BN, Trevaskis B, Guerinot ML, Udvardi MK, Puppo A, Day DA (2002) GmZIP1 encodes a symbiosis specific zinc transporter in soybean. J Biol Chem 277(7):4738–4746

Nene YL (1966) Symptoms, cause and control of *Khaira* disease of paddy. Bull Ind Phytopathol Soc 3:97–191

Neue HU, Lantin RS (1994) Micronutrient toxicities and deficiencies in rice. In: Yeo AR, Flowers TJ (eds) Soil mineral stresses: approaches to crop improvement. Springer, Berlin, pp 175–200

Ptashnyk M, Roose T, Jones DL, Kirk GJD (2011) Enhanced zinc uptake by rice through phytosiderophore secretion: a modeling study. Plant Cell Environ 34:2038–2046

Reichman SM (2002) The responses of plants to metal toxicity: a review focusing on copper, manganese and zinc. Occasional paper no.14. Australian Minerals & Energy Environment Foundation, Melbourne

Rose TJ, Rose MT, Pariasca-Tanaka J, Heuer S, Wissuwa M (2011) The frustration with utilization: why have improvements in internal phosphorus utilization efficiency in crops remained so elusive? Front Plant Nutr 2:1–5

Rose TJ, Impa SM, Rose MT, Pariasca-Tanaka J, Mori A, Heur S, Johnson-Beebout SE, Wissuwa M (2013) Enhancing phosphorus and zinc acquisition efficiency in rice: a critical review of root traits and their potential utility in rice breeding. Ann Bot 112(2):331–345. doi:10.1093/aob/mcs217

Schützendübel A, Polle A (2002) Plant responses to abiotic stresses: heavy metal-induced oxidative stress and protection by mycorrhization. J Exp Bot 53(372):1351–1365

Shaul O, Hilgemann DW, de-Almeida-Engler J, Van Montagu M, Inzé D, Galili G (1999) Cloning and characterization of a novel Mg^{2+}/H^{+} exchanger. EMBO J 18:3973–3980

Song W-Y, Choic KS, Kimb DY, Geislera M, Park J, Vincenzettia V, Schellenberg M, Kim SH, Limd YP, Nohe EW, Leeb Y, Martinoia E (2010) *Arabidopsis* PCR2 is a zinc exporter involved in both zinc extrusion and long-distance zinc transport. Plant Cell 22(7):2237–2252

Suzuki M, Takahashi M, Tsukamoto T, Watanabe S, Matsuhashi S, Yazaki J, Kishimoto N, Kikuchi S, Nakanishi H, Mori S, Nishizawa NK (2006) Biosynthesis and secretion of mugineic acid family phytosiderophores in zinc-deficient barley. Plant J 48:85–97

Suzuki M, Tsukamoto T, Inoue H, Watanabe S, Matsuhashi S, Takahashi M, Nakanishi H, Mori S, Nishizawa NK (2008) Deoxymugineic acid increases Zn translocation in Zn-deficient rice plants. Plant Mol Biol 66:609–617

Tennstedt P, Peisker D, Böttcher C, Trampczynska A, Clemens S (2009) Phytochelatin synthesis is essential for the detoxification of excess zinc and contributes significantly to the accumulation of zinc. Plant Physiol 149(2):938–948

Van Breemen N, Castro RU (1980) Zinc deficiency in wetland rice along a toposequence of hydromorphic soils in the Philippines. II. Cropping experiment. Plant Soil 57:215–221

Waters BM, Uauy C, Dubcovsky J, Grusak MA (2009) Wheat (*Triticum aestivum*) NAM proteins regulate the translocation of iron, zinc, and nitrogen compounds from vegetative tissues to grain. J Exp Bot 60(15):4263–4274

Widodo B, Broadley MR, Rose T, Frei M, Pariasca-Tanaka J, Yoshihashi T, Thomson M, Hammond JP, Aprile A, Close TJ, Ismail AM, Wissuwa M (2010) Response to zinc deficiency of two rice lines with contrasting tolerance is determined by root growth maintenance and organic acid exudation rates, and not by zinc-transporter activity. New Phytol 186(2):400–414

Williams LE, Mills RF (2005) P_{1B}-ATPases—an ancient family of transition metal pumps with diverse functions in plants. Trends Plant Sci 10:491–502

Yoshida S, Tanaka A (1969) Zinc deficiency of the rice plant in calcareous soils. Soil Sci Plant Nutr 15:75–80

12 Manganese (Mn) Uptake

Abstract

Mn deficiency is rarely observed since its cellular requirement is low. Mn is a component of photosynthetic proteins and enzymes. Mn is a cofactor of about 35 enzymes. Mn^{2+} uptake by roots is biphasic and consists of (1) an initial rapid reversible and non-metabolic process and (2) a slow second phase. The gene families involved in Mn transport include cation/H^+ antiporters, Nramps, the ZIP family and the CDF family.

12.1 Occurrence of Mn and Soil Reactions

Earth's crust contains an average of 1,000 μg g^{-1} of Mn. Mn is found in most of the Fe–Mg rocks. The total Mn in soils varies between 20 and 3,000 μg g^{-1} with an average of about 600 μg g^{-1}. The available (DTPA) Mn in acid soils of India has been reported to be 0.2–950 ppm. Maximum available Mn has been reported from Kashmir valley (Sahu and Mitra 1997).

Mn in the form of its oxides and hydroxides is coated on soil particles along with iron oxides and other constituents. Mn can exist in various oxidation states (0, II, III, IV, VI and VII). In biological systems, Mn occurs preferably in the oxidation states of II, III and IV (Guest et al. 2002). Divalent Mn^{2+} is the most available form present in soil solution. Concentration of Mn^{2+} decreases 100-fold for each unit increase in pH (Lindsay 1981). The concentration of Mn^{2+} in soil solution is controlled primarily by MnO_2 and is within 1 ppm.

$$MnO_2 + 4H^+ + 2e^- \rightleftarrows Mn^{2+} + 2H_2O$$

About 90 % of solution Mn^{2+} exists as organic complexes. Since organic matter (OM) is negatively charged, it absorbs most of the Mn^{2+} ions. However, Mn^{2+} absorbed by OM is released by exchange with H^+ from the roots (Bradl 2004). Organic acids excreted by roots in anionic form can react with oxides of Mn (MnOx) and form chelates with Mn (Ryan et al. 2001). Acidification and complexation with organic acids (citric acid) are primary mechanisms through which micronutrients including Mn are mobilised in the rhizosphere (Neumann and Romheld 2001). Solubility of Mn^{2+} increases under reducing conditions. In acid soils with pH less than 5.0, increased Mn^{2+} concentration may become toxic (Lindsay 1979).

G.N. Mitra, *Regulation of Nutrient Uptake by Plants: A Biochemical and Molecular Approach*,
DOI 10.1007/978-81-322-2334-4_12,

12.2 Mn Content of Plants

Manganese (Mn) is an essential micronutrient for plants. Similar to other micronutrients, Mn shows toxicity symptoms in plants beyond a threshold concentration. This concentration varies with plant species and their genotypes. Mn deficiency is rarely observed since its cellular requirement is low. Typical concentration of Mn in plants is in the range of 20–500 μg g^{-1}.

12.3 Functions of Mn in Plants

12.3.1 Mn Deficiency

Mn deficiency is observed in plants grown on calcareous and alkaline soils due to oxidation and immobilisation of Mn^{2+}. Further, high concentration of Fe may also cause Mn deficiency due to their negative interaction. Symptoms of Mn deficiency include yellowing of young leaves of dicotyledonous plants or development of grey specs in the mature leaves of cereals (Marshner 1995). In *A. thaliana*, the root pattern and development of root hairs are altered. Mn-deficient roots of *Arabidopsis* have an elevated Fe concentration, decreased expression of Fe-deficiency-induced genes and increased expression of ferritin genes (Yang et al. 2007).

Sensitivity of some of the crops to Mn deficiency is given in Table 12.1. Some of the crops have been repeated under different groups due to varietal differences within the same crop and environmental factors.

Table 12.1 Sensitivity of crops to Mn deficiency

Sensitivity to Mn deficiency	Name of the crops
Highly sensitive	Potato, radish, soybean, tomato, wheat, apples, citrus, cucumber, grapes, lettuce, oats, onion, sugar beet, peas
Medium sensitivity	Rice, wheat, barley, oats, corn, sweet corn, tomato, cabbage, cauliflower, rye, Sweet corn
Low sensitivity	Rye, soybean, barley, blueberries, cotton, field beans, corn, turfgrass

12.3.2 Mn Toxicity

In plants, Mn toxicity causes chlorosis and brown speckles on mature leaves and necrosis, which results in reduced yield (Marshner 1995). Mn toxicity symptoms such as chlorosis of leaves (inter-venial and marginal) and necrotic leaf spots have been reported for various crops: canola (Moroni et al. 2003), clover (Rosas et al. 2007), rye grass (Mora et al. 2009) and in leaves of barley and cowpea (Demirevska-Kepova et al. 2004; Führs et al. 2008). Deficiency of K, Ca, Mg, Fe and Si intensifies Mn toxicity (Abou et al. 2002). All the symptoms of Mn toxicity are caused due to its effects on photosynthesis of plants (Millaleo et al. 2010).

12.3.3 Biochemical Functions of Mn in Plants

Mn is a component of photosynthetic proteins and enzymes. Its deficiency especially in chloroplast affects water splitting mechanism of photosystem II (PSII), which provides electrons for photosynthetic electron transport (Buchanan et al. 2000). A group of four Mn atoms (Mn cluster) is associated with oxygen-evolving complex (OEC) bound to the reaction centre protein (DI) of PSII (water–plastoquinone oxidoreductase). Ferreira et al. (2004) have identified five metal ions in the Mn cluster; four are Mn and one Ca. The reaction involves the stepwise catalysis of a four electron oxidation ($2H_2O$ to O_2), one electron at a time, but without release of reactive oxygen intermediates. The enzyme advances from S_0 state through S_1, S_2 and S_3 to the S_4 states by sequential loss of a single electron. The enzyme on reaching the S4 state returns to the S_0 state and O_2 is evolved. Mn is useful in the redox reaction because it can exist in several stable oxidation states and can act as a catalyst involving multi-step sequential electron transfer (Merchant 2005).

Mn is a cofactor of about 35 enzymes such as Mn-superoxide dismutase, Mn catalase, pyruvate carboxylase, phosphoenol pyruvate carboxykinase, etc. Mn is essential for synthesis of chlorophyll; ATP synthesis; biosynthesis of fatty acids, acyl-lipids and proteins; and synthesis of tyrosine and plant secondary products, flavonoids and lignin. It is involved in RuBP carboxylase reactions (Pfeffer et al. 1986; Houtz et al. 1988; Ducic and Polle 2005), biosynthesis of isoprenoids (Lidon et al. 2004) and assimilation of nitrate (Ducic and Polle 2005). Mn is involved in some of the essential metabolic processes, such as photosynthesis, respiration, activation of hormones (IAA through IAA oxidase, Burnell 1988) and synthesis of amino acids (Millaleo et al. 2010).

12.4 Mechanism of Mn Uptake by Plants

Uptake of Mn^{2+} by roots according to Millaleo et al. (2010) is a biphasic process. The initial rapid reversible and non-metabolic process is caused by absorption of Mn^{2+} by negatively charged cell wall constituents of root-cell apoplastic spaces (Humphries et al. 2007). The second phase, which involves transport into the symplast, is slow since it depends on plant metabolism and carrier proteins for transport across plasma membrane. Xylem transport from roots to the above-ground parts is carried out by the transpiration stream (Marshner 1995). Experiments with the use of ^{54}Mn (Page and Feller 2005; Page et al. 2006) on wheat and white lupine plants (*Lupinus albus*) show that 7 days after labelling phase, almost all ^{54}Mn moves to the youngest fully expanded leaves and only a small fraction to the other leaves. Mn accumulation is found in the periphery of old leaves. Roots release Mn rapidly into the xylem to reach the photosynthetically active leaves through the transpiration stream. Mn tends to accumulate primarily in shoots rather than in roots of plants.

12.4.1 Mn Transporters

The gene families involved in Mn transport include (1) cation/H^+ antiporters (Sect. 9.4.2.6), (2) Nramps (Sect. 9.4.2.3), (3) the ZIP family (Sect. 9.4.2.5), (4) the CDF family (Sect. 9.4.2.4) and (5) P-type ATPases (Sect. 9.4.2.1).

12.4.1.1 Cation/H^+ Antiporters

The *Arabidopsis* AtCAX1 is a vacuolar high-affinity Ca^{2+}/H^+ antiporter. AtCAX2 has low affinity for Ca^{2+} and possibly transports Mn^{2+} and Cd^{2+} across the tonoplast (Hirschi et al. 1996, 2000).

12.4.1.2 Nramps

In *Arabidopsis*, out of six Nramp genes, five (AtNramp1–4 and AtNramp6) have been characterised at the molecular level (Curie et al. 2000; Thomine et al. 2000; Cailliatte et al. 2009). Heterologous expressions of AtNramp1, AtNramp3 and AtNramp4 from *Arabidopsis* in yeast mutants indicate that these proteins can transport Fe, Mn and Cd (Curie et al. 2000; Thomine et al. 2000). AtNramp1 in *Arabidopsis* acts as a Mn transporter for high-affinity Mn uptake by the roots from the soil under conditions of Mn deficiency with Mn concentration less than 1 μM (Cailliatte et al. 2010). A low-affinity system operates when Mn concentration is higher. Expression of AtNramp1 is restricted to root plasma membrane and upregulated due to Mn deficiency. Overexpression of Nramp1 in plants enhances growth and increases Mn content of the plants under Mn-deficient conditions (Cailliatte et al. 2010). Recent characterisation of OsNramps5 indicates its involvement in transport and uptake of Mn, Fe and Cd by rice (Ishimaru et al. 2012). MbNramp1 in a fruit tree (*Malus baccata*) is involved in Fe, Mn and Cd transport (Xiao et al. 2008).

12.4.1.3 The ZIP Family

Recently 11 members of ZIP family of *Arabidopsis* have been studied in detail by Milner et al. (2013). They report from yeast complementation studies that, possibly, ZIP7 can transport Zn, Mn and Fe; ZIP1 and ZIP2 transport Zn and Mn; ZIP3, ZIP11 and ZIP12 transport Zn alone; ZIP5, ZIP6 and ZIP9 transport Mn alone; and none can transport Cu. AtZIP1 does not have a major role in Zn uptake. Due to its localisation in the vacuole, it probably remobilises Zn and Mn from the vacuole to the cytosol. AtZIP2, localised in the plasma membrane, is probably not involved in root Zn or Mn uptake from the soil. It translocates Mn (possibly Zn) from root to shoot. IRT1, a member of ZIP family, is a high-affinity Fe^{2+} transporter under Fe-deficient conditions but also transports a number of other cations including Mn (Eide et al. 1996; Vert et al. 2002). In a Mn-efficient genotype of barley, expression of IRT1 has been found to be about 40 % greater suggesting existence of an efficient Mn uptake system (Pedas et al. 2008).

12.4.1.4 The CDF Family

The proteins of this family are involved in efflux of transitional metal cations, Zn^{2+}, Cd^{2+}, Co^{2+}, Ni^{2+} or Mn^{2+}, from cytoplasm to outside of the cell or into subcellular compartments to maintain metal homeostasis and tolerance to their toxic effects (Paulsen and Saier 1997; Eide 1998; van der Zaal et al. 1999; Hall and Williams 2003; Hanikenne et al. 2005).

References

Abou M, Symeonidis L, Hatzistavrou E, Yupsanis T (2002) Nucleolytic activities and appearance of a new DNase in relation to nickel and manganese accumulation in *Alyssum múrale*. J Plant Physiol 159:1087–1095

Bradl H (2004) Adsorption of heavy metal ions on soils and soils constituents. J Colloid Interface Sci 277:1–18

Burnell J (1988) The biochemistry of manganese in plants. In: Graham RD, Hannam RJ, Uren NJ (eds) Manganese in soil and plants. Kluwer Academic Publishers, Dordrecht, pp 125–137

Cailliatte R, Lapeyre B, Briat JF, Mari S, Curie C (2009) The NRAMP6 metal transporter contributes to cadmium toxicity. Biochem J 422:217–228

Cailliatte R, Schikora A, Briat J-F, Mari S, Curie C (2010) High-affinity manganese uptake by the metal transporter NRAMP1 is essential for Arabidopsis growth in low manganese conditions. Plant Cell 22(3):904–917

Curie C, Alonso JM, Le Jean M, Ecker JR, Briat JF (2000) Involvement of NRAMP1 from *Arabidopsis thaliana* in iron transport. Biochem J 347:749–755

Demirevska-Kepova K, Simova-Stoilova L, Stoyanova Z, Holzer R, Feller U (2004) Biochemical changes in barley plants after excessive supply of copper and manganese. Environ Exp Bot 52:253–266

Ducic T, Polle A (2005) Transport and detoxification of manganese and copper in plants. Braz J Plant Physiol 17:103–112

Eide DJ (1998) The molecular biology of metal ion transport in *Saccharomyces cerevisiae*. Annu Rev Nutr 18:441–469

Eide D, Broderius M, Fett J, Guerinot ML (1996) A novel iron-regulated metal transporter from plants identified by functional expression in yeast. Proc Natl Acad Sci U S A 93:5624–5628

Ferreira KN, Iverson TM, Maghlaoui K, Barber J, Iwata S (2004) Architecture of the photosynthetic oxygen-evolving center. Science 303:1831–1838

Führs H, Hartwig M, Buitrago L, Heintz D, Van Dorsselaer A, Braun H, Horst W (2008) Early manganese-toxicity response in *Vigna unguiculata* L. – a proteomic and transcriptomic study. Proteomics 8:149–159

Guest C, Schulze D, Thompson I, Huber D (2002) Correlating manganese X-ray absorption near-edge structure spectra with extractable soil manganese. Soil Sci Soc Am J 66:1172–1181

Hall JL, Williams LE (2003) Transition metal transporters in plants. J Exp Bot 54(393):2601–2613

Hanikenne M, Motte P, Wu MCS, Wang T, Loppes R, Matagne RF (2005) A mitochondrial half-size ABC transporter is involved in cadmium tolerance in *Chlamydomonas reinhardtii*. Plant Cell Environ 28(7):863–873

Hirschi KD, Zhen R-G, Cunningham KW, Rea PA, Fink GR (1996) CAX1, an H^+/Ca^{2+} antiporter from *Arabidopsis*. Proc Natl Acad Sci U S A 93:8782–8786

Hirschi KD, Korenkov VD, Wilganowski NL, Wagner GJ (2000) Expression of *Arabidopsis CAX2* in tobacco, altered metal accumulation and increased manganese tolerance. Plant Physiol 124:125–134

Houtz RL, Nable RO, Cheniae GM (1988) Evidence for effects on the *in vivo* activity of ribulose-biphosphate carboxylase/oxygenase during development of Mn toxicity in tobacco. Plant Physiol 86:1143–1149

Humphries J, Stangoulis J, Graham R (2007) Manganese. In: Barker A, Pilbeam D (eds) Handbook of plant nutrition. Taylor and Francis, Boca Raton, pp 351–366

Ishimaru Y, Takahashi R, Bashir K et al (2012) Characterizing the role of rice NRAMP5 in manganese, iron and cadmium transport. Sci Rep 2:286

Lidon FC, Barreiro M, Ramalho J (2004) Manganese accumulation in rice: implications for photosynthetic functioning. J Plant Physiol 161:1235–1244

Lindsay WL (1979) Solubilities of common zinc minerals in soils, chemical equilibria in soils. John Wiley and Sons, New York

Lindsay WL (1981) Chemistry in soil environment. ASA, Madison

Marshner H (1995) Mineral nutrition of higher plants. Academic, London, pp 313–323

Merchant S (2005) The light reactions: a guide to recent acquisition for the picture gallery. Plant Cell 17(3):648–663

Millaleo R, Reyes-Diaz M, Ivanov AG, Mora ML, Alberdi M (2010) Manganese as essential and toxic element for plants: transport, accumulation and resistance mechanisms. J Soil Sci Plant Nutr 10(4):470–481

Milner MJ, Seamon J, Craft F, Kochian LV (2013) Transport properties of members of the ZIP family in plants and their role in Zn and Mn homeostasis. J Exp Bot 64(1):369–381

Mora M, Rosas A, Ribera A, Rengel R (2009) Differential tolerance to Mn toxicity in perennial ryegrass genotypes: involvement of antioxidative enzymes and root exudation of carboxylates. Plant Soil 320:79–89

Moroni J, Scott B, Wratten N (2003) Differential tolerance of high manganese among rapeseed genotypes. Plant Soil 253:507–519

Neumann G, Romheld V (2001) The release of root exudates as affected by the plants physiological status. In: Pinto R, Varanini Z, Nannipieri P (eds) The rhizosphere: biochemistry and organic substances at the soil-plant interface. Marcel Dekker, New York, pp 41–93

Page V, Feller U (2005) Selective transport of zinc, manganese, nickel, cobalt and cadmium in the root system and transfer to the leaves in young wheat plants. Ann Bot 96:425–434

Page V, Weisskopf L, Feller U (2006) Heavy metals in white lupin: uptake, root-to-shoot transfer and redistribution within the plant. New Phytol 171:329–341

Paulsen IT, Saier MH Jr (1997) A novel family of ubiquitous heavy metal ion transport proteins. J Membr Biol 156:99–103

Pedas P, Ytting CK, Fuglsang AT, Jahn TP, Schjoerring JK, Husted S (2008) Manganese efficiency in barley: identification and characterization of the metal ion transporter HvIRT1. Plant Physiol 148:455–466

Pfeffer PE, Tu S, Gerasimowicz WV, Cavanaugh JR (1986) In vivo 3IP NMR studies of corn root tissue and its uptake of toxic metals. Plant Physiol 80:77–84

Rosas A, Rengel Z, Mora M (2007) Manganese supply and pH influence growth, carboxylate exudation and peroxidase activity of ryegrass and white clover. J Plant Nutr 30:253–270

Ryan P, Delhaize E, Jones D (2001) Function and mechanism of organic anion exudation from plant roots. Annu Rev Plant Physiol Plant Mol Biol 52:527–560

Sahu SK, Mitra GN (1997) Acid soils of India. Publication and Information Division, Indian Council of Agricultural Research, New Delhi

Thomine S, Wang R, Ward JM, Crawford NM, Schroeder JI (2000) Cadmium and iron transport by members of a plant metal transporter family in Arabidopsis with homology to Nramp genes. Proc Natl Acad Sci U S A 97:4991–4996

Van der Zaal BJ, Neuteboom LW, Pinas JE, Chardonnens AN, Schat H, Verkleij JAC, Hooykaas PJJ (1999) Over-expression of a novel *Arabidopsis* gene related to putative zinc-transporter genes from animals can lead to enhanced zinc resistance and accumulation. Plant Physiol 119:1047–1055

Vert G, Grotz N, Dedaldechamp F, Gaymard F, Guerinot ML, Briat JF, Curie C (2002) IRT1, an *Arabidopsis* transporter essential for iron uptake from the soil and for plant growth. Plant Cell 14:1223–1233

Xiao H, Yin L, Xu X, Li T, Han Z (2008) The iron-regulated transporter, MbNRAMP1, isolated from *Malus baccata* is involved in Fe, Mn and Cd trafficking. Ann Bot 102:881–889

Yang et al (2007) Manganese uptake and transportation as well as antioxidant response to excess manganese in plants. College of Plant Science, Agricultural Division, Jilin University, Changchun 130062, China 2007 Dec, 33(6):480–488 (Article in Chinese, English Abstract)

13 Copper (Cu) Uptake

Abstract

An average of 9 μg g^{-1} of copper is present in the soil. Cu deficiency is rarely observed in plants since its requirement is low. A wide range of gene families and proteins have been identified, which are involved in Cu transport and Cu homeostasis, such as COPT1, AtHMA6/PAA1, AtHMA8/PAA2, AtHMA7/RAN1, AtHMA5 and possibly YSL1 and YSL3. To protect Cu from improper interactions with other cellular constituents, Cu is chelated with nicotianamine (CuNA) and transported within the xylem sap from root to shoot. CuCCH (copper chaperone) complex is involved in inserting Cu into the active sites of Cu-dependent enzymes.

13.1 Occurrence of Cu and Soil Reactions

Copper content of earth's crust is about 50–70 μg g^{-1}. Total Cu content of soils varies between 1 and 40 μg g^{-1} with an average of 9 μg g^{-1}. Copper (Cu^{2+}) is chemically adsorbed to the surface of clays and Fe, Al and Mn oxides. It is one of the divalent cations, which is most strongly adsorbed to oxides of Fe and Al, forming Cu-O-Fe and Cu-O-Al surface bonds. It also forms such bonds with clay minerals. In soil solution, the dominant ion at pH below 7.0 is Cu^{2+} and above pH 7.0 Cu $(OH)_2$. The solubility of Cu^{2+} decreases with increase in pH:

$$Cu^{2+} + 2H_2O \rightleftarrows Cu(OH)_2 + 2H^+$$

Effects of pH on Cu uptake differ among crop plants. In rape (*Brassica napus* L.) and tomato, Cu concentrations in plants have been reported to be higher in acidic soils as compared to calcareous soils (Chaignon et al. 2003; Cornu et al. 2007). Accumulation of Cu in maize is the same in acidic and calcareous soils (Brun et al. 2001). No correlation has been found between Cu uptake and pH in durum wheat (*Triticum turgidum* L. var *durum*) grown on copper-contaminated soils (Michaud et al. 2007).

Copper concentration in soil solution is within a range of 10^{-6} to 10^{-9}. About 98 % of Cu in soil solution is complexed with low molecular weight organic molecules (Marshner 1995). Concentrations of free Cu ion and Cu chelates are very low in normal soils and are governed by soil properties and crop management practices. Microbial and root activity in the rhizosphere may influence Cu availability and uptake by plants. Under Fe-deficient conditions, roots of *graminaceous* monocots secrete mugineic acid (MA) family phytosiderophores to the

G.N. Mitra, *Regulation of Nutrient Uptake by Plants: A Biochemical and Molecular Approach*,
DOI 10.1007/978-81-322-2334-4_13, © Springer India 2015

rhizosphere, which form chelates with copper and increase Cu uptake in calcareous soils (Chaignon et al. 2002).

13.2 Copper (Cu) Content of Plants

Copper (Cu) is an essential micronutrient required by plants for their metabolism. Copper deficiency is rarely observed in plants. Copper concentration in plant tissues is about 1–5 μg g^{-1} of dry weight (Marshner 1995) and in leaves 5–20 μg g^{-1} of dry weight (Baker and Senef 1995). However, there is considerable variation among plant species and their varieties.

Toxicity of Cu is observed beyond a threshold value, which differs among different species of plants and their genotypes. Threshold values have been reported for groundnut (shoot), 230 mg kg^{-1} (Borkert et al. 1998), soybean (shoot) 140 mg kg^{-1} (Borkert et al. 1998), rice (whole plant) 35 mg kg^{-1} (Borkert et al. 1998), rice (shoot) <20 mg kg^{-1}, wheat (shoot), 75 mg kg^{-1} (Wheeler and Power 1995) and black gram (leaves) 67 mg kg^{-1} (Kalyanaraman and Sivagurunathan 1993).

13.3 Functions of Copper

13.3.1 Cu Deficiency

Cu deficiency may affect plant productivity at concentration below 5 μg g^{-1} of plant dry weight. Deficiency symptoms include chlorosis of young leaves, which starts from the tip and moves downwards along leaf margins, in corn young leaves turn yellow and are stunted in growth, and in many vegetable crops, flowering does not take place (Marshner 1995).

Cu deficiency occurs in sensitive crops grown under adverse soil conditions such as highly leached coarse-textured soils, alkaline soils (solubility of Cu^{2+} decreases with increase in pH) and interactions with other nutrients. High Fe and Zn content and heavy application of N and P fertilisers interfere with Cu uptake by plants.

13.3.1.1 Crops Sensitive to Cu Deficiency

Sensitivity to Cu of some of the crop plants under Cu-deficient conditions is given in Table 13.1.

13.3.2 Cu Toxicity

Cu toxicity occurs in soils manured with urban sewage and sludge, in soils enriched with mining dusts from Cu mining area, in some saline soils (Mitra et al. 2002) and in soils growing crops sprayed repeatedly with Cu-containing pesticides. Visual symptoms include chlorosis of leaves, stunted growth of plants and discoloured roots.

13.3.3 Biochemical Functions of Cu in Plants

Cu is involved in photosynthetic and respiratory electron transport chain, in cell wall lignification, vitamin C metabolism, ethylene perception, carbohydrate metabolism, nitrogen fixation, fatty acid desaturation and hydroxylation, protection from oxidative stress, pathogen defence and biogenesis of molybdenum cofactor. Cu is an integral part of plastocyanin, a protein involved in photosynthetic electron transport, and of cytochrome c oxidase involved in respiratory electron transport chain. Deficiency of Cu reduces plastocyanin biosynthesis, which affects PSI electron transport (Shikanai et al. 2003). Cu-deficient chloroplasts have decreased PSII activity due to disintegration of thylakoid membranes and modification of PSII acceptor site (Heneriques 1989;

Table 13.1 Sensitivity of crops to Cu deficiency

Sensitivity to Cu deficiency	Name of the crops
Highly sensitive	Rice, wheat, carrots, onion, beets, spinach, citrus, flax
Medium sensitivity	Barley, oats, corn, sweet corn, tomato, cabbage, cauliflower, cucumber, radishes
Low sensitivity	Rye, soybean, canola, rape seed, beans, grapes, apples

Droppa et al. 1987). Several enzymes need Cu ion as a cofactor such as polyphenol, ascorbate, diamine and laccase oxidases. The detoxification of superoxide radicals is carried out by the enzyme Cu/Zn superoxide dismutase. Cu is involved in oxidative phosphorylation, iron mobilisation, signalling transcription and protein trafficking machinery (Yruela 2005; Pilon et al. 2006; Puig et al. 2007).

Several *Arabidopsis* genes are upregulated due to Cu deficiency, such as genes of COPT1, COPT2, ZIP2 transporters, FRO3-metal reductases, CCH chaperones and chloroplastic Fe-SODs (Himelblau et al. 1998; Sancenon et al. 2003; Abdel-Ghany et al. 2005; Mukherjee et al. 2006).

Transcriptional regulation control is the primary response to Cu deficiency. The Cu-responsive elements, CuREs, located in the 5′-upstream region of Cu deficiency-induced genes with the critical GTAC core sequence, are responsible for transcriptional activation of genes under Cu-deficient conditions (Quinn et al. 2000, 2002). The *Arabidopsis* genome contains 17 SPL (squamosa protein-like) proteins with a well-conserved DNA-binding SBP domain (squamosa promoter-binding protein) (Birkenbihl et al. 2005). Expression of miR398s is induced due to Cu deficiency with concurrent downregulation of CSD1 and CSD2. There is simultaneously upregulation of Fe-SOD (FSD), which takes over dismutase function (Sunkar et al. 2006; Cuypers et al. 2011; Gielen et al. 2012). Such regulation is carried out by SPL7 (squamosa promoter-binding protein-like 7), which directly binds GTAC motifs of both FSD and miR398b/c promoters and upregulates their expression. This results in positive regulation of FSDs and negative regulation of CSDs (Abdel-Ghany and Pilon 2008; Yamasaki et al. 2009).

Redox properties of Cu contribute primarily to its toxicity. Redox reactions between Cu^{2+} and Cu^{+} can catalyse production of highly toxic hydroxyl radicals (HO′), which fragment Cu/Zn SOD (Casano et al. 1997), causing damage to cell membranes, nucleic acids, proteins and other biomolecules (Halliwell and Gutteridge 1984). Damages to cell membrane cause ion leakage such as K^{+} ion leakage from the excised roots of *Agrostis tenuis* Sibth (Wainwrighst and Woolhouseh 1977) and roots of wheat (Quartacci et al. 2001). Oxidative damage caused by Cu toxicity interferes with various cellular processes such as photosynthesis, pigment synthesis and a number of other metabolic processes resulting in stunted growth of plants (Marshner 1995; Kupper et al. 2003; Yruela 2005, 2009). Cu toxicity causes destruction of thylakoid structure in chloroplast and significant modification of protein and lipid composition of thylakoid membranes (Maksymiec 1997; Shingles et al. 2004).

13.4 Mechanism of Cu Uptake by Plants

Under physiological conditions, copper exists as Cu^{+} and Cu^{2+} forms. The Cu^{2+} form is often bound by N in histidine side chain and Cu^{+} preferably to S in cysteine or methionine.

Both deficiency and toxicity of Cu adversely affects crucial physiological processes in plants. It is essential that Cu concentrations in tissues and cells need to be controlled within a narrow physiological range. This involves uptake of Cu from soil, transport to different parts of the plants and regulation of its concentration in tissues, cells and intracellular organelles. A wide range of gene families and proteins have been identified, which are involved in Cu homeostasis.

13.4.1 Copper Transporter Proteins (COPT)

The COPT family of transporters have been identified in plants through sequence homology with eukaryotic copper transporter, Ctr, and functional complementation studies in yeast (Puig and Thiel 2002; Puig et al. 2007). The *Arabidopsis* genome contains six genes encoding COPT transporters from COPT1 to COPT6. The well-characterised COPT1 is a high-affinity transporter specific for Cu^{+} ion (Sancenon et al. 2003). COPT1 plays a significant physiological role in Cu acquisition by roots under Cu-deficient conditions (Yruela

2009). COPT1 transporters, possibly located in the plasma membrane, allow transport of Cu from exterior into cytoplasm. Their transport ability is stimulated by extracellular K^+ ion. The COPT1 gene is highly expressed in the root tips, stomata, embryos, trichomes and pollen, and its expression is negatively regulated by Cu (Sancenon et al. 2003; Yruela 2009).

All members of COPT family have the predicted three TMs (transmembrane domains) and most of them contain N-terminus methionine and histidine-rich putative metal-binding domains. An extracellular methionine residue located approximately 20 amino acids before TM1 and an MxxxxM motif within TM2 are possibly involved in metal coordination and are essential for Cu acquisition and transport (Puig and Thiel 2002; Klomp et al. 2003).

13.4.2 P-type ATPases

Arabidopsis has eight P_{1B}-ATPases, four of which belong to Cu^{2+} cluster and the other four to Zn^{2+} cluster (Baxter et al. 2003; Lee et al. 2007).

AtHMA6/PAA1 (P_{1B}-type ATPase of *Arabidopsis*1), which belongs to Cu^{2+} cluster, is involved in Cu transport system in chloroplast and delivery of cofactor to stomatal Cu/Zn superoxide dismutase (Shikanai et al. 2003). AtHMA8/PAA2 transports Cu into the thylakoid lumen to supply plastocyanin (Abdel-Ghany et al. 2005). AtHMA5 is involved in transmembrane transport of Cu and also interacts with Cu metallochaperones (CCH) (Williams et al. 2000; Andres-Colas et al. 2006). AtHMA7/RAN1 is possibly associated with the delivery of Cu ions to ethylene receptors (Hirayama et al. 1999).

AtHMA1, which phylogenetically falls in Zn cluster, delivers Cu ions to the stroma for chloroplast superoxide dismutase activity (Seigneurin-Benny et al. 2006).

A family of nine proteins belonging to P_{1B}-type ATPases has been identified in rice, whereas their numbers in barley is 10 (Williams and Mills 2005). In rice phylogenetic analysis indicates that OsHMA1 to OsHMA3 belong to Zn cluster and OsHMA4 to OsHMA9 belong to Cu cluster.

13.4.3 The ZIP Family

The members of ZIP family are involved in influx of Zn, Fe, Mn and Cu from outside the cell or from subcellular compartments into the cytoplasm with variable substrate range and specificity. The ZIP family (Sect. 9.4.2.5) has been classified into four subfamilies based on sequence conservation (Gaither and Eide 2001). ZIPs from plants and fungi come under subfamily I. ZIP transporters have been identified from a number of plants mainly dicots (Grotz and Guerinot 2006). These are involved in transport of various metal ions such as Mn^{2+}, Fe^{2+}/Fe^{3+}, Cd^{2+}, Co^{2+}, Cu^{2+}, Ni^{2+} and especially Zn^{2+}. There are 15 ZIP genes in *Arabidopsis* (Mäser et al. 2001). Recently 11 members of ZIP family of *Arabidopsis* have been studied in detail by Milner Milner et al. (2013). They report from yeast complementation studies that none can transport Cu.

13.4.4 The NRAMPs

The Nramp genes (Sect. 9.4.2.3) have been identified in several plant species.

In *Arabidopsis,* out of six Nramp genes, five (AtNramp1–4 and AtNramp6) have been characterised at the molecular level (Curie et al. 2000; Thomine et al. 2000; Cailliatte et al. 2009). Nramps are involved in Cu transport in yeast and could possibly have similar function in plants (Yruela 2005).

13.4.5 The YSL Transporters

The transport activity and specificity of YS1 transporters have been examined from maize (ZmYS1), barley (HvYS1) and rice (OsYSL15). Expression of ZmYS1, HvYS1 and OsYSL15 is strongly upregulated under Fe-deficient conditions (Curie et al. 2009). ZmYS1 and OsYSL15 have broader transporter capacities with possibilities of transporting Ni^{2+}, Zn^{2+} and Cu^{2+}-PS complexes. *Arabidopsis* has eight YSL transporters and three of them AtYSL1, AtYSL2 and AtYSL3 have been characterised (Curie et al.

2009). It has been suggested that AtYSL1 and AtYSL3 are possibly involved in Cu delivery from vascular tissues to seeds along with iron (Yruela 2009).

13.4.6 Nicotianamine (NA)

NA (Sect. 10.4.2.2) forms stable complexes with Fe^{2+}, Co^{2+}, Zn^{2+}, Ni^{2+} and Cu^{2+} *in vitro* at an increasing order of affinity (Curie et al. 2009). The stability of metal complexes is at a maximum with pH 6.5. Overexpression of barley NAS gene in transgenic tobacco resulted in increased concentration of Fe, Zn and Cu in leaves and flowers and enhanced Fe and Zn content of pollen and seeds (Takahashi et al. 2003). The CuNA complex is stable at xylem pH (5–6) and probably acts as a Cu chelator in the xylem sap for its transport from root to shoot. NA-metal complexes are possibly involved in phloem loading and unloading processes for Cu, Fe and Zn (Curie et al. 2009).

13.4.7 CCH (Copper Chaperone)

The low solubility and high activity of Cu^{+} inside the cell need to be protected from improper interactions with other cellular constituents (Hall and Williams 2003; Yruela 2009). Copper chaperones belong to a family of metal receptors, which are cytosolic, soluble, low molecular weight proteins involved in inserting Cu into the active sites of Cu-dependent enzymes (O'Halloran and Culotta 2000; Huffman and O'Halloran 2001). *Arabidopsis* has two homologues of yeast antioxidant 1p (ATX1) Cu chaperone, CCH and ATX1 (Himelblau et al. 1998; Andres-Colas et al. 2006). Similar to ATX1 metal chaperones, CCH has typical lysine residues, a βαββαβα-fold structure and an MxCxxC Cu-binding motif in the N terminus (Pufahl et al. 1997). However, CCH of higher plants, in contrast to ATX1 of non-plants, have an exclusive C-terminus domain (Andres-Colas et al. 2006). Higher levels of expression of CCH have been found in *Arabidopsis* stem and vascular cells that lack nuclei (Yruela 2009). Expression of CCH increases in oxidative stress, senescence and Cu deficiency. It has been suggested that the extra C-terminus domain of CCH is involved in symplastic Cu transport through plasmodesmata associated with nutrient mobilisation in senescing leaves.

13.4.8 CCS (Copper Chaperone for Cu/Zn Superoxide Dismutase)

CCS (copper chaperone for Cu/Zn superoxide dismutase) genes (homologous to yeast *Ccs1p/Lys7p*) encode proteins, which deliver Cu to Cu/Zn SOD by a protein–protein interaction. These have been identified in tomato, LeCCS (Zhu et al. 2000); *Arabidopsis thaliana, AtCCS* (Wintz and Vulpe 2002); potato, StCCS (Trindade et al. 2003); and soybean, GmCCS (Yruela 2009). In *Arabidopsis,* AtCCS has predicted chloroplastic targeting sequence, but it is localised in both plastids and cytosol, which indicates its involvement in Cu transport to Cu/Zn SOD enzymes of both chloroplast and cytosol. It has been reported that Cu deficiency downregulates expression of both AtCCS and Cu/Zn SODs (Abdel-Ghany et al. 2005; Yruela 2009).

13.4.9 miRNA and siRNA

MicroRNA (miRNA) and siRNA (see Sect. 3.4.9) have been reported to be involved in response to heavy metal stress (both deficiency and sufficiency) in plants. miRNAs regulate various biological processes by negatively controlling the expression of corresponding genes either by (1) post-transcriptional cleavage of target mRNA or inhibition of its translation or (2) transcriptionally by methylation of target DNA (Gielen et al. 2012).

Expressions of two closely related Cu/Zn superoxide dismutases (cytosolic CSD1 and chloroplastic CSD2) transcripts, which can detoxify Cu-induced oxidative stress, are upregulated by Cu sufficiency. There is also a downregulation of transcription of miR398, which otherwise would have cleaved mRNA of CSD1 and CSD2. This results in post-transcriptional accumulation of mRNA of CSD1 and CSD2 (Sunkar et al. 2006). Expression of all the three miR398s (miR398$_a$, miR398$_b$, miR398c) is downregulated in *Arabidopsis*, when exposed to excess of Cu. Expression of miR398s is induced due to Cu deficiency with concurrent downregulation of CSD1 and CSD2. Simultaneously Fe-SOD (FSD) is upregulated, which takes over dismutase function (Sunkar et al. 2006; Cuypers et al. 2011; Gielen et al. 2012). Such regulation is carried out by SPL7 (squamosa promoter-binding protein-like 7), which directly binds GTAC motifs of both FSD and miR398b/c promoters and upregulates their expression. This results in positive regulation of FSDs and negative regulation of CSDs (Abdel-Ghany and Pilon 2008; Yamasaki et al. 2009). A conserved KIN17, curved DNA-binding domain protein, assembles with SPL7 to adapt *Arabidopsis* growth and development to limiting copper availability (Garcia-Molina et al. 2014).

References

Abdel-Ghany SE, Pilon M (2008) MicroRNA-mediated systemic down-regulation of copper protein expression in response to low copper availability in *Arabidopsis*. J Biol Chem 283:15932–15945

Abdel-Ghany SE, Burkhead JL, Gogolin KA, Andres-Colas N, Bodecker JR, Puig S, Penarrubia L, Pilon M (2005) AtCCS is a functional homolog of the yeast copper chaperone Ccs1/Lys7. FEBS Lett 579:2307–2312

Andres-Colas N, Sancenon V, Rodriguez-Navarro S, Mayo S, Thiele DJ, Ecker JR, Puig S, Penarrubia L (2006) The *Arabidopsis* heavy metal P-type ATPase HMA5 interacts with metallochaperones and functions in copper detoxification of roots. Plant J 45:225–236

Baker DE, Senef JP (1995) In: Alloway BJ (ed) Heavy metals in soils. Blackie Academic and Professional, London, pp 179–205

Baxter I, Tchieu J, Sussman MR, Boutry M, Palmgren MG, Gribskov M, Harper JF, Axelsen KB (2003) Genomic comparison of P-type ATPase ion pumps in *Arabidopsis* and rice. Plant Physiol 132:618–628

Birkenbihl RP, Jach G, Saedler H, Huijser P (2005) Functional dissection of the plant specific SBP domain: overlap of the DNA binding and nuclear localization domains. J Mol Biol 352:585–596

Borkert CM, Cox FR, Tucker MR (1998) Zinc and copper toxicity in peanut, soybean, rice, and corn in soil mixtures. Commun Soil Sci Plant Anal 29:2991–3005

Brun LA, Maillet J, Hinsinger P, Pepin M (2001) Evaluation of copper availability to plants in copper contaminated vine yard soils. Environ Pollut 111:293–302

Cailliatte R, Lapeyre B, Briat JF, Mari S, Curie C (2009) The NRAMP6 metal transporter contributes to cadmium toxicity. Biochem J 422:217–228

Casano LM, Gomez LD, Lascano HR, Gonzales CA, Trippi VS (1997) Inactivation and degradation of CuZn–SOD by active oxygen species in wheat chloroplasts exposed to photo-oxidative stress. Plant Cell Physiol 38:433–440

Chaignon V, Di Malta D, Hinsinger P (2002) Fe-deficiency increases Cu acquisition by wheat cropped in a copper contaminated vine yard soil. New Phytol 154:121–130

Chaignon V, Sanchez-Neira I, Hermann P, Jaillard B, Hinsnger P (2003) Copper bioavailability and extractability as related to chemical properties of contaminated soils from vine growing area. Environ Pollut 123:229–238

Cornu JY, Staunton S, Hinsinger P (2007) Copper concentration in plants and in the rhizosphere as influenced by the iron status of tomato (*Lycopersicon esculentum* L). Plant Soil 292:63–77

Curie C, Alonso JM, Le Jean M, Ecker JR, Briat JF (2000) Involvement of NRAMP1 from Arabidopsis thaliana in iron transport. Biochem J 347:749–755

Curie C, Cassin G, Couch D, Divol F, Higuchi K, Jean ML, Misson J, Schikora A, Czernic P, Mari S (2009) Metal movement within the plant: contribution of nicotianamine and yellow stripe 1-like transporters. Ann Bot 103(1):1–11

Cuypers A, Smeets K, Ruytinx J, Opdenakker K, Keunen E, Remans T, Horemans N, Vanhoudt N, van Sanden S, van Belleghem F et al (2011) The cellular redox state as a modulator in cadmium and copper responses in *Arabidopsis thaliana* seedlings. J Plant Physiol 168:309–316

Droppa M, Masojidek J, Rozsa Z, Wolak A, Horvath LI, Farkas T, Horvath G (1987) Characteristics of Cu deficiency-induced inhibition of photosynthetic electron transport in spinach chloroplasts. Biochim Biophys Acta 891:75–84

Gaither LA, Eide DJ (2001) Eukaryotic zinc transporters and their regulation. Biometals 14:251–270

Garcia-Molina A, Xing S, Huijser P (2014) A conserved KIN17 curved DNA-binding domain protein assembles with Squamosa promoter-binding protein like7 to adapt *Arabidopsis* growth and development to limiting copper availability. Plant Physiol 164(2): 828–840

Gielen H, Remans T, Vangronsveld J, Cuypers A (2012) MicroRNAs in metal stress: specific roles or secondary responses? Int J Mol Sci 13(2):15826–15847

Grotz N, Guerinot ML (2006) Molecular aspects of Cu, Fe and Zn homeostasis in plants. Biochim Biophys Acta 1763:595–608

Hall JL, Williams LE (2003) Transition metal transporters in plants. J Exp Bot 54(393):2601–2613

Halliwell B, Gutteridge JMC (1984) Oxygen toxicity, oxygen radicals, transitional metals and diseases. Biochem J 219:1–14

Heneriques FS (1989) Effect of copper deficiency on photo synthetic apparatus of sugar beet (Beta vulgaris L.). J Plant Physiol 135:453–458

Himelblau E, Mira H, Lin SJ, Culotta VC, Penarrubia L, Amasino RM (1998) Identification of functional homolog of the yeast copper homeostasis gene ATX1 from *Arabidopsis*. Plant Physiol 117:1227–1234

Hirayama T, Kieber JJ, Hirayama N, Kogan M, Guzman P, Nourizadeh S, Alonso JM, Dailey WP, Dancis A, Ecker JR (1999) RESPONSIVE-TO-ANTAGONIST1, a Menkes/Wilson disease-related copper transport, is required for ethylene signaling in Arabidopsis. Cell 97:383–393

Huffman DL, O'Halloran TV (2001) Function, structure and mechanisms of intracellular copper trafficking proteins. Annu Rev Biochem 70:677–701

Kalyanaraman SB, Sivagurunathan P (1993) Effect of cadmium, copper, and zinc on the growth of blackgram. J Plant Nutr 16:2029–2042

Klomp AE, Juijn JA, van der Gun LT, van der Berg IE, Berger R, Klomp LW (2003) The N-terminus of human copper transporter (hCTR1) is localized extracellularly and interacts with itself. Biochem J 370:881–889

Kupper H, Setlik I, Setlikova E, Ferimazova N, Spiller M, Kupper FC (2003) Copper induced inhibition of photosynthesis: limiting steps of in vivo copper chlorophyll formation in *Scenedesmus quadricauda*. Funct Biol 30:1187–1196

Lee S, Kim Y-Y, Lee Y, An G (2007) Rice P_{1B}-type heavy-metal ATPase, OsHMA9, is a metal efflux protein. Plant Physiol 145:831–842

Maksymiec M (1997) Effect of copper on cellular processes in higher plants. Photosynthetica 34:321–342

Marshner H (1995) Mineral nutrition of higher plants. Academic, London, pp 313–323

Mäser P, Thomine S, Schroeder JI et al (2001) Phylogenetic relationships within cation transporter families of *Arabidopsis*. Plant Physiol 126:1646–1667

Michaud AM, Bravin MN, Galleguillos M, Hinsinger P (2007) Copper uptake and phytotoxicity as assessed in situ for durum wheat (*Triticum turgidum durum* L.) cultivated in copper contaminated former vine yard. Plant Soil 298:99–111

Milner MJ, Seamon J, Craft E, Kochian LV (2013) Transport properties of members of the ZIP family in plants and their role in Zn and Mn homeostasis. J Exp Bot 64(1):369–381

Mitra GN, Misra UK, Sahu SK (2002) Macro and micronutrient status of soils of Orissa. IFFCO, Kolkata

Mukherjee I, Campbell NH, Ash JS, Connolly EL (2006) Expression profiling of the *Arabidopsis* Ferric chelate reductase (FRO) gene family reveals differential regulation by iron and copper. Planta 223:1178–1190

O'Halloran TV, Culotta VC (2000) Metal chaperones: an intracellular shuttle service for metal ions. J Biol Chem 275:25057–25060

Pilon M, Abdel-Ghany SE, Cohu CM, Gogolin KA, Ye H (2006) Copper cofactor delivery in plant cells. Curr Opin Plant Biol 9:1–8

Pufahl RA, Singer CP, Peariso KL, Lin SJ, Schmidt PJ, Fahrni CJ, Culotta VC, Penner-Hahn JE, O'Halloran TV (1997) Metal ion chaperone function of soluble Cu(I) receptor Atx1. Science 278:853–856

Puig S, Thiel DJ (2002) Molecular mechanisms of copper uptake and distribution. Curr Opin Chem Biol 6:171–180

Puig S, Andres-Colas N, Garcia-Molina A, Penarrubia L (2007) Copper and iron homeostasis in *Arabidopsis*: response to metal deficiencies, interactions and biotechnological applications. Plant Cell Environ 30:271–290

Quartacci MF, Cosi E, Navari-Izzo F (2001) Lipids and NADPH-dependent superoxide production in plasma membrane-vesicles from roots of wheat grown under copper deficiency or excess. J Exp Bot 52:77–84

Quinn JM, Barrako P, Eriksson M, Merchant S (2000) Coordinate copper and oxygen responsive Cyt6 and Cpx1 expression in *Chlamydomonas* is mediated by the same element. J Biol Chem 275:6080–6089

Quinn JM, Eriksson M, Moseley JL, Merchant S (2002) Oxygen responsive gene expression *Chlamydomonas reinhardtii* through a copper sensing signal transduction pathway. Plant Physiol 128:463–471

Sancenon V, Puig S, Mira H, Thiele DJ, Penarubia L (2003) Identification of copper transporter family in *Arabidopsis thaliana*. Plant Mol Biol 51:577–587

Seigneurin-Benny D, Gravota A, Auroy P, Mazard C, Kraut A, Finazzi G, Grunwald D, Rappaport F, Vavasseur A, Joyrd J, Richaud P, Rolland NJBC (2006) HMA1 a new Cu-ATPase of chloroplast envelop is essential for growth under adverse light conditions. J Biol Chem 28:2882–2892

Shikanai T, Müller-Moulé P, Munekage Y, Niyogi KK, Pilon M (2003) PAA1, a P-type ATPase of *Arabidopsis*, functions in copper transport in chloroplasts. Plant Cell 15:1333–1346

Shingles R, Wimmers LE, McCarty RE (2004) Copper transport across pea thylakoid membranes. Plant Physiol 135(1):145–151

Sunkar R, Kapoor A, Zhu JK (2006) Posttranscriptional induction of two Cu/Zn superoxide dismutase genes in *Arabidopsis* is mediated by down regulation of miR398 and important for oxidative stress tolerance. Plant Cell 18:2051–2065

Takahashi M, Terada Y, Nakai I, Nakanishi H, Yoshimura E, Mori S, Nishizawa N (2003) Role of nicotianamine in the intracellular delivery of metals and plant reproductive development. Plant Cell 15:1263

Thomine S, Wang R, Ward JM, Crawford NM, Schroeder JI (2000) Cadmium and iron transport by members of a plant metal transporter family in *Arabidopsis* with homology to Nramp genes. Proc Natl Acad Sci U S A 97:4991–4996

Trindade LM, Horvath BM, Bergervoet MJE, Visser RJF (2003) Isolation of a gene coding copper chaperone for copper/Zinc superoxide dismutase and characterisation of its promoter in potato. Plant Physiol 133:618–629

Wainwrighst J, Woolhouseh W (1977) Some physiological aspects of copper and zinc tolerance in *Agrostis tenuis* Sibth: Cell elongation and membrane damage. J Exp Bot 28:1029–1036

Wheeler DM, Power IL (1995) Comparison of plant uptake and plant toxicity of various ions in wheat. Plant Soil 172:167–173

Williams LE, Mills RF (2005) P_{1B}-ATPases—an ancient family of transition metal pumps with diverse functions in plants. Trends Plant Sci 10:491–502

Williams LE, Pittman JK, Hall JL (2000) Emerging mechanisms for heavy metal transport in plants. Biochim Biophys Acta 1465:104–126

Wintz H, Vulpe C (2002) Plant copper chaperones. Biochem Soc Trans 30:732–735

Yamasaki H, Hayashi M, Fukazawa M, Kobayashi Y, Shikanai T (2009) *SQUAMOSA* promoter binding protein-like7 is a central regulator for copper homeostasis in *Arabidopsis*. Plant Cell 21: 347–361

Yruela I (2005) Copper in plants. Braz J Plant Physiol 17:145–146

Yruela I (2009) Copper in plants: acquisition, transport and interactions. Funct Plant Biol 36(5):409–430

Zhu H, Shipp E, Sanchez RJ, Liba A, Stine JE, Hart PJ, Gralla EB, Nersissian AM, Valentine JS (2000) Cobalt (II) binding to human and tomato copper chaperone for superoxide dismutase: implication for the metal ion transfer mechanism. Biochemistry 39:5413–5421

Boron (B) Uptake

14

Abstract

Boron deficiency generally occurs at a B concentration of < 20 μg g^{-1} in mature leaf tissues of most of the crops. Boron is essential for cell wall structure and functions. Borate cross-linking of RG-II is correlated with the ability of vascular plants to maintain upright growth. Boric acid is charge neutral and is the major chemical form in which boron is taken up by plants. NIP5;1 and BOR1 are the genes involved in B uptake under conditions of B deficiency. *Japonica* rice cultivars are more tolerant to higher levels of B than *indica* rice cultivars.

14.1 Occurrence of Boron and Soil Reactions

Boron occurs on the earth's crust in most of the igneous rocks at a concentration less than 10 μg g^{-1}. The primary B mineral in soil is Tourmaline, a relatively insoluble borosilicate. Total B concentration in soils is mostly around 7–80 μg g^{-1}. About 95 % of soil B is not available to plants. About 0.1 μg g^{-1} of B in soil solution is considered adequate for most of the monocots (Havlin et al. 2007).

Soil texture, moisture content, pH, organic matter status and interaction with other nutrients are some of the factors which affect B availability in soil. Well-drained, coarse-textured soils are generally low in B and respond to B application especially by B-loving plants. Low moisture content reduces release of B from soil organic matter and impedes diffusion and mass flow to the rhizosphere of plants. Availability of B decreases with increase in pH above pH 6.3–6.5. Liming acid soils may cause temporary B deficiency. Boron exists primarily as boric acid, $B(OH)_3$, in soil solution. Boric acid is a weak Lewis acid with a pKa of 9.24. At a soil pH >9.0, boric acid forms $B(OH)_4^-$ ion (Woods 1996):

$$B(OH)_3 + H_2O \rightleftarrows B(OH)_4 + H^+$$

H_3BO_3 is the preferred form in which roots absorb B. Boric acid forms esters with alcohols in a pH-dependant manner. The most stable borate esters are formed with *cis*-diols on furanoid ring such as ribose and apiose (Power and Woods 1997).

High Ca^{2+} concentration in alkaline soils or recently over-limed soils negatively affects B availability. Application of higher doses of K fertilisers may also affect B availability.

G.N. Mitra, *Regulation of Nutrient Uptake by Plants: A Biochemical and Molecular Approach*,
DOI 10.1007/978-81-322-2334-4_14, © Springer India 2015

14.2 Boron (B) Content of Plants

Boron (B) is a non-metal micronutrient recognised as essential to plants more than 90 years ago. The range of available boron concentration in soils between boron deficiency and toxicity to plants is narrow. Tolerance to B deficiency as well as toxicity differs among plant species and their genotypes. It has been reported that boron deficiency affects productivity of 132 crops in 80 countries (Takano et al. 2002). Boron has long been considered as essential to plants and not to microorganisms and animals (Marshner 1995). However, boron-containing compounds have been found in bacteria (Dembitsky et al. 2002; Chen et al. 2002; Amin et al. 2007).

Boron concentration in monocots varies between 6 and 18 μg g^{-1} and in dicots from 20 to 60 μg g^{-1}. Boron deficiency generally occurs at a B concentration of <20 μg g^{-1} in mature leaf tissues of most of the crops.

14.3 Functions of B in Plants

14.3.1 B Deficiency

Boron is not readily translocated from older to younger leaves. The first visual symptom is cessation of terminal bud growth and death of young leaves. Young leaves become pale green in colour. The leaves have twisted appearance. There is rotting of fruits, tubers or roots. There is breakdown of internal tissues leading to formation of darkened areas called black hearts. In citrus fruits, there is uneven thickness of peel, lumpy fruits and gummy deposit in the fruits. Hollow heart in groundnut is caused by B deficiency due to its negative effect on Ca translocation, which causes inhibition of cell wall formation and cell division. Plants are more sensitive to boron deficiency in the reproductive stage than in vegetative stage. Significant yield reductions have been reported due to boron deficiency at the reproductive stage without expression of any symptom at the vegetative stage (Dell and Huang 1997).

Table 14.1 Names of crops sensitive to B deficiency

Sensitivity to B deficiency	Names of the crops
Highly sensitive	Cauliflower, rape seed, peanut, sugar beet, canola, turnip, apple, broccoli
Medium sensitivity	Cabbage, carrot, cotton, radish, spinach, tomato, grapes, lettuce
Low sensitivity	Wheat, corn, oat, barley, bean, pea, potato, sorghum, soybean, onion

Boron deficiency in pulse crops causes reduced seed germination and seedling vigour.

14.3.2 Crops Sensitive to B Deficiency

A list of crops sensitive to B deficiency is given in Table 14.1.

14.3.3 Boron Toxicity

The highest concentration of B in soil is generally found in marine evaporates and in marine argillaceous sediments (Erd 1980). A relatively high concentration of B has been found in coastal saline soils (Mitra et al. 2002). Sea water contains 4.5–5 mg L^{-1} of B. It has also been reported that many cereal-growing regions of Australia, North Africa and Western Asia suffer from B toxicity (Patterson et al. 2007).

14.3.3.1 Boron Content of Irrigation Water and Crop Sensitivity

Irrigation water has been classified into three groups according to their B content (USDA 1954; Keren and Bingham 1985) and crop sensitivity as given in Table 14.2.

14.3.3.2 Visual Symptoms of B Toxicity

Visual symptoms of B toxicity generally consist of chlorotic leaves with necrotic patches often in the margins and tips of older leaves (Nable et al. 1997). These patches have relatively higher B concentrations as compared to surrounding leaf

Table 14.2 Sensitivity of crops to B content in irrigation water

B content of irrigation water	B sensitivity of crops	Names of crops affected
<0.3 mg L^{-1}	Sensitive crops	Bean (*Phaseolus vulgaris*), apple, avocado, orange
1–2 mg L^{-1}	Semi-tolerant crops	Oat, maize, potato, sweet pea, tomato, radish
2–4 mg L^{-1}	Tolerant crops	Carrot, alfalfa, sugar beet (*Beta vulgaris*), onion, palm (*Phoenix canariensis*), date palm, turnip

tissues (Oerth and Roth 1969). B toxicity in fruits manifests itself in the form of gummy nuts. Internal necrosis, bark necrosis and stem die back (Brown and Hu 1996). Boron concentration in root tissues generally remains low and roots appear not to be affected by B toxicity (Oerth and Roth 1969; Nable et al. 1997). In several cereals and legume, B concentrations in the shoots of susceptible species have been found to be much higher than the resistant ones (Chhipa and Lal 1990; Paul et al. 1992a, b).

14.3.4 Biochemical Functions of B in Plants

Boron is essential for cell wall structure and functions. Primary cell wall of higher plants consist of cellulose, hemicelluloses (xyloglucan and arabinoxylan) and pectic polysaccharides, which consist of galacturonic acid-rich polysaccharides that form a hydrated matrix in which cellulose–hemicellulose network is embedded. The major components of pectic matrix are homogalacturonan (HG), rhamnogalacturonan I (RG-I) and rhamnogalacturonan II (RG-II). Boron forms cross-links in pectic polysaccharides through borate-diol bonding of two rhamnogalacturonan II (RG-II) molecules in the cell wall. RG-II is present in primary cell wall of *angiosperms, gymnosperms, lycophytes and pteridophytes*. RG-II is composed of at least 12 different glycosyl residues linked together by more than 20 glycosidic bonds. The glycosyl sequence of RG-II is conserved in vascular plants. RG-II consists of two homogalacturonan chains and is covalently cross-linked by a borate diester bond (Fig. 14.1) to form a dimer. Such dimer formation is required for the three-dimensional network of pectins in the cell wall. This network provides the mechanical strength to the cell wall and is required for normal plant growth and development. Deficiency of boron results in reduced borate cross-linking and changes in cell wall properties of plants (O'Neill et al. 2004). According to Matsunaga et al. (2004), borate cross-linking of RG-II is correlated with the ability of vascular plants to maintain upright growth. Pabst et al. (2013) report from their studies that rhamnogalacturonan II structure shows variation in the side chain monosaccharide composition and methylation status within and across different plant species and constitute a family rather than a unique structure. RG-II probably is the only polysaccharide containing boron, which is the component of a primary cell wall.

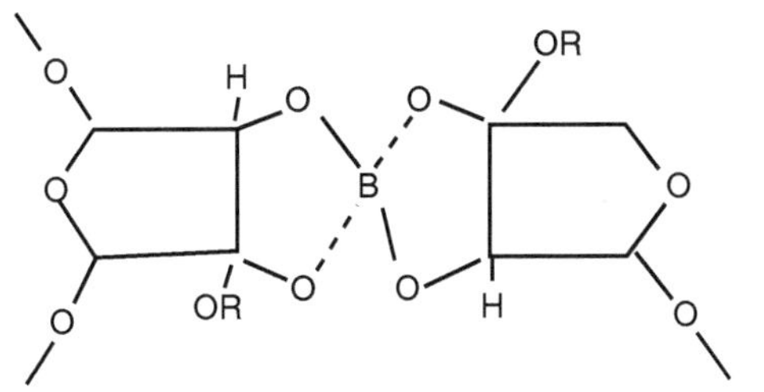

Fig 14.1 Boron-diol diester bond of RG-II

14.4 Mechanism of Boron Uptake by Plants

Boron is primarily present in soil solution as boric acid (H_3BO3) or as a borate. Boric acid is charge neutral and is the major chemical form in which boron is taken up by plants (Marshner 1995). Boric acid is permeable through lipid bilayer. It was thought prior to 1990 that B uptake by plants is through passive transport without any support of protein transporters. In root tip, where casparian strips are not fully developed, solutes can get into the xylem by apoplastic flow, an important pathway for Ca^{2+} transport to shoots (White 2001). Casparian strips are hydrophobic lipid layers (suberin) present in the cell wall between the endodermal cells, which block apoplastic flow of solutes into the stele. Nutrient uptake by plants through most of its root length involves transport through plasma membrane and casparian strips twice, once getting into the cell and then export out of the cell into the xylem. Two types of protein transporters are required for such symplastic flux, one for influx and another for efflux of solutes (Miwa and Fujiwara 2010).

14.4.1 Boron Transporters

Two types of B transporters have been identified in *Arabidopsis thaliana*, which are involved in B uptake under conditions of B deficiency, NIP5;1 and BOR1.

14.4.1.1 NIP 5;1 (Nodulin-26-Like Intrinsic Proteins)

NIP5;1 is a boric acid channel protein, which facilitates B influx into the root cells under B-limiting conditions (Takano et al. 2006). NIP5;1 is a member of MIP (major intrinsic protein) superfamily, which includes aquaporins (see Sect.1.3). NIP5;1 gene is upregulated under B-limiting conditions. NIP5;1 protein is localised in the plasma membrane. NIPs among all plant MIPs have been reported to transport a number of small uncharged molecules such as glycerol and urea in addition to water (Wallace and Roberts 2005; Takano et al. 2006). NIP5;1 also transports other substances in addition to boric acid and water (Takano et al. 2006). Studies have indicated that NIP5;1 is expressed in shoot organs such as stem nodes and leaves (Schmid et al. 2005).

14.4.1.2 NIP 6;1

NIP 6;1 has close similarity with NIP 5;1. It facilitates rapid transport of boric acid across membranes but is completely impermeable to water. Boron deficiency results in NIP 6;1 transcript accumulation in shoots but not in roots. NIP 6;1 is a boric acid channel required for proper distribution of B particularly among young developing tissues. It has been proposed that NIP 6;1 is involved in transfer of boric acid from xylem to phloem in the nodal regions (Tanaka et al. 2008).

14.4.1.3 BOR1

BOR1 is the first boron transporter identified in the biological system. BOR1 is homologous with bicarbonate transporter in animals. The BOR1 gene in *Arabidopsis* was identified by genetic mapping and complementation studies. It encodes a plasma membrane protein, which is involved in boron efflux from cells (Takano et al. 2002). BOR1 is involved in xylem loading of boron against concentration gradient under boron-deficient conditions and protects shoots from boron deficiency. Expression of BOR1 of *A. thaliana* is regulated post-transcriptionally. BOR1 accumulation is high especially in the plasma membrane under boron-deficient conditions. When B concentration is high, BOR1 expression is negligible. Excess BOR1 is incorporated into endosomes and transported back to the vacuole for degradation (Takano et al. 2005). This boron-regulated mechanism of BOR1 expression prevents unnecessary accumulation of boron in shoots and prevents B toxicity.

14.4.1.4 OsBOR1

OsBor1 in rice is a plasma membrane-localised efflux transporter of B and is required for normal growth of rice plant under B-deficient conditions. Disruption of OsBOR1 reduces B uptake and xylem loading in rice. The accumulation of OsBOR1 transcripts is higher in roots than in shoots and is independent of B deprivation. OsBOR1 is required both for root uptake and xylem loading in rice (Nakagawa et al. 2007).

14.4.1.5 BOR4

BOR4 has been identified. It is an efflux transporter, which is expressed under high boron conditions. It is localised to the distal side of epidermis in the roots. Constitutive expression of BOR4 in plants makes them tolerant to high B content (Miwa and Fujiwara 2009).

Under conditions of boron deficiency, the coordinated expression of NIP5;1 for boron acquisition and BOR1 for translocation to xylem appears to be an efficient mechanism for the movement of B from soil to shoot of plants. BOR1 probably generates a concentration gradient between root cells and the rhizosphere to facilitate NIP5;1, a boric acid channel, to take up B into the root cells (Miwa and Fujiwara 2010). BOR1 is not only involved in xylem loading but also in distribution of B within shoots.

14.4.2 Genetic Manipulation to Improve Tolerance to B Deficiency and Toxicity

14.4.2.1 Boron Deficiency

It has been reported that overexpression of BOR1 enhances shoot growth under B-deficient conditions but has no effect on root growth (Miwa and Fujiwara 2010). Overexpression of NIP5;1 enhances root elongation in low B conditions (Kato et al. 2009). Overexpression of both BOR1 and NIP5;1 in *Arabidopsis thaliana* enhances root and shoot growth, and the plant becomes significantly tolerant to B deficiency (Miwa and Fujiwara 2010).

14.4.2.2 Boron Toxicity

Overexpression of BOR4, a BOR1 paralogue in *Arabidopsis thaliana*, has been reported to be very effective in making the plants tolerant to relatively higher concentration of B (Miwa et al. 2007). HvBOR2/Bot1 in barley, TaBOR2 in wheat and AtBOR4 of *Arabidopsis* efflux excess B from plant cells and make them tolerant to B toxicity (Reid 2007). It has been reported that tolerant cultivars of barley and wheat show necrosis of leaves at higher leaf-B concentrations. However, leaf protoplasts contain lower B concentration. Boron transporters probably efflux toxic boron out of the cell to the apoplast and protect the cytoplasm of the leaf cells from B toxicity (Reid and Fitxpstrick 2009).

14.4.2.3 Crop Tolerance to B

It has been reported that *Japonica* rice cultivars are more tolerant to higher levels of B than *indica* rice cultivars. However, B concentrations in their roots and shoots are not significantly different (Ochiai et al. 2008).

Boron-tolerant barley varieties have been reported to contain elevated levels of three enzymes involved in phytosiderophore production such as iron-deficiency sensitive proteins IDS2, IDS3 and a methyl thio-ribose kinase (Patterson et al. 2007). There appears to be a potential link between Fe, B and the siderophore hydroxy mugineic acid. Boron-tolerant barley variety 'Sahara' has been found to accumulate more B under Fe-deficient conditions as compared to Fe-replete condition. However, in the B-sensitive barley variety 'Clippre', B uptake is not affected by Fe status of the plant (Patterson et al. 2007).

References

Amin SA, Kupper FC, Green DH, Harris WR, Carrano CJ (2007) Boron binding by a siderophore isolated from marine bacteria associated with the toxic *dinoflagellate Gymnodinium catenatum*. J Am Chem Soc 129:478–479

Brown PH, Hu H (1996) Phloem mobility of species dependent evidence for phloem mobility in sorbitol rich species. Ann Bot 77:497–505

Chen X, Schauder S, Potier N et al (2002) Structural identification of a bacterial quorum-sensing signal containing boron. Nature 415:545–549

Chhipa BR, Lal P (1990) A comparative study on the effect of soil boron on yield, yield attributes and nutrient uptake by susceptible and tolerant varieties of wheat. An Edafol Agrobiol 48:489–492

Dell B, Huang LB (1997) Physiological response of plants to low boron. Plant Soil 193:103–120

Dembitsky VM, Smoum R, Al-Quntar A, Ali HA, Pergament I, Srebnik M (2002) Natural occurrence of boron-containing compounds in plants, algae and microorganisms. Plant Sci 163:931–942

Erd RC (1980) The minerals of boron. In: Thomson R (ed) Supplement to Mellor's comprehensive treatise on Inorganic and theoretical chemistry, vol V. Longman, New York, pp 7–71

Havlin JL, Tisdale SL, Beaton JD, Nelson WL (2007) In soil fertility and fertilizers. Prentice Hall, New Delhi

Kato Y, Miwa K, Takano J, Wada M, Fujiwara T (2009) Highly boron deficiency-tolerant plants generated by enhanced expression of NIP5;1, a boric acid channel. Plant Cell Physiol 50:58–66

Keren R, Bingham FT (1985) Boron in soil and water. Adv Soil Sci 1:230–276

Marshner H (1995) Mineral nutrition of higher plants. Academic, London, pp 313–323

Matsunaga T, Ishii T, Matsumoto S, Higuchi M, Darvill A, Albersheim P, O'Neill MA (2004) Occurrence of the primary cell wall polysaccharide rhamnogalacturonan II *Pteridophytes, Lycophytes* and *Bryophytes*. Implication for the evolution of vascular plants. Plant Physiol 134(1):339–351

Mitra GN, Misra UK, Sahu SK (2002) In Macro and micronutrient status of soils of Orissa. IFFCO India, Calcutta

Miwa K, Fujiwara T (2009) Boron transport in plants: coordinated regulation of transporters. In: The proceedings of international plant nutrition colloquium XVI. UC, Davis

Miwa K, Fujiwara T (2010) Boron transport in plants: coordinated regulation of transporters. Ann Bot 105(7): 1103–1108

Miwa K, Takano J, Omori H, Seki M, Shinozaki K, Fujiwara T (2007) Plants tolerant of high boron levels. Science 318:1417

Nable RO, Bannuelos GS, Paull G (1997) Boron toxicity. Plant Soil 193:181–198

Nakagawa Y, Hanaoka H, Kobayashi M, Miyoshi K, Miwa K, Fujiwara T (2007) Cell-type specificity of the expression of Os *BOR1*, a rice efflux boron transporter gene, is regulated in response to boron availability for efficient boron uptake and xylem loading. Plant Cell 19:2624–2635

O'Neill MA, Ishii T, Albersheim P, Darvill AG (2004) Rhamnogalacturonan II: structure and function of a borate cross-linked cell wall pectic polysaccharide. Annu Rev Plant Physiol Plant Mol Biol 55:109–139

Ochiai K, Uemura S, Shimizu A, Okumoto Y, Matoh T (2008) Boron toxicity in rice (*Oryza sativa* L.), Quantitative trait locus (QTL) analysis of tolerance to boron toxicity. Theor Appl Genet 117(1):125–133

Oerth JJ, Roth JA (1969) Boron nutrition of sugar beet, cotton and soybean. Agron J 61:191–195

Pabst M, Fischl RM, Brecker L, Morelle W, Fauland A, Köfeler H, Altmann F, Léonard R (2013) Rhamnogalacturonan II structure shows variation in the side chains monosaccharide composition and methylation status within and across different plant species. Plant J 76:61–72. doi:10.1111/tpj.12271

Patterson J, Ford K, Cassin A, Natera S, Bacic A (2007) Increased abundance of proteins involved in phytosiderophore production in boron-tolerant barley. Plant Physiol 144(3):1612–1631

Paul JG, Nable RO, Rathjen AJ (1992a) Physiological and genetic control of the tolerance of wheat to high concentrations of boron and implications for plant breeding. Plant Soil 146:251–260

Paul JG, Nable RO, Lake AWH, Materne MA, Rathjen AJ (1992b) Response of annual medics (*Medicago Spp.*) and field peas (*Pisum sativum*) to concentrations of boron, genetic variation and the mechanism of tolerance. Aust J Agric Res 43:203–213

Power P, Woods WG (1997) The chemistry of boron and its speciation in plants. Plant Soil 193:1–13

Reid R (2007) Identification of boron transporter genes likely to be responsible for tolerance to boron toxicity in wheat and barley. Plant Cell Physiol 48:1673–1678

Reid R, Fitxpstrick K (2009) Influence of leaf tolerance mechanisms and rain on boron toxicity in barley and wheat. Plant Physiol 151(1):413–420

Schmid M, Davison TS, Henz SR, Pape UJ, Demar M, Vingron M, Schölkopf B, Weigel D, Lohmann J (2005) A gene expression map of *Arabidopsis thaliana* development. Nat Genet 37:501–506

Takano J, Noguchi K, Yasumori M, Kobayashi M, Gajdos Z, Miwa K, Hayashi H, Yoneyama T, Fujiwara T (2002) *Arabidopsis* boron transporter for xylem loading. Nature 420(6913):337–340

Takano J, Miwa K, Yuan L, von Wirén N, Fujiwara T (2005) Endocytosis and degradation of BOR1, a boron transporter of *Arabidopsis thaliana*, regulated by boron availability. Proc Natl Acad Sci U S A 102:12276–12281

Takano J, Wada M, Ludewig U, Schaaf G, von Wirén N, Fujiwara T (2006) The *Arabidopsis* major intrinsic protein NIP5;1 is essential for efficient boron uptake and plant development under boron limitation. Plant Cell 18:1498–1509

Tanaka M, Wallace IS, Takano J, Roberts DM, Fujiwara T (2008) NIP6;1 is a boric acid channel for preferential transport of boron to growing shoot tissues in *Arabidopsis*. Plant Cell 20(10):2860–2875

USDA (1954) Diagnosis and improvement of saline and alkali soils, US Salinity Laboratory, hand book, no. 60, US Department of Agriculture

Wallace IS, Roberts DM (2005) Distinct transport selectivity of two structural subclasses of the nodulin-like intrinsic protein family of plant aquaglyceroporin channels. Biochemistry 44:16826–16834

White PJ (2001) The pathways of calcium movement to the xylem. J Exp Bot 52:891–899

Woods WG (1996) Review of possible boron speciation relating to its essentiality. J Trace Elem Exp Med 9:153–163

15 Molybdenum (Mo) Uptake

Abstract

All forms of life contain molybdenum enzymes, which are involved in global cycling of C, S and N. Molybdenum (Mo) is taken up by plants as molybdate (MoO_4^{2-}), which is then used for synthesis of pterin-based Mo cofactor (Moco). All Mo enzymes are activated by Moco except nitrogenase. The Moco-containing enzymes found so far in plants consist of (1) nitrate reductase, NR; (2) sulphite oxidase, SO; (3) xanthine dehydrogenase, XDH; and (4) aldehyde oxidase. A member of group 5 sulphate transporter, Sultr 5;2 is probably an intracellular transporter involved in Mo metabolism in *Arabidopsis* and is named as MOT1. MOT1 is specific for Mo and allows plant to take up Mo from the scarce resource of Mo in soil.

15.1 Occurrence of Molybdenum (Mo) and Soil Reactions

The average molybdenum content of the earth's crust is about 2 μg g^{-1} and of soil 0.2–5 μg g^{-1}. Mo is found in various forms in soil, such as MoS_2, Fe_2 $(MoO_4)_3$, $CaMoO_4$ and $PbMoO_4$. The major form in which Mo is found in both acidic and alkaline soils is $CaMoO_4$. Molybdenum takes various forms in soil solution such as MoO_4^{2-}, $HMoO_4^-$ and H_2MoO_4 depending upon pH. Availability of Mo increases with increase in soil pH. Liming of acid soils decreases Mo availability. Mo is strongly adsorbed by Fe/Al oxides and may form insoluble compounds, which become unavailable to plants.

$$2Fe^{3+} + 3MoO_4^{2-} \rightleftarrows Fe_2\left(MoO_4\right)_3$$

A critical limit of 0.1 μg g^{-1}has been fixed for available (acid ammonium oxalate extractable) Mo deficiency in soil (AICRPM 1981). Soils containing less than this are likely to suffer from Mo deficiency.

15.2 Molybdenum Content of Plants

Molybdenum (Mo) is an essential micronutrient required by all living organisms including plants. Plants containing <0.2 μg g^{-1} are likely to show Mo deficiency. However, this concentration may vary in different plant species. Mo toxicity is rare under field conditions.

G.N. Mitra, *Regulation of Nutrient Uptake by Plants: A Biochemical and Molecular Approach*, DOI 10.1007/978-81-322-2334-4_15, © Springer India 2015

15.3 Functions of Mo in Plants

15.3.1 Mo Deficiency

Phosphorus increases Mo availability and SO_4^{2-} depresses Mo uptake by plants. Application of sulphate fertilisers may cause Mo deficiency in plants grown on Mo-deficient soils.

The common symptoms of Mo deficiency consist of yellowing of leaves, stunted growth, leaf rolling, curling and scorching of leaf margins (Gupta 1997). Mo deficiency symptoms are similar to Fe deficiency with interveinal chlorosis of leaves since Mo has a role in Fe translocation (Havlin et al. 2007).

15.3.2 Sensitivity of Crops to Mo Deficiency

Crops differ in their sensitivity to Mo deficiency. A list of some of the crops with different sensitivity to Mo deficiency is given in Table 15.1.

15.3.3 Mo Toxicity

Mo toxicity rarely occurs under field conditions. Forage crops with high molybdenum content may occur in wet, high-pH and high-organic-matter soils. Cattle consuming such forage may suffer from a disease called molybdenosis. This disease is caused by an imbalance in Mo and Cu in their diet if Mo content is more than 5 μg g^{-1}. Mo toxicity causes stunted growth and bone deformation in animals, which may be cured by oral feeding of Cu.

Table 15.1 Sensitivity of different crops to Mo deficiency

Sensitivity to Mo deficiency	Names of crops
High sensitivity	Legumes, cauliflower, onion, rapeseed, spinach
Mild sensitivity	Beet, cabbage, cotton, potato, tomato, radish
Low sensitivity	Carrot, barley, corn, wheat, sugar beet, beans, apple

15.3.4 Biochemical Functions of Mo

Molybdenum Cofactor (Moco)

Molybdenum (Mo) is taken up by plants as molybdate (MoO_4^{2-}), which is then used for synthesis of pterin-based Mo cofactor (Moco). Mo in Moco is covalently bound to two S atoms of a unique tricyclic pterin moiety known as molybdopterin (Schwarz and Mendel 2006).

Moco is highly conserved in eukaryotes, eubacteria and archaebacteria (Schwarz et al. 2000). All forms of life contain Mo enzymes, which are involved in global cycling of C, S and N. All of these enzymes are activated by Moco except nitrogenase. Moco becomes unstable when it is dissociated from the protein part of the enzymes (Basu and Burgmayer 2011).

Pterin

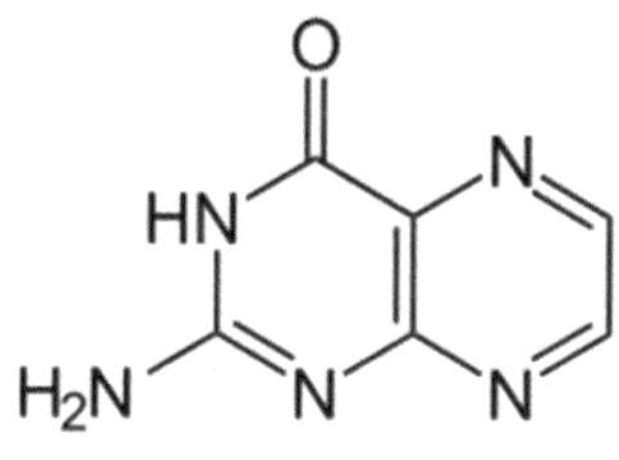

Structure of *Pterin* (en.wikipedia.org/wiki)

Pterin belongs to a larger family of bicyclic N-heterocycles known as pteridines. The pteridine ring systems consist of a pyrazine ring joined to a pyrimidine ring at 5, 6 positions. Pterin refers to a pteridine molecule where there is substitution of an amino group at position 2 and a keto group at position 4. (Numbering is done in a clockwise manner with N atom of pyrimidine ring assigned number 1 and the C atom in para-position numbered 4. Nos.5 and 6 of pyrimidine ring are fused with pyrazole ring. In the fused ring, Nos. 5, 6, 7 and 8 are assigned to N and C atoms of pyrazole ring.) Pterin undergoes keto–enol tautomerism, but keto form is more stable. A number of pterin derivatives generally with substitution in the 6th position have been isolated from different organisms (Basu and Burgmayer 2011).

Moco-Containing Enzymes

The Moco-containing enzymes found so far in plants consist of (1) nitrate reductase, NR; (2) sulphite oxidase, SO; (3) xanthine dehydrogenase, XDH; and (4) aldehyde oxidase, AO (Schwarz and Mendel 2006; Ide et al. 2010). These are divided into two groups:

1. NR and SO
2. XDH and AO

Nitrate reductase (NR): NR is localised in the cytosol and involved in the reduction of nitrate to nitrite.

Sulphite oxidase (SO): SO is located in peroxisomes and oxidises sulphite into sulphate.

Xanthine dehydrogenase (XDH): XDH is required for purine degradation in nitrogen metabolism and can generate reactive oxygen species (ROS) (Nakagawa et al. 2007).

Aldehyde oxidase: AO is involved in synthesis of abscisic acid (ABA), auxins and glucosinolates (Ibdah et al. 2009; Ide et al. 2010).

Moco is sulphurated before insertion into XDH and AO by Moco sulphurase, ABA3 (Bittner et al. 2001). Moco-binding proteins (MoBPs) involved in the distribution of Moco within the cell have been isolated from *Arabidopsis thaliana* (Kruse et al. 2010). MoBPs are localised in cytosol and undergo protein–protein contact with both Moco donor protein CNX1 and acceptor protein nitrate reductase for transfer of Moco, which indicates that MoBPs are involved in cellular distribution of Moco in *Arabidopsis* (Kruse et al. 2010).

Biosynthesis of Moco

In plants CNX genes are involved in synthesis of Moco, whereas four genes in humans, mocs1, mocs2, mocs3 and geph, are involved.

The first step in Moco biosynthesis in plants (Mendel and Hansch 2002) involves conversion of guanosine triphosphate (GTP) into cyclic pyranopterin mono phosphate (cPMP) also known as precursor Z. This step is catalysed by CNX2 and CNX3 proteins.

(i) conversion of GTP to cPMP

CNX2, CNX3

↓

GTP → **cPMP** (precursor Z)
Guanosine triphosphate → Cyclic pyranopterin mono phosphate

In the second step, two sulphur atoms are incorporated into precursor Z (cPMP). This reaction is catalysed by the enzyme MPT synthase, a heterodimeric complex of two small CNX7 and two large CNX6 subunits that convert precursor Z into MPT (molybdopterin). The S is bound to the C-terminus of CNX7 as thiocarboxylate. After the MPT synthase has transferred the two S atoms to precursor Z, it has to be re-sulphurated by the sulphurase CNX5 to reactivate the enzyme for the next reaction cycle of precursor Z conversion. Donor of S is a cysteine residue in the protein.

(ii) Incorporation of 2S atoms into precursor Z and convert it into MPT

Sulphurase (CNX5)(S) → resulphurisation → MPT synthase [2CNX7(S) + 2CNX6]

↓ +2S

cPMP (Precursor Z) → **MPT** (Molybdo pterin)

The third step involves transfer of Mo to MPT by CNX1 for synthesis of Moco. CNX1 is a two domain protein, where N-terminal domain is essential for generating an activated form of Mo, which is incorporated into the C-terminal domain with the bound MPT. The dithiolene group of molybdopterin coordinates Mo to form Moco. CNX1 is essential to stabilise Moco.

(iii) Transfer of Mo to MPT and synthesis of Moco

CNX1(Mo)

↓+Mo

MPT → **Moco**

Molybdo pterin **Molybdenum cofactor**

Structure of Moco

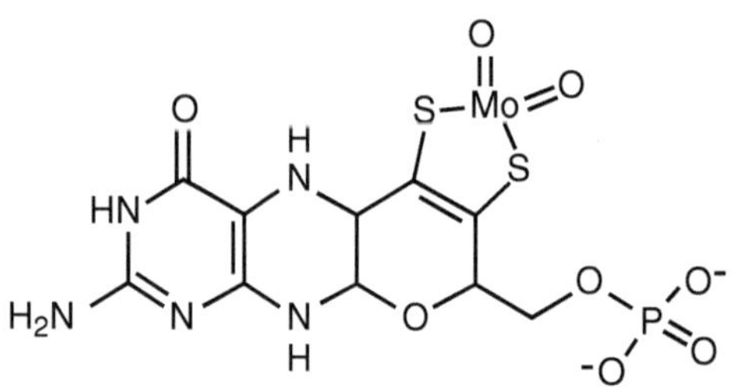

Nitrogenase and Mo

Nitrogenase is the enzyme involved in biological nitrogen fixation by microorganisms called diazotrophs. There are four known types of nitrogenases coded by unique sets of genes with different combinations of metals at the active sites (Seefeldt et al. 2009).

The nitrogenase enzyme, which has a MoFe cofactor at the active site and occurs in most of the diazotrophs, has been widely studied. Mo nitrogenase has two component enzymes, (i) a dinitrogen reductase and (ii) a dinitrogenase:

(i) Dinitrogen reductase is an Fe protein or NifH (a γ_2 homodimer, encoded by *nifH* gene).
(ii) Dinitrogenase is a MoFe protein or NifDK (a $\alpha_2\beta_2$ tetramer, coded by *nifK* and *nifD* genes, respectively).

The Fe protein (dinitrogen reductase) (M.W.$\approx$64000) a γ_2 homodimer has a binding site for one molecule of MgATP/MgADP buried within each subunit and a single [4Fe-4S] cluster, which bridges the two subunits. The Fe protein has therefore two bound molecules of MgATP and a [4Fe-4S] in a 1+ oxidation state.

The MoFe protein, a $\alpha_2\beta_2$ hetero tetramer (MW$\approx$250,000), contains two pairs of metalloclusters, named as P cluster [8Fe-7S] and FeMo cofactor [7Fe-Mo-9S-homocitrate-X]. One of each cluster is enclosed within a αβ-unit. The MoFe protein has therefore two catalytic units. Mo is bound by (R)-homocitrate by its two hydroxyl and two carboxyl groups in a coordination complex, which consists of a transition metal S framework.

Following steps are suggested to take place for reduction of N_2 to $2NH_3$.

The Fe protein with two bound molecules of MgATP and a [4Fe-4S] in a 1+ oxidation state transiently contacts MoFe protein. During this contact, the 2MgATP molecules from Fe protein are hydrolysed to 2MgADP molecules, and a single electron is transferred from the Fe protein [4Fe-4S] cluster into MoFe protein, where reduction of N_2 to $2NH_3$ takes place (Seefeldt et al. 2009). The oxidised Fe protein with 2MgADP then dissociates from the MoFe protein and regenerated in two steps. The 2MgADP is replaced by 2MgATP, and the oxidised $[4Fe\text{-}4S]^{2-}$ cluster is reduced to 1+ oxidation state by the common electron donors, the reduced ferredoxin or flavodoxin.

The mechanism and various intermediate steps involved in the reaction $N_2+3H_2\rightarrow 2NH_3$ are yet to be elucidated, although the proteins involved in the reaction have been sequenced and exhaustive crystallographic studies have been made on the metalloclusters (Rees et al. 2005; Seefeldt et al. 2009; Huergo et al. 2012; Duval et al. 2013). The overall reaction can be written as follows:

$$N_2 + 8H^+ + 16MgATP + 8e^- = 2NH_3 + H_2 + 16MgADP + 16Pi$$

The conversion of N_2 to $2NH_3$ is an energy expensive process and requires hydrolysis of 16 ATPs (since 2ATPs are required for transfer of one electron, transfer of 8 electrons as required by the equation needs 16 ATPs). To avoid energy wastage, the diazotrophs have both transcriptional and post-translational mechanisms to shut down nitrogen fixation, when ammonium is available in the environment (Huergo et al. 2012).

15.4 Mechanism of Molybdenum Uptake by Plants

15.4.1 Sultr5;2 (MOT1)

The uptakes of Mo and Se (selenium) are probably through sulphate uptake pathway (Shinmachi et al. 2010). A member of group 5 sulphate transporter, Sultr 5;2, is probably an intracellular transporter involved in Mo (molybdenum) metabolism in *Arabidopsis* and is named as MOT1 (Tomatsu et al. 2007; Baxter et al. 2008). In *Arabidopsis* the gene family of sulphate transporters consists of 14 isoforms, which can be subdivided into five groups. Alignment and phylogenetic analysis of the first four groups of *Arabidopsis* and rice sulphate transporter proteins indicate that all have 12 transmembrane spanning domains and a STAS (sulphate transporter and anti-sigma factor antagonist) domain at the carboxy-terminal (Aravind and Koonin 2000). The fifth group involved in Mo transport is more diverse but closely related, with two smaller proteins but lacks the C-terminal STAS domain (Hawkesford 2003), which is important for transport activity (Shibgaki Shibagaki and Grossman 2004). Groups 1, 2 and 3 of MOT1 are located in the plasma membrane and group 4 in the tonoplast (Kataoka et al. 2004).

MOT1 is expressed in all tissues of wheat (Shinmachi et al. 2010).

MOT1 is a high-affinity molybdate transporter specific for Mo and allows plant to take up Mo from the scarce resource of Mo in soil (Tomatsu et al. 2007; Baxter et al. 2008). Mo deficiency affects N and S metabolism in a manner different from N and S deficiency (Tomatsu et al. 2007; Baxter et al. 2008). Under conditions of Mo deficiency, Mo and its transporter MOT1 play an important role in primary metabolism of C, N and S (Ide et al. 2010).

15.4.2 Seeds as a Source of Mo

It has been reported that large seeds in some of the plants can complement Mo requirement of plants in addition to soil resource. Mo content of 3.639 ± 0.751 μg $seed^{-1}$ in common bean plants has been reported to be adequate to complement its Mo requirement from soil (Vieira et al. 2011).

15.4.3 Interaction with Other Nutrients

Studies with rice seedlings indicate that Mn^{2+}, Zn^{2+}, Cu^{2+}, Cl^- or SO_4^{2-} reduced MoO_4^{2-} uptake, but Fe^{2+} has a positive effect (Kannan and Ramani 1978). There is a significant accumulation of phosphate in plants due to Mo deficiency. This is caused by induction of PHO;H1 (a member of PHO1 family) and its expression in roots of Mo-deficient plants. PHO1;H1 is involved in phosphate acquisition and induced by phosphate deficiency (Stefanovic et al. 2007). Phosphate deficiency has been reported to enhance Mo uptake in tomato plants (Heuwinkel et al. 1992).

Forage crops with high molybdenum may occur in wet, high-pH and high-organic-matter soils. Cattle consuming such forage may suffer from a disease called molybdenosis. This disease is caused by an imbalance in Mo and Cu in their diet if Mo content is more than 5 μg g^{-1}. Mo toxicity causes stunted growth and bone deformation in animals, which may be cured by oral feeding of Cu.

References

AICRPM (1981) Annual report, 1981 of all India coordinated research project on micro and secondary nutrients. Indian Institute of Soil Science, Bhopal

Aravind L, Koonin EV (2000) The STAS domain: a link between anion transporters and antisigma-factor antagonists. Curr Biol 10:53–55

Basu P, Burgmayer SJN (2011) Pterin chemistry and its relationship to the molybdenum cofactor. Coord Chem Rev 255(9–10):1016–1038

Baxter I, Muthukumar B, Park HC, Buchner P, Lahner B, Danku J, Zhao K, Lee J, Hawkesford MJ, Guerinot ML et al (2008) Variation in molybdenum content across broadly distributed populations of *Arabidopsis thaliana* is controlled by a mitochondrial molybdenum transporter (MOT1). PLoS Genet 4:1–13

Bittner F, Oreb M, Mendel RR (2001) ABA3 is a molybdenum cofactor sulfurase required for activation of aldehyde oxidase and xanthine dehydrogenase in *Arabidopsis thaliana*. J Biol Chem 276:40381–40384

Duval S, Danyal K, Shaw S, Lytle AK, Dean DR, Hoffmab BM, Antony E, Seefeldt C (2013) Electron transfer precedes ATP hydrolysis during nitrogenase catalysis. PNAS 110(41):16414–16419

Gupta UC (1997) Symptoms of molybdenum deficiency and toxicity in crops. In: Gupta UC (ed) Molybdenum in agriculture. Cambridge University Press, Cambridge, pp 160–170

Havlin JL, Tisdale SL, Beaton JD, Nelson WL (2007) Soil fertility and fertilisers. Prentice Hall (India), New Delhi

Hawkesford MJ (2003) Transporter gene families in plants: the sulphate transporter gene family-redundancy or specialization? Physiol Plant 117:155–165

Heuwinkel H, Kirkby EA, Le Bot J, Marschner H (1992) Phosphorus deficiency enhances molybdenum uptake by tomato plants. J Plant Nutr 15:549–568

Huergo L, Pedrosa FO, Muller-Santos M, Chubatsu LS, Monteiro RA, Merrick M, Souza EM (2012) P_{II} signal transduction proteins: pivotal players in post-translational control of nitrogenase activity. Microbiology 158(1):176–190

Ibdah M, Chen YT, Wilkerson CG, Pichersky E (2009) An aldehyde oxidase in developing seeds of Arabidopsis converts benzaldehyde to benzoic acid. Plant Physiol 150:416–423

Ide Y, Kusano M, Oikawa A, Fukushima A, Tomatsu H, Saito K, Hirai MY, Fujiwara T (2010) Effects of molybdenum deficiency and defects in molybdate transporter MOT1 on transcript accumulation and nitrogen/sulphur metabolism in *Arabidopsis thaliana*. J Exp Bot 62(4):1483–1499

Kannan S, Ramani S (1978) Studies on molybdenum absorption and transport in bean and rice. Plant Physiol 62(2):179–181

Kataoka T, Hayashi N, Yamaya T, Takahashi H (2004) Root-to-shoot transport of sulfate in *Arabidopsis*: evidence for the role of SULTR3;5 as a component of low-affinity sulfate transport system in the root vasculature. Plant Physiol 136:4198–4204

Kruse T, Gehl C, Geisler M, Lehrke M, Ringel P, Hallier S, Hänsch R, Mendel RR (2010) Identification and biochemical characterization of molybdenum cofactor-binding proteins from *Arabidopsis thaliana*. J Biol Chem 285:6623–6635

Mendel RR, Hansch R (2002) Molybdo enzymes and molybdenum cofactor in plants. J Exp Bot 53(375):1689–1698

Nakagawa A, Sakamoto S, Takahashi M, Morikawa H, Sakamoto A (2007) The RNAi-mediated silencing of xanthine dehydrogenase impairs growth and fertility and accelerates leaf senescence in transgenic Arabidopsis plants. Plant Cell Physiol 48:1484–1495

Rees DC, Tezcan FA, Haynes CA, Walton MY, Andrade S, Einsle O, Howard B (2005) Structural basis of biological nitrogen fixation. Phil Trans R Soc A 363(1829):971–984

Schwarz G, Mendel RR (2006) Molybdenum cofactor biosynthesis and molybdenum enzymes. Annu Rev Plant Physiol Plant Mol Biol 57:623–647

Schwarz G, Schulze J, Bittner F, Eilers T, Kuper J, Bollmann G, Nelich A, Brinkmann H, Mendel RR (2000) The Molybdenum cofactor biosynthetic protein Cnx1 complements molybdate-repairable mutants, transfers molybdenum to the metal binding pterin and is associated with cytoskeleton. Plant Cell 12(12):2455–2471

Seefeldt LC, Hoffman BM, Dean DR (2009) Mechanism of Mo-dependent nitrogenase. Annu Rev Biochem 78:701–722

Shibagaki N, Grossman AR (2004) Probing the function of STAS domains of the Arabidopsis sulfate transporters. J Biol Chem 279:30791–30799

Shinmachi F, Buchner P, Stroud JL, Parmar S, Zhao F-J, McGrath SP, Hawkesford MJ (2010) Influence of sulfur deficiency on the expression of specific sulfate transporters and the distribution of sulfur, selenium, and molybdenum in wheat. Plant Physiol 153(1):327–336

Stefanovic A, Ribot C, Rouached H, Wang Y, Chong J, Belbahri L, Delessert S, Poirier Y (2007) Members of the PHO1 gene family show limited functional redundancy in phosphate transfer to the shoot, and are regulated by phosphate deficiency via distinct pathways. Plant J 50:982–994

Tomatsu H, Takano J, Takahashi H, Watanabe-Takahashi A, Shibagaki N, Fujiwara T (2007) An *Arabidopsis thaliana* high-affinity molybdate transporter required for efficient uptake of molybdate from soil. Proc Natl Acad Sci U S A 104:18807–18812

Vieira RF, Paula TJ, Pires AA, Carneiro JES, da Rocha GS (2011) Common bean seed complements molybdenum uptake by plants from soil. Agron J 103(6):1843–1848

16 Nickel (Ni) Uptake

Abstract

Nickel deficiency is rarely encountered under field conditions since its requirement by plants is extremely low (nanogram level). Ni deficiency results in accumulation of toxic concentration of urea in the leaves due to depression of urease activity. Nickel acts as a cofactor of enzyme urease and is essential for conversion of urea into NH_4^+ for use by plant tissues. AtIRT1 (iron-regulated transporter1), a member of ZIP family involved in high-affinity iron uptake by roots of *Arabidopsis*, has also been suggested to transport Ni.

16.1 Occurrence of Nickel (Ni) and Soil Reactions

Nickel (Ni) constitutes about 3 % of the earth's crust. Total concentration of Ni in soil is around 5–500 mg kg^{-1} with an average of 50 mg kg^{-1}. Arable soils contain 3–1,000 mg $kg^{-1}$of Ni (Liu et al. 2012). Nickel becomes available to plants in the form of Ni^{2+} ions. Nickel is readily oxidised in soil and becomes unavailable to plants above pH 6.7.

16.2 Nickel (Ni) Content of Plants

Nickel (Ni) was recognised as an essential plant micronutrient in 1987 (Dixon et al. 1975; Brown et al. 1987), but was accepted as such only in 2004 since its requirement in nanogram level could be estimated only after development of modern instruments (Liu et al. 2012). Nickel is the 17th element added to the list of essential plant nutrient (Liu 2001).

The original work of Brown et al. (1987), which established Ni as an essential micronutrient for temperate cereal crops, reported that barley plants with 40–80 nanogram (ng) of Ni per gram of dry weight showed significant reduction in germination, 50 % reduction in grain yield and low seedling vigour as compared to plants grown with adequate Ni concentration. In cowpea 0.01–0.14 μg g^{-1} of Ni concentration in plant has been reported to be deficient and 0.22–10.3 μg g^{-1} of Ni as adequate. In soybean 0.02–0.04 μg $g^{-1}$of Ni concentration in plants is considered as deficient (Walker et al. 1985; Eskew et al. 1984; Brown et al. 1987).

G.N. Mitra, *Regulation of Nutrient Uptake by Plants: A Biochemical and Molecular Approach*,
DOI 10.1007/978-81-322-2334-4_16, © Springer India 2015

16.3 Functions of Ni in Plants

Nickel deficiency causes delayed nodulation and reduced N fixation by leguminous plants. Adequate levels of Ni and urease are important for N metabolism of legumes at the reproductive phase (Brown et al. 1987; Brown 2006). Ni deficiency is rarely encountered under field conditions since its requirement by plants is extremely low.

16.3.1 Visual Symptoms of Ni Deficiency

Deficiency symptoms for Ni include leaf tip necrosis. Since Ni is involved in nitrogen metabolism and N fixation, visual symptoms closely resemble N deficiency. A characteristic Ni-deficiency symptom 'mouse ear' develops in pecan plants (Brown et al. 1987).

16.3.2 Nickel Deficiency-Induced Toxicity

Nickel is a constituent of enzyme urease (Dixon et al. 1975), which is present in a large number of plants. It has been observed in several plants that Ni deficiency results in accumulation of toxic concentration of urea in the leaves due to depression of urease activity. Nickel acts as a cofactor of enzyme urease and is essential for conversion of urea into NH_4^+ for use by plant tissues:

$$\underset{\text{Urea}}{NH_2CONH_2} + 3H_2O \underset{\substack{\text{Urease}\\(\text{Ni})}}{\rightarrow} 2NH_4^+ + CO_2 + 2OH^-$$

In Ni-deficient pecan plants, lactic acid has been found to increase by 3.2-fold and oxalic acid by 2.4-fold as compared to Ni-sufficient plants. This causes disruption of organic acid metabolism and results in manifestation of toxic symptoms such as leaf bronzing, chlorosis and leaf tip necrosis (Graham et al. 1985).

16.3.3 Nickel Sufficiency-Induced Toxicity

Nickel concentration >10 µg g^{-1} is generally considered toxic to sensitive plants (Liu et al. 2012). Divalent cations like Co^{2+}, Ni^{2+} and Zn^{2+} can displace Mg^{2+} from its position in ribulose 1, 5-diphosphate carboxylase/oxygenase, which results in loss of its activity (Wildner and Henkel 1979; van Assche and Clijsters 1986). It has been observed that plants exude histidine and citrate, which can chelate Ni and prevent Ni toxicity to plants (Salt et al. 2000). Nickel hyperaccumulating plants, *Alyssum* species (*Brassicaceae*), generally have the capacity to produce free histidine (Krämer et al. 1996). When Ni-hyperaccumulating plants of *A. lesbiacum* are exposed to higher concentration of Ni, there is proportionate increase in histidine concentration in its xylem sap. This does not happen in the non-accumulator *A. montanum.* However, when *A. Montanum* is supplied with exogenous histidine along with Ni, there is an influx of Ni to the xylem, and it shows tolerance to Ni (Kerkeb and Krämer 2003). However, none of the three genes involved in biosynthesis of histidine are regulated by Ni (Persans et al. 1999). Transgenic plants with twofold increase in histidine content give 10-fold higher biomass productions in the presence of toxic concentration of Ni as compared to wild plants. There appears to be a mechanistic link between histidine and tolerance to Ni toxicity (Wycisk et al. 2004). Other amino acids such as aspartic acid and alanine are possibly involved in complexationofNiinxylemofNi-hyperaccumulator *Stackhousia tryonii* (Bhatia et al. 2005).

Exposure to Ni and Zn in toxic concentrations has been reported to cause severe depletion of glutathione (GSH) in pigeon pea (Rao and Sresty 2000). Nickel is a poor inducer of phytochelatins (PCs) and has low binding affinity to PCs (Vatamaniuk et al. 1999).

Vacuole is not a major site of Ni accumulation since Ni/H^+ antiporter or any other nucleotide-dependent Ni pumps have not been found to operate in the tonoplast of the roots of oat and barley leaves (Brune et al. 1995; Hall 2002).

It is suggested that Ni has probably a role in synthesis of phytoalexin and plant disease resistance (Graham et al. 1985; Liu et al. 2012). It has been observed that application of Ni to roots of cowpea deficient in Ni (0.03 μg $g^{-1}$of Ni in dry roots) reduces fungal infection by 50 % (Brown 2006).

16.4 Mechanism Nickel Uptake by Plants

16.4.1 AtIRT1

AtIRT1 (iron-regulated transporter1), a member of ZIP family (Eide et al. 1996), is possibly the major transporter involved in high-affinity iron uptake by roots of *Arabidopsis* (Connolly et al. 2002; Vert et al. 2002). It has been suggested that AtIRT1 also transports Ni (Nishida et al. 2011). In hydroponic cultures, excess exposure to Ni results in Fe accumulation and increased expression of AtIRT1 in roots of *Arabidopsis* possibly due to induction AtIRT1 expression by excess Ni. Under Fe-deficient conditions, there is accumulation of Ni in roots with increased expression of AtIRT1 (Nishida et al. 2011). Exposure to Ni also induces expression of FRO2, a ferric chelate reductase and FIT (Fer-like iron deficiency-induced transcription factor) (Nishida et al. 2012) (see Sect. 10.4.5).

16.4.2 The YSL Transporters

Yellow stripe-like (YSL) transporters are so named due to formation of yellow-striped leaves in a mutant phenotype of maize, when the gene encoding these transporters is disrupted (von Wirén et al. 1994; Curie et al. 2001). YS1 found in *Poaceae* roots is a proton-coupled symporter of Fe (III)–PS complexes (Schaaf et al. 2004). The transport activity and specificity of YS1 transporters have been examined from maize (ZmYS1), barley (HvYS1) and rice (OsYSL15). Expressions of ZmYS1, HvYS1 and OsYSL15 are strongly upregulated under Fe-deficient conditions (Curie et al. 2009). ZmYS1 and OsYSL15 have broader transporter capacities with possibilities of transporting Ni (II)–PS, Zn(II)–PS and Cu(II)–PS complexes (see Sect. 10.4.3.5).

16.4.3 The CAX Family (Cation/H^+ Antiporters)

The CAX proteins are divalent cation/H^+ antiporters involved in cation influx into the vacuole. The *Arabidopsis* AtCAX4 is expressed in root apex and lateral root primordia. Expression of AtCAX4 is induced by increasing Ni^{2+} and Mn^{2+} levels and decreasing Ca^{2+} levels (see Sect. 5.4.3.1). This indicates that AtCAX4 is a cation/H^+ antiporter, which is involved in root growth under heavy metal stress conditions (Mei et al. 2009).

16.4.4 The NRAMPs

These proteins are proton/metal symporters and have broad spectrum of divalent metal cation substrate, such as Fe^{2+}, Mn^{2+}, Cd^{2+}, Co^{2+}, Cu^{2+}, Ni^{2+} and Pb^{2+} (Gunshin et al. 1997; Nevo and Nelson 2006) (see section 'Nramp genes in plants').

16.4.5 The Cation Diffusion Facilitators (CDFs) Family

The CDFs are involved in efflux of transitional metal cations, Zn^{2+}, Cd^{2+}, Co^{2+}, Ni^{2+} or Mn^{2+}, from cytoplasm to outside of the cell or into sub-cellular compartments to maintain metal homeostasis and tolerance to their toxic effects (Paulsen and Saier 1997; Hall and Williams 2003) (see Sect. 9.4.2.4).

16.4.6 Nickel Transport Within Plant

Studies with **^{63}Ni** on wheat plants showed that there was quick movement of **^{63}Ni** from labelled part of roots to newly developed roots and transiently to expanding younger leaves. There was also rapid redistribution of **^{63}Ni** from older to younger leaves indicating a high mobility through phloem (Page and Feller 2005).

16.4.7 Interaction of Ni with Other Plant Nutrients

There is a negative interaction between Ni and other nutrients such as Zn, Cu, Mn, Fe, Ca or Mg (Liu et al. 2012). Excess concentration of any of these nutrients alone or with other nutrients as cited above may cause Ni deficiency in soil.

References

Bhatia NP, Walsh KB, Baker AJM (2005) Detection and quantification of ligands involved in nickel detoxification in the herbaceous Ni hyper accumulator *Stackhousia tryonii* Bailey. J Exp Bot 56:1343–1349

Brown PH (2006) Nickel. In: Barker AV, Pilbeam DJ (eds) Handbook of plant nutrition. CRC Press Taylor & Francis Group, Boca Raton, pp 395–410

Brown PH, Welch RM, Cary EEE (1987) Nickel: a micronutrient essential for higher plants. Plant Physiol 85:801–803

Brune A, Urbach W, Dietz K-J (1995) Differential toxicity of heavy metals is partly related to a loss of preferential extraplasmic compartmentation: a comparison of Cd-, Mo-, Ni- and Zn-stress. New Phytol 129:403–409

Connolly EL, Fett JP, Guerinot ML (2002) Expression of the IRT1 metal transporter is controlled by metals at the levels of transcript and protein accumulation. Plant Cell 14:1347–1357

Curie C, Panaviene Z, Loulergue C, Dellaporta SL, Briat JF, Walker EL (2001) Maize yellow stripe1 encodes a membrane protein directly involved in Fe(III) uptake. Nature 409:346

Curie C, Cassin G, Couch D, Divol F, Higuchi K, Jean ML, Misson J, Schikora A, Czernic P, Mari S (2009) Metal movement within the plant: contribution of nicotianamine and yellow stripe 1-like transporters. Ann Bot 103(1):1–11

Dixon NE, Gazzola C, Blakel RL, Zerner YB (1975) Jack-Bean urease (E.C.3.5. 1.5.3.) a metallo-enzyme, a simple biological role for nickel. J Am Chem Soc 97:4131–4133

Eide D, Broderius M, Fett J, Guerinot ML (1996) A novel iron-regulated metal transporter from plants identified by functional expression in yeast. Proc Natl Acad Sci U S A 93:5624–5628

Eskew DL, Welch RM, Norvell WA (1984) Nickel in higher plants: further evidence for an essential role. Plant Physiol 76:691–693

Graham RD, Welch RM, Walker CD (1985) A role of nickel in the resistance of plants to rust. In: Proceedings of the 3rd Australian agron conference, Hobart

Gunshin H, Mackenzie B, Berger UV, Gunshin Y, Romero MF, Boron WF, Nussberger S, Gollan JL, Hediger MA (1997) Cloning and characterization of a mammalian proton-coupled metal-ion transporter. Nature 388:482–488

Hall JL (2002) Cellular mechanisms for heavy metal detoxification and tolerance. J Exp Bot 53(366):1–11

Hall JL, Williams LE (2003) Transition metal transporters in plants. J Exp Bot 54(393):2601–2613

Kerkeb L, Krämer U (2003) The role of free histidine in xylem loading of nickel in *Alyssum lesbiacum* and *Brassica juncea*. Plant Physiol 131:716–724

Krämer U, Cotter-Howells JD, Charnock JM, Baker AJM, Smith JAC (1996) Free histidine as a metal chelator in plants that accumulate nickel. Nature 379:635–638

Liu GD (2001) A new essential mineral element – nickel. Plant Nutr Fertil Sci 7(1):101–103

Liu GE, Simonne H, Li Y (2012) Nickel nutrition in plants, HS1191, One of a series of the horticultural sciences, Florida Cooperative Extension Service, Institute of Food and Agricultural Sciences, University of Florida. Original publication date June 2011

Mei H, Cheng NH, Zhao J, Park S, Escareno RA, Pitman JK, Hirschi KD (2009) Root development under metal stress in *Arabidopsis thaliana* requires the H+/cation antiporter CAX4. New Phytol 183(1):95–105. doi:10.1111/j.1469-8137.2009.02831.x. Epub 2009 Apr 8

Nevo Y, Nelson N (2006) The NRAMP family of metal-ion transporters. Biochim Biophys Acta 1763:609–620

Nishida S, Tsuzuki C, Kato A, Aisu A, Yoshida J, Mizuno T (2011) AtIRT1, the primary iron uptake transporter in the root, mediates excess nickel accumulation in *Arabidopsis thaliana* plant. Cell Physiol 52(8):1433–1442

Nishida S, Aisu A, Mizuno T (2012) Induction of IRT1 by Nickel-induced iron-deficient response in *Arabidopsis*. Plant Signal Behav 73:329–331

Page V, Feller U (2005) Selective transport of zinc, manganese, nickel, cobalt and cadmium in the root system and transfer to the leaves in young wheat plants. Ann Bot 96:425–434

Paulsen IT, Saier MH Jr (1997) A novel family of ubiquitous heavy metal ion transport proteins. J Membr Biol 156:99–103

Persans MW, Yan X, Patnoe JM, Krämer U, Salt DE (1999) Molecular dissection of the role of histidine in Ni hyper accumulation in *Thlaspi goesingense*. Plant Physiol 121:1117–1126

Rao KVM, Sresty TVS (2000) Antioxidative parameters in the seedlings of pigeon pea (*Cajanus cajan* (L.) Millspaugh) in response to Zn and Ni stresses. Plant Sci 157:113–128

Salt DE, Kato N, Krämer U, Smith RD, Raskin I (2000) The role of root exudates in nickel hyper accumulation and tolerance in accumulator and non accumulator species of *Thlaspi*. In: Terry N, Banuelos G (eds) Phytoremediation of contaminated soil and water. Lewis Publishers, Boca Raton, pp 189–200

Schaaf G, Ludewig U, Erenoglu BE, Mori S, Kitahara T, von Wiren N (2004) ZmYS1 functions as a proton-coupled symporter for phytosiderophore and nicotianamine-chelated metals. J Biol Chem 279:9091

van Assche F, Clijsters H (1986) Inhibition of photosynthesis in *Phaseolus vulgaris* by treatment with toxic concentration of zinc: effect on ribulose-1,5-bisphosphate carboxylase/oxygenase. J Plant Physiol 125:355–360

Vatamaniuk OK, Mari S, Lu YP, Rea PA (1999) AtPCS1, a phytochelatin synthase from *Arabidopsis*: isolation and *in vitro* reconstitution. Proc Natl Acad Sci U S A 96:7110–7115

Vert G, Grotz N, Dédaldéchamp F, Gaymard F, Guerinot M, Briat J-F, Curie C (2002) IRT1, an *Arabidopsis* transporter essential for iron uptake from the soil and for plant growth. Plant Cell 14:1223–1233

von Wirén N, Mori S, Marschner H, Römheld V (1994) Iron inefficiency in maize mutant ys1 (*Zea mays* L. cv yellow-stripe) is caused by a defect in uptake of iron phytosiderophores. Plant Physiol 106(1):71–77

Walker CD, Graham RD, Madison JT, Cary EE, Welch RM (1985) Effects of nickel deficiency on some nitrogen metabolites in cowpeas, *Vigna unguiculata*. Plant Physiol 79:474–479

Wildner GF, Henkel J (1979) The effect of divalent metal ion on the activity of Mg^{2+}-depleted ribulose-1,5-bisphosphate oxygenase. Planta 146:223–228

Wycisk K, Kim EJ, Schroeder JI, Krämer U (2004) Enhancing the first enzymatic step in the histidine biosynthesis pathway increases the free histidine pool and nickel tolerance in *Arabidopsis thaliana*. FEBS Lett 578:128–134

17 Chloride (Cl^-) Uptake

Abstract

Chlorine (Cl_2) in the form of chloride Cl^- ion is an essential plant nutrient. Chloride constitutes an important part of nutrient management for cops like coconut and oil palm. The use of chloride fertilisers suppresses a number of diseases in different crops. Chloride is a major osmotically active ion in the vacuole and is involved in turgor and osmoregulation. Electrophysiological studies have identified several anion channels, which operate in different cells and tissues of plants. They transport Cl^- along with other anions. CLC genes belong to the family of voltage-gated chloride channels and are generally expressed in endo-membranes of all tissues.

17.1 Occurrence of chloride (Cl^-) and Soil Reactions

Broyer et al. (1954) established chlorine (Cl_2) in the form of chloride (Cl^-) ion as an essential plant micronutrient in 1954 from their studies on tomato plants. Chloride concentration in the earth's crust is around 0.02–0.05 %. It occurs in igneous and metamorphic rocks. Sea water contains about 3.5 % of NaCl; Na^+, 1.08 %; Cl^-, 1.94 %; SO_4^{2-}, 0.27 %; Mg^{2+}, 0.13 %; Ca^{2+}, 0.04 %; and K^+, 0.04 % (Munns 2002). Chloride is added to soil along with potassic fertiliser KCl, which constitutes 92 % of world potassium consumption in agriculture.

Chlorine exists as chloride (Cl^-) ion in soil solution and is reported to be in the range of 0.5 ppm in acid soils and more than 6,000 ppm in saline and sodic soils. Annual atmospheric wet deposition of Cl_2 in USA has been reported to be <0.5 to >10 kg Cl_2 ha^{-1} (USNADP 2002). Chloride is highly soluble and leached to lower depths in porous soils. Chloride is not adsorbed by organic matter or clay and is not readily precipitated from solution. Most of the chloride salts are readily soluble in water. Chloride content of arable soils may increase due to impeded drainage, high water table, capillary rise Cl^- from lower depths and irrigation water with high Cl^- content.

17.2 Chloride Content of Plants

The amount of chloride taken by plants is about 20 kg to 80 kg ha^{-1} depending on plant type and density in the field (Fixen 1993). Concentration of chloride in higher plants is usually 0.2–2.0 % but may go up to 10 % in saline soils (Fixen 1993).

G.N. Mitra, *Regulation of Nutrient Uptake by Plants: A Biochemical and Molecular Approach*, DOI 10.1007/978-81-322-2334-4_17, © Springer India 2015

17.3 Functions of Chloride in Plants

Chloride (Cl^-) is involved in osmotic adjustments (cell turgor and stomatal movements and regulation of water loss), transport of nutrients in plants (Ca, Mg and K) and photosynthesis, increases yield of cereal crops, regulates function of several enzymes and reduces disease attack. Chloride constitutes an important part of nutrient management for crops like coconut and oil palm.

17.3.1 Chloride Deficiency

Chloride-deficient leaves show wilting, chlorosis, necrosis and an unusual bronze discolouration (Broyer et al. 1954). Chloride deficiency causes characteristic leaf spot in wheat (Engel et al. 1997). Application of chloride fertilisers has been reported to suppress physiological disorders such as wheat leaf spot and hollow heart and brown centre of potato.

17.3.2 Chloride and Disease Resistance

The use of chloride fertilisers suppresses a number of diseases in different crops as given in Table 17.1.

Table 17.1 Suppression of different diseases in crops by application of chloride fertilisers

Name of the crops	Diseases suppressed by chloride
Barley	Common root rot, *Fusarium* caused root rot, blotch rot
Coconut	Grey leaf spot
Corn	Stalk rot
Pearl millet	Downy mildew
Rice	Stem rot, sheath blight
Wheat	Common root rot, stripe rust, leaf rust, *Septoria*

Table 17.2 Sensitivity of some of the crops to chloride deficiency

Sensitivity to Cl^- deficiency	Name of the crops
Highly sensitive	Legumes, avocado, tobacco, peach, lettuce
Medium sensitivity	Wheat, potato, cotton, soybean, oats
Low sensitivity	Tomato, corn, barley, sugar beet, spinach

17.3.3 Sensitivity of Crops to Chloride Deficiency

Sensitivity of some of the crops to Cl^- deficiency is given in Table 17.2.

17.3.4 Chloride Toxicity

It has been reported that Cl^- can be toxic to plants if its concentration exceeds 4–7 mg g^{-1} of dry weight for chloride-sensitive species and 15–50 mg g^{-1} for chloride-tolerant plants (Xu et al. 2000). Crops grown on saline soils accumulate both Na^+ and Cl^- ions in their shoots. While Na^+ toxicity has been widely studied, the toxic effects of Cl^- as an independent ion have scarcely been reported. While both Na^+ and Cl^- are toxic at higher concentration, some plants can regulate Na^+ uptake better than Cl^- (Munns and Tester 2008). It has been recently reported (Tavakkoli et al. 2011) from soil- and solution-based studies that barley genotypes with varying susceptibility to salinity show different exclusion mechanisms for Na^+ and Cl^-. High concentrations of Na^+ reduce uptake of Ca^{2+} and K^+ and reduce stomatal conductance, resulting in reduced photosynthesis. High Cl^- concentration causes chlorophyll degradation and reduces actual quantum yield of PSII electron transport. There is a difference in response to salinity between soil and solution culture studies. Capacity to control Cl^- exclusion from shoots has been correlated with salt tolerance in many species of plants (Teakle et al. 2007; Teakle and Tyerman 2009).

17.3.5 Biochemical Functions of Chloride in Plants

Minimum concentration of chloride in plant tissue essential for biochemical reactions is about 100 mg kg^{-1} of dry weight (Fixen 1993). Chloride is a major osmotically active ion in the vacuole and is involved in turgor and osmoregulation (White and Broadly 2001).

Chloride is reported to act as an inhibitor of malic enzyme (Wedding 1989). Chloride acts as an activator of a number of enzymes. Chloride-dependent α-amylases, angiotensin-converting enzymes (ACE) and photosystem II are activated by bound chloride (Pokhrel et al. 2011). In these enzymes, positively charged arginine and lysine residues constitute the chloride binding sites. The mechanism of chloride activation of ACE- and chloride-dependent α-amylases consists of (i) correctly positioning catalytic or other residues involved in stabilising enzyme–substrate complex and (ii) fine-tuning of pKa of a catalytic residue (Pokhrel et al. 2011). Chloride binding in photosystem II is essential for photosynthetic water oxidation. Recent structural studies indicate that there is at least one chloride binding site in the vicinity of oxygen evolving complex (OEC). The absence of chloride may cause a conformational shift of D1:D61 and makes it an inefficient proton acceptor during the S-state cycle (Pokhrel et al. 2011; Rivalta et al. 2011). Recent data indicate that chloride transfer across plasma membrane is important for pollen germination and pollen tube growth (Tavares et al. 2011).

17.4 Mechanism of Chloride Uptake by Plants

17.4.1 Active and Passive Uptake

Influx of Cl^- into the symplasm appears to be active, whereas efflux is passive since there is a gradient for passive efflux of anions from cytoplasm to the external medium (Teakle and Tyerman 2009). As Cl^- LATs increase, influx of Cl^- into the cytoplasm passive Cl^- efflux starts to maintain cellular chloride homeostasis. Chloride efflux is stimulated by increased concentration of external Cl^- and may reach 90 % of influx in barley (Britto et al. 2004). According to White and Broadly (2001), chloride uptake by roots is through a symplastic pathway, and the nature of chloride fluxes depends on the chloride concentration in the roots. At low chloride concentration in soil solution, chloride uptake through plasma membrane of root cells is through active influx, whereas passive influx takes place under more saline conditions. Both active and passive Cl^- transports occur through tonoplast. Electrophysiological studies indicate the presence of $Cl^-/2H^+$ symporter in the plasma membrane of root hair cells, and Cl^-/nH^+ antiporter mediates chloride influx across tonoplast.

17.4.2 Chloride Channels and Transporters

Channels and transporters have been placed in two classes for K^+ transport, Class 1 consisting of low-affinity systems effective at 1 mM and above and Class 2 consisting of high-affinity systems effective at K^+ concentration in the micromolar range (Véry and Sentenac 2003). There are however channels which operate in high-affinity ion uptake and transporters operating in both high- and low-affinity ion uptake.

Electrophysiological studies have identified several anion channels, which operate in different cells and tissues of plants. They transport other anions along with Cl^-.

17.4.2.1 Outward Rectifying Depolarisation-Activated Anion Channels (ORDAACs)

Outward rectifying depolarisation-activated anion channels (ORDAACs) have been reported to be present in the protoplasts of epidermis and cortical cells of roots in wheat, maize, *Arabidopsis* and lupin (Skerret and Tyerman 1994; Pineros and Kochian 2001; Diatloff et al. 2004; Zhang et al. 2004). These are twice as permeable to NO_3^- than Cl^-.

17.4.2.2 Inward Rectifier Anion Channel (IRAC)

An inward rectifier anion channel (IRAC)-type channel with high permeability to citrate over chloride has been identified from lupin roots especially under low Pi conditions (Zhang et al. 2004). It is involved in efflux of citrate from roots to mobilise external Pi in conjunction with a proton pump that activates the channel by hyperpolarisation.

Protoplasts of barley xylem parenchyma have 3 types of anion channels (Köhler and Raschke 2000). These consist of:

1. An inwardly rectifying anion channel (X-IRAC), activated at highly negative hyperpolarisation and equally permeable to both NO_3^- and Cl^-
2. A quick activating anion channel (X-QUAC), possibly involved in anion efflux at hyperpolarised potential in conjunction with outward cation rectification, more selective to NO_3^- than Cl^- in both maize and barley (Köhler and Raschke 2000; Gilliham and Tester 2005)
3. A slow-activating anion channel (X-SLAC) (Teakle and Tyerman 2009)

17.4.3 Genes Involved in Chloride Transport

17.4.3.1 Chloride Channel Gene (CLC)

Unlike plant genes of chloride channels, CLCs have been well characterised in animals, bacteria and yeast (Hechenberger et al. 1996). Genes of CLCs from plants have been cloned from tobacco (Lurin et al. 1996), *Arabidopsis* (Hechenberger et al. 1996), rice (Nakamura et al. 2006) and soybean (Li et al. 2006). All of them belong to the family of voltage-gated chloride channel and are generally expressed in endo-membranes of all tissues. There are seven members of this family in *Arabidopsis* (AtCLCa, b, c, d, e, f and g) and rice (OsCLC, 1–7). In *Arabidopsis* AtCLCa cprotein is expressed in the tonoplast; AtCLCd, trans-Golgi; AtCLCf, *cis*-Golgi; and AtCLCe, thylakoid of chloroplast (Marmagne et al. 2007; Teakle and Tyerman 2009; de Angeli et al. 2009). In *Arabidopsis,* AtCLCa, c, e and f have been genetically linked to NO_3^+ accumulation (Harada et al. 2004). AtCLCa has been found to be primarily a nitrate transporter, a $2NO_3^-/1H^+$ exchanger, which accumulates primarily NO_3^- in the vacuole (de Angeli et al. 2009). Electrophysiological studies on tobacco CLCNt1 expressed in *Xenopus oocytes* indicate that Cl^- transport is 33 % lower than NO_3^- and SO_4^{2-} and also malate and glutamate less than 33 % of NO_3^- (Lurin et al. 1996). OsCLC1 and OsCLC2 in rice are located in tonoplast (Nakamura et al. 2006). In soybean both GmCLC1 and GmNHX1 (a Na^+/H^+ antiporter) are located in the tonoplast. GmCLC1 accumulates more chloride in the vacuole (Li et al. 2006).

It has been reported that ATP inhibits NO_3^+ influx into vacuole by AtCLCa to a maximum of 60 %. This is caused by interaction of ATP with C-terminal domain of AtCLCa, which is composed of two cystathionine β-synthetase motifs, found in all eukaryotic CLC family (de Angeli et al. 2009).

Recent data indicate that Cl^- transport across plasma membrane is important for pollen germination and pollen tube growth (Tavares et al. 2011).

Structural Studies on CLC

Hechenberger et al. (1996) cloned cDNAs of four CLCs of *Arabidopsis,* AtCLCa, AtCLCb, AtCLCc and AtCLCd, from ESTs (expressed sequence tags) and reported that the open reading frames of the four cDNA clones code for proteins consisting of 775 amino acids (AA) for AtCLCa, 780 AA for AtCLCb, 779 AA for AtCLCc and 792 AA for AtCLCd. All of the four have molecular mass of about 85 kDa. Both AtCLCa and AtCLCb have 87 % identical protein sequence, but only 53 % identical with AtCLCc and 48 % with AtCLCd. All the four show about 30 % identities with mammalian proteins CLC6 and CLC7. AtCLCc and CLC-Nt1 from tobacco have 75 % identical AA sequence. CLC proteins have up to 12 transmembrane domains with N-glycosylation sites (4 in AtCLCa and b and one site in AtCLCd and e) present intracellularly.

CLCs are homodimers; each monomer forms an anion transport pathway and has key residues for ion binding and selectivity. Most CLCs have large C-termini cytosolic domain containing two cystathionine β-synthetase motifs that are crucial for their activity in plant cell (Bogusz 2012).

17.4.3.2 Cation Chloride Cotransporters (CCC) (2A.30)

CCC family of proteins occurs in animals, plants, fungi and bacteria. They are KCl/NaCl symporters. The NaCl/KCl symporters are specifically inhibited by bumetanide, and NaCl symporters specifically by thiazide. The generalised transport reaction of CCC symporters is

$$\{Na^{+} \text{or } K^{+} + Cl^{-}\}(\text{out}) \rightleftarrows \{Na^{+} \text{or } K^{+} + Cl^{-}\}(\text{in})$$

Most of the CCC family proteins have been characterised in higher animals. Only a few have been studied in plants. CCC family proteins are usually large with 1,000–1,200 aminoacyl residues and have large hydrophilic C- and N-termini. They possess putative 12 transmembrane domains. The protein forms dimers in solution. The CCCs have a conserved structural scaffold consisting of transmembrane transport domain and a cytoplasmic regulatory domain (transport classification data base).

Arabidopsis thaliana cDNA clone encoding a CCC transporter protein (confirmed by its inhibition of transport activity by bumetanide) has been named as AtCCC. It has been found to be a bona fide Na^+ or K^+, Cl^- cotransporter. AtCCC is preferentially expressed in the root and shoot vasculature at the xylem symplast boundary root tips, trichomes, leaf hydathodes, leaf stipules and anthers. AtCCC is involved in long-distance Cl^- transport and Cl^- homeostasis (Colmenero-Flores et al. 2007).

cDNAs encoding CCC transporters in rice have been cloned (Kong et al. 2011) and named as OsCCC1. It is induced by KCl in the roots and shoots and expressed at a higher level in the leaves and root tips. OsCCC1 is located in the plasma membrane of rice and onion. OsCCC1 plays a significant role in K^+ and Cl^- homeostasis and development of rice plant.

17.4.3.3 Other Transporters Involved in Chloride Channels

A number of anion channels and associated proteins have been identified, which are involved in transport of specific anions but may also transport chloride to a lesser extent.

Al^{3+}-Activated Malate Transporter

Plants respond to Al^{3+} toxicity by exuding di- and tri-carboxylic acids, which form complexes with Al^{3+} at the surface of the roots and prevent it from entering into the root cells. Several genes from ALMT and MATE family encode transporter proteins, which transport organic acids across plasma membrane in response to Al toxicity. A gene isolated from wheat (TaALMT1) encodes a protein with six putative transmembrane domains and is located in the root plasma membrane (Motoda et al. 2007). It is expressed in response to extracellular Al^{3+} (5 μM) but can mediate ion transport in the absence of Al^{3+}. It can also transport Cl^- in a malate/chloride ratio of 30:1. It transports SO_4^{2-} and NO_3^- to a lesser extent. TaALMT can mediate large anion influx as well (Pineros et al. 2008b). Aluminium-activated malate transporters have also been identified in *Arabidopsis,* AtALMT1 (Hoekenga et al. 2006); maize, ZmALMT1 (Pineros et al. 2008a); and rape, BuALMT1 (Ligaba et al. 2006), which mediate Al-activated malate exudation and provide Al tolerance to plants.

NRTs

The nitrate transporters NRTs (Sect. 2.4.1) may be involved in Cl^- transport (about 20 % of NO_3^-) (Pouliquin et al. 2000; Brumos et al. 2009).

17.4.4 Chloride and Salt Tolerance

There are differences between Na^+ and Cl^- ions with respect to salt tolerance of plants. There is no correlation between concentrations of Cl^- ion in shoot or root and salt tolerance in wheat (Kinraide 1999; Husain et al. 2004; Plett and Møller 2010), though there is a negative correlation with concentration of Na^+ ion. Similar correlations are found in rice (Lin and Kao 2001).

However, in soybean, Cl^- concentration in leaf is negatively correlated with salt tolerance and no correlation with Na^+ concentration (Luo et al. 2005). According to Teakle and Tyerman (2009), salt tolerance of a plant depends on its capacity to minimise concentrations of both Na^+ and Cl^- in the cytoplasm, since either of them is toxic at higher concentration. (Please also read the paragraph on chloride toxicity.)

References

Bogusz A (2012) The ClC chloride channels and their role in plant cell. Postepy Biochem 58(1):54–60 [Article in Polish]

Britto DT, Ruth TJ, Lapi S, Kronzucker HJ (2004) Cellular and whole-plant chloride dynamics in barley: insights into chloride nitrogen interactions and salinity responses. Planta 218:615–622

Broyer TC, Carlton AB, Johnson CM, Stout PR (1954) Chlorine-a micronutrient element for higher plants. Plant Physiol 29(6):526–532

Brumos J, Colmenero-Flores JM, Conesa A, Izquierdo P, Sanchez G, Iglesias DJ, Lopez-Climent MF, Gomez-Cadenas A, Talon M (2009) Membrane transporters and carbon metabolism implicated in chloride homeostasis differentiate salt stress responses in tolerant and sensitive Citrus rootstocks. Funct Integr Genomics 9:293–309

Colmenero-Flores JM, Martinez G, Gamba G, Vazquez N, Iglesias DJ, Brumos J, Talon M (2007) Identification and functional characterization of cation-chloride co-transporters in plants. Plant J 50:278–292

de Angeli A, Monachello D, Ephritikhine G, Franchisse J-M, Thomine S, Gambale F, Barbier-Brygoo G (2009) CLC mediated anion transport in plant cells. Philos Trans R Soc B 364(1514):195–201

Diatloff E, Roberts M, Sanders D, Roberts SK (2004) Characterization of anion channels in the plasma membrane of *Arabidopsis* epidermal root cells and the identification of a citrate-permeable channel induced by phosphate starvation. Plant Physiol 136:4136–4149

Engel RE, Bruchner DE, Mathre DE, Brumfield SKZ (1997) A chloride deficient leaf spot syndrome of wheat. Soil Sci Soc Am J 61:176–184

Fixen PE (1993) Crop response to chloride. Adv Agron 50:107–150

Gilliham M, Tester M (2005) The regulation of anion loading to the maize root xylem. Plant Physiol 137:819–828

Harada H, Kuromori T, Hirayama T, Shinozaki K, Leigh RA (2004) Quantitative trait loci analysis of nitrate storage in *Arabidopsis* leading to an investigation of the contribution of the anion channel gene, AtCLC-c, to variation in nitrate levels. J Exp Bot 55:2005–2014

Hechenberger M, Schwappach B, Fischer WN, Frommer WB, Jentsch TJ, Steinmeyer K (1996) A family of putative chloride channels from *Arabidopsis* and functional complementation of a yeast strain with a *CLC* gene disruption. J Biol Chem 271:33632–33638

Hoekenga OA, Maron LG, Pineros MA et al (2006) AtALMT1 which encodes a malate transporter, is identified as one of several genes critical for aluminium tolerance in *Arabidopsis*. Proc Natl Acad Sci U S A 103:9738–9743

Husain S, von Caemmerer S, Munns R (2004) Control of salt transport from roots to shoots of wheat in saline soil. Funct Plant Biol 31:1115–1126

Kinraide TB (1999) Interactions among Ca2+, Na+ and K+ in salinity toxicity: quantitative resolution of multiple toxic and ameliorative effects. J Exp Bot 50:1495–1505

Köhler B, Raschke K (2000) The delivery of salts to the xylem, three types of anion conductance in the plasmalemma of the xylem parenchyma of roots of barley. Plant Physiol 122:243–254

Kong XQ, Gao XH, Sun W, An J, Zhao YX, Zhang H (2011) Cloning and functional characterization of a cation-chloride cotransporter gene, OsCCC1. Plant Mol Biol 75(6):567–578

Li WYF, Wong FL, Tsai SN, Phang TH, Shao G, Lam HM (2006) Tonoplast-located *GmCLC1* and *GmNHX1* from soybean enhance NaCl tolerance in transgenic bright yellow (BY)-2 cells. Plant Cell Environ 29:1122–1137

Ligaba A, Katsuhara M, Ryan PR, Shibasaka M, Matsumoto H (2006) The *BnALMT1* and *BnALMT2* genes from rape encode aluminium-activated malate transporters that enhance the aluminium resistance of plant cells. Plant Physiol 142:1294–1303

Lin CC, Kao CH (2001) Relative importance of Na+, Cl–, and abscisic acid in NaCl induced inhibition of root growth of rice seedlings. Plant Soil 237:165–171

Luo Q, Bingjun Y, Liu Y (2005) Differential selectivity to chloride and sodium ions in seedlings of *Glycine max* and *G. Soja* under NaCl stress. J Plant Physiol 162:1003–1012

Lurin C, Geelen D, BarbierBrygoo H, Guern J, Maurel C (1996) Cloning and functional expression of a plant voltage dependent chloride channel. Plant Cell 8:701–711

Marmagne A, Vinauger-Douard M, Monachello D, de Longevialle A-F, Charo C, Allot M, Rappaport F, Wollman F-A, Brugoo HB, Ephritikhine G (2007) Two members of the *Arabidopsis* CLC (chloride channel) family, AtCLCe and AtCLCf, are associated with thylakoid and Golgi membranes, respectively. J Exp Bot 58(12):3385–3393. doi:10.1093/jxb/erm187

Motoda H, Sasaki T, Kano Y, Ryan P, Delhaize E, Matsumoto H, Yamamoto Y (2007) The membrane topology of ALMT1, an aluminium-activated malate transport protein in wheat (*Triticum aestivum*). Plant Signal Behav 2:467–472

Munns R (2002) Comparative physiology of salt and water stress. Plant Cell Environ 25:239–250

Munns R, Tester M (2008) Mechanisms of salinity tolerance. Annu Rev Plant Biol 59:651–681

Nakamura A, Fukuda A, Sakai S, Tanaka Y (2006) Molecular cloning, functional expression and subcellular localization of two putative vacuolar voltage-gated chloride channels in rice (*Oryza sativa* L.). Plant Cell Physiol 47:32–42

Pineros MA, Kochian LV (2001) A patch-clamp study on the physiology of aluminium toxicity and aluminum tolerance in maize, Identification and characterization of Al^{3+}-induced anion channels. Plant Physiol 125:292–305

Pineros MA, Cancado GMA, Kochian LV (2008a) Novel properties of the wheat aluminium tolerance organic acid transporter (TaALMT1) revealed by electrophysiological characterization in *Xenopus oocytes*: functional and structural implications. Plant Physiol 147:2131–2146

Pineros MA, Cancado GMA, Maron LG, Lyi SM, Menossi M, Kochian LV (2008b) Not all ALMT1-type transporters mediate aluminium-activated organic acid responses: the case of ZmALMT1 – an anion-selective transporter. Plant J 53:352–367

Plett DC, Møller IS (2010) Na^+ transport in plants: what we know and would like to know. Plant Cell Environ 33(4):612–626

Pokhrel R, McConnell IL, Brudvig GW (2011) Chloride regulation of enzyme turnover: application to the role of chloride in photosystem II. Biochemistry 50(14):2725–2734

Pouliquin P, Boyer JC, Grouzis JP, Gibrat R (2000) Passive nitrate transport by root plasma membrane vesicles exhibits an acidic optimal pH like the H^+-ATPase. Plant Physiol 122:265–273

Rivalta I, Amin M, Luber S, Vassiliev S, Pokhrel R, Umena Y, Kawakami K, Shen JR, Kamiya N, Bruce D, Brudvig GW, Gunner MR, Batista VS (2011) Structural-functional role of chloride in photosystem II. Biochemistry 50(29):6312–6315

Skerret M, Tyerman SD (1994) A channel that allows inwardly directed fluxes of anions in protoplasts derived from wheat roots. Planta 192:295–305

Tavakkoli E, Fatehi F, Coventry S, Rengasamy P, McDonald GK (2011) Additive effects of Na+ and Cl– ions on barley growth under salinity stress. J Exp Bot 62(6):2189–2203

Tavares B, Domingos P, Dias PN, Feijo JA, Bicho A (2011) The essential role of anionic transport in plant cells: the pollen tube as a case study. J Exp Bot 62(7):2273–2298

Teakle NL, Tyerman SD (2009) Mechanisms of Cl(–) transport contributing to salt tolerance. Plant Cell Environ 33(4):566–589

Teakle N, Flowers T, Real D, Colmer T (2007) *Lotus tenuis* tolerates the interactive effects of salinity and water logging by 'excluding' Na^+ and Cl^- from the xylem. J Exp Bot 58:2169–2180

USNADP (2002) US National Atmospheric Deposition Program, Washington, DC

Véry A-A, Sentenac H (2003) Molecular mechanisms and regulation of K^+ transport in higher plants. Annu Rev Plant Biol 54:575–603

Wedding RT (1989) Malic enzyme of higher plants. Plant Physiol 90(2):362–371

White PJ, Broadly MR (2001) Chloride in soils and its uptake and movement within the plant: a review. Ann Bot 88(6):967–988

Xu G, Magen H, Tarchitzky J, Kafkafi U (2000) Advances in chloride nutrition. Adv Agron 68:96–150

Zhang WH, Ryan PR, Tyerman SD (2004) Citrate-permeable channels in the plasma membrane of cluster roots from white lupin. Plant Physiol 136:3771–3783

Sodium (Na) Uptake

18

Abstract

While Na^+ is not an essential nutrient for all plants, it is essential for halophytes, which accumulate salt in vacuoles to maintain turgor and growth. A few of the C_4 plants (except corn and sorghum) need Na^+ essentially for specific functions, such as in the concentration of CO_2. High-affinity Na^+ uptake probably occurs in most of the land plants. Na^+ can be beneficial to plants under conditions of K^+ deficiency. Na^+ can undertake osmotic functions, reduce the total K^+ requirements and improve growth when the lack of K^+ is a limiting factor.

18.1 Occurrence of Na and Soil Reactions

Sodium is the sixth most abundant element and is about 2.8 % of the earth's crust. Rain water contains 86 μM L^{-1} of Na^+, which increases with decreasing distance from the sea and is affected by wind direction. Sea water contains 470 mM L^{-1} of Na^+ (Munns 2002). The concentration of NaCl in sea water is about 3.5 % of NaCl (Na^+, 1.08 %; Cl^-, 1.94 %; SO_4^{2-}, 0.27 %; Mg^{2+}, 0.13 %; Ca^{2+}, 0.04 % and K^+, 0.04 %).

Soils are classified (USDA 1954) according to their electrical conductivity of the saturated extract of soil (EC_{se}), pH and exchangeable sodium percentage (ESP) into (1) saline soils, (2) sodic soils and (3) saline sodic soils as given in Table 18.1.

Nonsaline normal soils contain 0.1–1 % of Na with very little exchangeable Na^+. Most of the soils from arid or semiarid regions are rich in Na, and it exists as NaCl, Na_2SO_4 and Na_2CO_3. Coastal saline soils are rich in NaCl. Sea water contains 3.5 % NaCl.

Saline soils can be reclaimed by leaching out salts with fresh water with no significant increase in soil pH. There are also salt-tolerant crops and salt-tolerant varieties within a crop, which can be grown on saline soils. With adequate crop management practices, getting a reasonable yield is possible.

Sodic soils are rich in Na, which disperses soil colloids and are toxic to most of the plants. Saline sodic soils with pH less than 8.5 and ESP >15 become sodic soil after the neutral salts are leached out. The pH increases and they become unsuitable for crop growth.

Soil solution contains about 0.5–5 ppm of Na^+ in temperate regions. Exchangeable Na^+, which is utilised by plants, is far less as compared to other cations. Sugar beet responds to Na^+ application, when exchangeable Na^+ is less than 0.05 meq/100 g of soil (Havlin et al. 2007).

G.N. Mitra, *Regulation of Nutrient Uptake by Plants: A Biochemical and Molecular Approach*, DOI 10.1007/978-81-322-2334-4_18, © Springer India 2015

Table 18.1 Classification of saline soils into saline, sodic and saline sodic soils

Type of soils	Salinity parameter		
	EC_{SE} (dS/m)	Soil pH	ESP (%)
(i) Saline soils	>4.0	<8.5	<15
(ii) Sodic soils	<4.0	>8.5	>15
(iii) Saline sodic	>4.0	<8.5	>15

Table 18.2 Sodium uptake capacity of some of the crops

Na^+ uptake capacity	Names of the crops
High	Spinach, sugar beet
Medium	Coconut, cotton, cabbage, oats
Low	Barley, rape, wheat, millet
Very low	Maize, soybean

18.2 Na^+ Content of Plants

Leaves of different plants may contain from 0.01 to 10 % of Na^+. Many C_4 plants have a higher requirement of Na^+ than C_3 plants. In sugar beet, a C_4 plant, Na^+ concentration in leaf tip may increase up to 10 %. Table 18.2 gives the Na uptake capacity of some of the crops.

18.3 Functions of Na in Plants

18.3.1 Beneficial Effects of Na Application

While Na^+ is not an essential nutrient for all plants, it is essential for *halophytes*, which accumulate salt in vacuoles to maintain turgor and growth. A few of the C_4 plants (except corn and sorghum) need Na^+ essentially for specific functions, such as in the concentration of CO_2 (Subbarao et al. 2003). Na^+ can be beneficial to plants under conditions of K^+ deficiency (Maathuis 2014). Na^+ can undertake osmotic functions, reduce the total K^+ requirements and improve growth when the lack of K^+ is a limiting factor. A near-complete replacement of K^+ by Na^+ in its osmotic function is possible (Shabala and Mackay 2011; Gattward et al. 2012). High-affinity Na^+ uptake probably occurs in most of the land plants (Haro et al. 2010). Beneficial effects of sodium especially at a low concentration (<1 mM) in the growth medium on a number of crop plants such as spinach, beet and sugar beet have been reported. Application of Na^+ in mM range as listed by Kronzucker et al. (2013) has been reported to increase in fresh and dry weight of beet (*Beta vulgaris* L.) (El-Seikh et al. 1967), increase in leaf area, water content and dry weight of barley (*Hordeum vulgare* l.) (Subbarao et al. 1999) and improve photosynthesis, water use efficiency and mineral nutrition in cacao tree (*Theobroma cacao*) (Gattward et al. 2012). In plants belonging to the family of *Chenopodiaceae* including crops like spinach, beet and sugar beet, K^+ depletion is not essential to derive benefits from Na^+ application. These benefits can be derived in the presence of significant concentration of K^+ in the growth medium (Subbarao et al. 2003).

18.3.2 Effect of Na Application on Some of the Quality Parameters

Improvement of some of the quality parameters due to addition of Na^+ has been reported, such as greener leaves, glossy leaves caused by increase in cuticular wax formation (Brownell and Crossland 1972) and improvement of taste and texture of crops (Zhang and Blumwald 2001). In sugar beet, Na^+ concentration in leaf tip may increase up to 10 %. Na^+ has effect on water relations and increases drought resistance of sugar beet. In Na-deficient soils beet leaves are dark green, thin and dull in hue (Havlin et al. 2007).

18.3.3 Na^+ Toxicity

Replacement of Na^+ for a significant part of cytoplasmic K^+ causes toxicity. The epidermal and cortical root cells of the plants absorb sufficient K^+ for the whole plant and exclude Na^+ uptake to keep it at the minimum. High Na^+ is toxic to plant roots especially under drought and causes dehydration of roots. Na^+ can replace Ca^{2+} in plasma membrane under sodic conditions and increase membrane permeability and transport of ions (Bresler et al. 1982).

The primary site of Na^+ toxicity is in the shoots, where Na^+ accumulates, disrupts metabolic processes and increases osmotic stress on the cells (Munns 2002). Some plants like grasses accumulate Na^+ in leaves, which results in necrosis of leaf tips and edges. Sudden high increase in concentration of Na^+ has osmotic consequences, disrupting membrane integrity of roots in crops (Britto et al. 2010) and shoots, as observed in rice (Flowers et al. 1991). Na^+ disrupts K^+ homeostasis and disrupts K^+ influx both at high- and low-affinity ranges especially at mM concentrations (Kronzucker et al. 2006).

Vacuole can tolerate a relatively higher concentration of Na^+ (Subbarao et al. 2003). Replacement of K^+ by Na^+ in the vacuole does not produce toxicity. Tissue tolerance results from effective sequestration of Na^+ in vacuole by transporters such as NHX, which does not harm cytosolic functions (Munns and Tester 2008).

(Na^+ toxicity due to saline conditions is not a part of this chapter.)

18.3.4 Biochemical Functions of Na in Plants

Generally Na^+ inhibits most enzymes activated by K^+ (about 60 in number) at concentrations above 100 mM and with high Na^+/K^+ ratio (Munns et al. 2006). A large number of monovalent cation-activated enzymes are found in both plants and animals. The structure of cation-binding sites and mechanism of activation have been elucidated for a number of enzymes such as pyruvate kinase, fructose-1,6-diphosphatase, S-adenosyl methionine synthetase, tryptophan synthetase, methylamine dehydrogenase and a catalytic RNA, through X-ray crystallography (Prasad et al. 2003). Two mechanisms have been suggested for enzyme activation (Suelter 1970). Either the monovalent cation forms a ternary complex with the enzyme and its substrate or activates the enzyme by allosteric modification. Monovalent cation-activated enzymes exhibit optimum activity in the presence of K^+ but reduced activity in the presence of Na^+ and Li^+ (Suelter 1970). In animals the clotting enzyme thrombin is allosterically activated by Na^+, and its activity is reduced by K^+ and Li^+. Unlike plants Na^+ is an essential nutrient for animals.

In C_4 plants Na^+ facilitates conversion of pyruvate to phosphoenolpyruvate (PEP), which occurs in mesophyll cells (Johnston et al. 1988). In the C_4 plant *Amaranthus tricolour*, Na^+ deficiency causes accumulation of the C_3 metabolites, alanine and pyruvate, whereas C_4 metabolites malate, aspartate and PEP decrease. In C_3 tomato none of these metabolites are affected due to Na^+ depletion (Johnston et al. 1988). The C_4 crops, corn and sorghum do not benefit from addition of Na^+ (Subbarao et al. 2003).

18.4 Mechanism of Sodium Uptake by Plants

18.4.1 High-Affinity Na Uptake

Na^+ can partially replace K^+ at low concentrations (<1 mM) and reduce potassium requirement of crops (Rodríguez-Navarro 2000; Subbarao et al. 2003). The decrease in cellular pH caused by reduced K^+ uptake can be ameliorated by Na^+ uptake (Carden et al. 2003). The scarce available Na^+ present in normal nonsaline soils can only be taken up by plants through a high-affinity uptake mechanism. High-affinity Na^+ uptake has been clearly demonstrated in only four species of plants, rice, wheat, barley and sunflower (Garciadeblás et al. 2003). Haro et al. (2010) studied Na^+ uptake by a number of plant species at very low concentrations, at which K^+ is generally taken up. They observed high-affinity Na^+ uptake by seven monocot and nine dicot species including two bryophytes *P. patens* and *R. fluitans* and concluded that high-affinity Na^+ uptake at low cation concentrations was probably a common feature of land plants. The plant species included in their study were K^+-starved seedlings (whole plants and excised roots) of rye (*Secale cereale*), sorghum (*S. bicolor*), maize (*Z. maise*), onion (*Allium cepa*), lentil (*Lens culinaris*), alfalfa (*Medicago sativa*), squash (*Cucurbita moschata*), melon (*Cucumis melo*), lettuce

(*Lactuca sativa*), chard (*Beta vulgaris*), carrot (*Daucus carota*), celery (*Apium graveolens*) and eggplant (*Solanum melongena*). They tested their capacity to deplete Na^+ from the growth medium at concentrations of 30 to ≤5 μM.

18.4.2 HKT Transporters and High-Affinity Na Uptake

HKT (high-affinity K^+ transporters) proteins of plants are part of the *Trk* superfamily of cation transporters and are topologically related to K^+ channels. All-known plant *HKT* genes contain two introns near the 3′ end. Plant HKT amino acid sequences group into two subfamilies, and the genes of subfamily one have longer introns than those of subfamily two (Platten et al.2006). Subfamily two exclusively contains monocot genes; subfamily one includes monocot genes and all-known *HKTs* from dicots. All of them contain a characteristic glycine or serine residue in the first pore-forming P loop (PA) that may be related to their functions.

According to Haro et al. (2010), high-affinity Na^+ uptake is probably mediated by HKT transporters in rice, species of *Triticeae* and *Aveneae* tribes of *Poaceae* family. Wheat HKT high-affinity K^+ transporter, which mediates Na^+-driven high-affinity K^+ uptake in yeast cells and low-affinity Na^+ uptake in *Xenopus* oocytes, was first suggested to be involved in high-affinity Na^+ uptake (Rubio et al. 1995). Subsequent studies in rice have shown that in K^+-starved plants, HKT transporters mediate high-affinity Na^+ transport without cotransport of K^+. Barium (Ba^{2+}) inhibits Na^+ uptake, but it has no effect on K^+ uptake by rice roots, which indicates specificity of HKT for Na^+ transport (Garciadeblás et al. 2003). There are nine HKT genes in rice in contrast to single gene in *Arabidopsis* (Platten et al. 2006) and probably more in other cereals (Huang et al. 2008). Rice HKT genes renamed as OsHKT2;1 and OsHKT2;2 are expressed in roots among other tissues (Kader et al. 2006). OsHKT2;1 is involved in high-affinity Na^+ transport under K^+-starved conditions and can partially replace K^+ (Horie et al. 2007). OsHKT2;2 catalyses Na^+-dependent K^+ uptake (Horie et al. 2001). In K^+-starved barley roots, HvHKT2;1, when heterologously expressed in yeast, mediates K^+ or Na^+ uniport, K^+/Na^+ symport or a combination of both, depending upon the construct from which the transporter is expressed (Haro et al. 2005; Craig Plett and Moller 2010). For example, HvHKT1 mediates K^+ or Na^+ uniport in yeast cells if the expression promoter is joined directly to the *HvHKT1* cDNA, but if a 59-nucleotide polylinker is inserted, it becomes a Na^+–K^+ symporter. Insertion of the 59-nt polylinker (uATGs) in the rice *HKT1* cDNA changes the mode of its Na+ uptake from K^+-insensitive to K^+-inhibitive. Both the modes of Na^+ uptake occur in rice plants (Banuelos et al. 2008).

It has been confirmed in rice (Garciadeblás et al. 2003), barley and wheat (Wang et al. 1998; Haro et al. 2005) that under K^+-starved conditions, expression of transcripts that encode HKT transporters significantly increases consistent with the hypothesis that high-affinity Na^+ uptake supplements K^+ uptake. Further in K^+-starved plants of rice, wheat and barley, Na^+ uptake is inhibited by addition of K^+ (Garciadeblás et al. 2003). In sunflower plants, Na^+ uptake is not K^+ sensitive (Garciadeblás et al. 2003).

All HKT transporters of subfamily 1 characterised so far have Ser-residue in their PA-loop, whereas all of subfamily 2 have Gly in this loop, with the exception of OsHKT2;1. While TaHKT2;1 of wheat and HvHKT2;1 of barley, which belong to cluster 1 of subfamily 2, have Gly-residues, the rice variety Pokkali, which has two HKT transporters, has Ser-residue in OsHKT2;1 and Gly-residue in OsHKT2;2 (Horie et al. 2001). High-affinity Na^+ uptake in sorghum is probably not mediated by an HKT transporter since K^+ deprivation does not cause transcript enhancement of any of its HKT genes (Haro et al. 2010).

In *Arabidopsis* AtHKT1;1 is a low-affinity Na^+ transporter, and a high-affinity Na^+ transporter appears to be absent in *Arabidopsis* genome (Mäser et al. 2002). AtHKT1;1 is localised in the stele (Sunarpi et al. 2005) and is involved in root accumulation of Na^+ and retrieval of Na^+ from the xylem, but it is not involved in root influx or recirculation in the phloem (Davenport et al. 2007; Craig Plett and Moller 2010).

No other transporter has been conclusively proved to be involved in high-affinity Na^+ uptake by plants.

References

Banuelos MA, Haro R, Fraile-Escanciano A, Rodriguez-Navarro A (2008) Effects of poly linker, uATGs on the function of grass HKT1 transporters expressed in yeast cells. Plant Cell Physiol 49:1128–1132

Bresler E, McNeal BL, Carter DL (1982) Saline and sodic soils- principles – dynamics – modelling, vol. 10, Advanced series in agricultural sciences. Springer, Berlin

Britto DT, Ebrahim-Abdebili S, Hamam AM, Coskun D, Kronzucker HJ (2010) ^{42}K analysis of sodium induced potassium efflux in barley: mechanism and relevance to salt tolerance. New Phytol 186:373–384

Brownell PF, Crossland CJ (1972) The requirement of sodium as a micronutrient by species having C_4 dicarboxylic photosynthetic pathway. Plant Physiol 49:794–797

Carden DE, Walker DJ, Flowers TJ, Miller AJ (2003) Single cell measurements of the contributions of cytosolic Na+ and K+ to salt tolerance. Plant Physiol 131:676–683

Craig Plett D, Moller IS (2010) Na^+ transport in glycophytic plants: what we know and would like to know. Plant Cell Environ 33:612–626

Davenport RJ, Muñoz-Mayor A, Jha D, Essah PA, Rus A, Tester M (2007) The Na+ transporter AtHKT1;1 controls retrieval of Na+ from the xylem in *Arabidopsis*. Plant Cell Environ 30:497–507

El-Seikh AM, Ulich A, Broyer TC (1967) Sodium and rubidium as possible nutrients for sugar beet plant. Plant Physiol 42:1202–1208

Flowers TJ, Hajibagheri MA, Yeo AR (1991) Ion accumulation in the cells walls of rice plants growing under saline conditions: evidence for the Oertli hypothesis. Plant Cell Environ 14:319–325

Garciadeblás B, Senn ME, Bañuelos MA, Rodríguez-Navarro A (2003) Sodium transport and HKT transporters: the rice model. Plant J 34:788–801

Gattward JN, Almeide AAF, Souza JO, Gomes FP, Kronzucker HJ (2012) Sodium – potassium synergism in *Theobroma cacao:* stimulation of photosynthesis, water use efficiency and mineral nutrition. Physiol Plant 146(3):350–362

Haro R, Bañuelos MA, Senn ME, Barrero-Gil J, Rodríguez-Navarro A (2005) HKT1 mediates sodium uniport in roots, pitfalls in the expression of HKT1 in yeast. Plant Physiol 139:1495–1506

Haro R, Bañuelos MA, Rodriguez-Navarro A (2010) High-affinity sodium uptake in land plants. Plant Cell Physiol 51(1):68–79

Havlin JL, Tisdale SL, Beaton JD, Nelson WL (2007) Soil fertility and fertilisers. Prentice Hall (India), New Delhi

Horie T, Yoshida K, Nakayama H, Yamada K, Oiki S, Shinmyo A (2001) Two types of HKT transporters with different properties of Na^+ and K^+ transport in *Oryza sativa*. Plant J 27:115–128

Horie T, Costa A, Kim TH, Han MJ, Horie R, Leung HY et al (2007) Rice OsHKT2;1 transporter mediates large Na^+ influx component into K+–starved roots for growth. EMBO J 26:3003–3014

Huang S, Spielmeyer W, Lagudah ES, Munns R (2008) Comparative mapping of HKT genes in wheat, barley, and rice, key determinants of Na^+ transport, and salt tolerance. J Exp Bot 59:927–937

Johnston M, Grof CPL, Brownell PF (1988) The effect of sodium nutrition on the sizes of intermediates of the C_4 photosynthetic pathway. Aus J Plant Physiol 15:749–760

Kader MA, Seidel T, Golldack D, Lindberg S (2006) Expressions of OsHKT1, OsHKT2, and OsVHA are differentially regulated under NaCl stress in salt-sensitive and salt-tolerant rice (*Oryza sativa* L.) cultivars. J Exp Bot 57:4257–4268

Kronzucker HJ, Szczerba MW, Moazami-Goudarzi M, Britto DT (2006) The cytosolic Na^+/K^+ ratio does not explain salinity induced growth impairment in barley-a dual tracer study using ^{42}K and ^{24}Na. Plant Cell Environ 29:2228–2237

Kronzucker HJ, Coskun D, Schulze LM, Wong LR, Britto DT (2013) Sodium as a nutrient and toxicant. Plant Soil 369:1–23

Maathuis FJM (2014) Sodium in plants: perception, signalling, and regulation of sodium fluxes. J Exp Bot 65(3):849. doi:10.1093/jxb/ert326

Mäser P, Hosoo Y, Goshima S, Horie T, Eckelman B, Yamada K, Bakker EP, Shinmyo A, Oiki S, Schroeder JL, Uozumi N (2002) Glycine residues in potassium channel-like selectivity filters determine potassium selectivity in four-loop-per-subunit HKT transporters from plants. Proc Natl Acad Sci 99(9):6428–6433

Munns R (2002) Comparative physiology of salt and water stress. Plant Cell Environ 25:239–250

Munns R, Tester M (2008) Mechanisms of salinity tolerance. Annu Rev Plant Biol 59:651–681

Munns R, James RA, Läuchli A (2006) Approaches to increasing the salt tolerance of wheat and other cereals. J Exp Bot 57(5):1025–1043

Platten JD, Cotsaftis O, Berthomieu P, Bohnert H, Davenport RJ, Fairbairn DJ, Horie T, Leigh RA, Lin HX, Luan S, Maser P, Pantoja O, Rodriguez-Navarro A, Schachtman DP, Schroeder JI et al (2006) Nomenclature for HKT transporters, key determinants of plant salinity tolerance. Trends Plant Sci 11:372–374

Prasad S, Wright KJ, Banerjee Roy D, Bush LA, Cantwell AM, Cera ED (2003) Redesigning the monovalent cation specificity of an enzyme. Proc Natl Acad Sci 100(24):13785–13790

Rodríguez-Navarro A (2000) Potassium transport in fungi and plants. Biochim Biophys Acta 1469:1–30

Rubio F, Gassmann W, Schroeder JI (1995) Sodium-driven potassium uptake by the plant potassium transporter HKT1 and mutations conferring salt tolerance. Science 270:1660–1663

Shabala S, Mackay AS (2011) Ion transport in halophytes. Adv Bot Res 57:151–159

Subbarao GV, Wheeler RM, Stutte GW, Levine LH (1999) How far can sodium substitute potassium in red beet? J Plant Nutr 22:1745–1761

Subbarao GV, Ito O, Berry WL, Wheeler RM (2003) Sodium – a functional plant nutrient. Crit Rev Plant Sci 22:391–416

Suelter CH (1970) Enzymes activated by monovalent cations. Science 168(3933):789–795

Sunarpi HT, Motoda J, Kubo M, Yang H, Yoda K, Horie R, Chan WY, Leung HY, Hattori K, Konomi M, Osumi M, Yamagami M, Schroeder JI, Uozumi N (2005) Enhanced salt tolerance mediated by AtHKT1 transporter-induced Na unloading from xylem vessels to xylem parenchyma cells. Plant J 44(6):928–938

USDA (1954) Diagnosis and improvement of saline and alkali soils, agriculture handbook no.60, US salinity laboratory, US Department of Agriculture, Washington, DC

Wang T-B, Gassmann W, Rubio F, Schroeder JI, Glass ADM (1998) Rapid up-regulation of HKT1, a high-affinity potassium transporter gene, in roots of barley and wheat following withdrawal of potassium. Plant Physiol 118:651–659

Zhang H-X, Blumwald E (2001) Transgenic salt-tolerant tomato plants accumulate salt in foliage but not in fruit. Nat Biotechnol 19:765–768

19 Silicon (Si) Uptake

Abstract

Silicon (Si) is the second most abundant element after oxygen and constitutes 27.7 % of the earth's crust. While the occurrence of pure silicon is rare, silicate minerals account for 90 % of the mass of the earth's crust. Silicon is a constituent of all plants, and its concentration in shoot may vary from 0.1 to 10 % of dry weight. Plants take up Si primarily as orthosilicic acid. LSi1 (low silicon 1) is a member of NIP2 subgroup of NIP subfamily of aquaporin-like proteins and functions as a Si-permeable channel. LSi6 is a homologue of LSi1 and is involved in xylem unloading of Si in rice.

19.1 Occurrence of Silicon (Si) and Soil Reactions

Silicon (Si) is the second most abundant element after oxygen and constitutes 27.7 % of the earth's crust. While the occurrence of pure silicon is rare, silicate minerals account for 90 % of the mass of the earth's crust. Si in soils ranges from 23 to 35 %. Major source of Si in soils is primary and secondary Si minerals. Sand and silt fraction of soil contain 90–95 % of quartz (SiO_2). Silicon occurs in soil solution primarily as H_4SiO_4 (orthosilicic acid) at a concentration of 0.1–0.6 mM and is taken up by plants in this form (Epstein 1994). Intensively weathered, high rain fall regions have low Si in soils, which are high in Al, low pH and low % base saturation. They have high P-fixing capacity due to high anion exchange capacity and high Fe/Al oxide content.

19.2 Silicon Content of Plants

Silicon is a constituent of all plants, and its concentration in shoot may vary from 0.1 to 10 % of dry weight (Epstein 1994; Ma and Takahashi 2002; Ma et al. 2007a). Takahashi et al. (1990) classified plants into three groups according to their silicon contents in the shoot as (1) silicon accumulators, containing more than 1 % of silicon, (2) silicon excluders with less than 0.5 % and (3) an intermediate type with 0.5–1 % in their shoot. Plants of the families *Poaceae*, *Equisetaceae* and *Cyperaceae* generally accumulate high concentrations of silicon (Si>4 %); plants of the families of *Cucurbitales*, *Urticales* and *Commelinaceae* accumulate intermediate concentrations of Si (2–4 %); and the rest of the species accumulate lower concentrations of Si in the order liverworts>horsetails>club

G.N. Mitra, *Regulation of Nutrient Uptake by Plants: A Biochemical and Molecular Approach*, DOI 10.1007/978-81-322-2334-4_19, mosses > mosses > *angiosperms* > *gymnosperms*>ferns (Hodson et al. 2005). Generally, dicots do not benefit from Si, since they cannot accumulate sufficient Si in their shoots (Ma 2004).

19.2.1 Rice

Rice is the greatest accumulator of Si and may contain 10–15 % of its dry weight when grown in the presence of sufficient Si fertilisers. *Japonica* rice varieties generally have higher concentrations of Si in their shoots than *indica* rice. Concentrations of Si in shoots of a number of *japonica* and *indica* rice varieties grown in nutrient solution containing 100 mg SiO_2 L^{-1} have been reported to be 117–171 mg g^{-1} (Ma et al. 2007a). There is variation in Si content of genotypes within the same species. Field experiments with 18 genotypes of rice showed Si concentration in a range of 18–41 mg g^{-1} (Deren et al. 2001). In a comparative study between a *japonica* (cv: Nipponbare) and *indica* (cv: Kasalath), it was observed that there was no significant difference in Si concentration in their roots. More than 99 % of Si was localised in their shoots (Ma et al. 2007a). However, *japonica* rice variety had a higher rate of silicon uptake as compared to *indica* rice variety since it had a higher root biomass (Ma et al. 2007a).

Different parts of the same rice plant may show different Si content. For example, Si content is 0.5 g kg^{-1} in polished rice, 50 g kg^{-1} in rice bran, 130 g kg^{-1} in rice straw, 230 g kg^{-1} in rice hulls and 350 g kg^{-1} in rice joints (found at the base of the grains) (Van Hoest 2006).

19.2.2 Other Plants

In 400 cultivars of barley (*Hordeum vulgare*), Si content in covered grains was in the range of 1.24–3.80 mg $^{-1}$ (Ma et al. 2003). Si content of sugarcane (*Saccharum officinarum*) grown in field has been reported to be in the range of 6.4–10.2 mg g^{-1} in the shoot (Deren et al. 2001). Si contents of wheat, oat, rye, barley, sorghum and corn are about10g kg^{-1} (Datnoff and Rodrigues 2005).

19.3 Functions of Silicon in Plants

Deficiency of Si in plants makes them structurally weak and susceptible to abnormalities in growth, development and reproduction. It is the only element which does not have any adverse effect when it accumulates in excess (Epstein 1999). Si deficiency symptom in rice includes soft droopy leaves, reduced photosynthetic efficiency due to mutual shading of leaves and reduced starch formation leading to incomplete grain filling.

19.3.1 The Beneficial Effects of Si

The beneficial effects of Si are expressed through its deposition in various tissues such as leaves, stems and hulls apart from the presence of soluble silica. The highest Si is deposited in the husk of rice and barley grains (Sangster et al. 2001).

Beneficial effects of Si application to rice plants include:

1. Increasing canopy photosynthetic efficiency by keeping leaves erect and compact.
2. Increasing resistance to fungi, bacteria and insects.
3. Reducing toxicity to heavy metals.
4. Improving water use efficiency by reducing cuticular transpiration.
5. Increasing resistance to lodging. Silicon strengthens the stem that prevents lodging in rice by increasing the thickness of culm wall and the size of the vascular bundle (Shimoyama 1958; Mitani and Ma 2005; Ma et al. 2007a).

19.3.2 Protection from Abiotic and Biotic Stress

Presence of Si in plants protects them from a number of biotic and abiotic stresses.

19.3.2.1 Si and Abiotic Stress

The abiotic stresses include salt stress, metal toxicity, drought stress, radiation damage, nutrient imbalance, high temperature and cold stress (Epstein 1999; Ma and Takahashi 2002; Ma 2004). Mineral stresses such as Mn and Al toxicity and phosphate deficiency are alleviated by Si.

Mn Toxicity

Leaf apoplastic Si has been reported to enhance Mn tolerance of cowpea (*Vigna unguiculata*). It is reported that Si-enhanced Mn^{2+} leaf tolerance in cowpea (Iwasaki et al. 2002a, b) and cucumber (Rogalla and Römheld 2002) is due to reduction in concentration of Mn^{2+} in leaf apoplastic washing fluid.

High concentrations of Mn in plants increase superoxide dismutase, catalase and ascorbate peroxidase activities but decrease concentrations of nonprotein thiols and glutathione, which results in accumulation of OH• (hydroxyl radical) and malondialdehyde. Addition of Si has been observed to significantly neutralise Mn-induced increase in OH• and malondialdehyde and enhance plant growth in rice (Li et al. 2012). It has also been reported that Si ameliorates Mn toxicity in cucumber by reducing hydroxyl radical accumulation in the leaf apoplast as well as free apoplastic Mn^{2+} (Maksimović et al. 2012).

Al Toxicity

Silicon has been reported to alleviate Al toxicity in conifers (Ryder et al. 2003), barley (Hammond et al. 1995), soybean (Baylis et al. 1994), maize (Barceló et al. 1993) and sorghum (Galvez et al. 1987; Wang et al. 2004). Various mechanisms have been suggested for alleviation of Al toxicity by Si.

1. Apoplast of root tips are the primary site of Al phytotoxicity. Al binds rapidly to the negatively charged binding sites of the cell wall and alters its properties. This affects root growth (Horst 1995). Treatment with Si results in esterification of cell wall component by Si and reduces binding sites on cell wall for Al (Corrales et al. 1997)
2. It has been suggested that Si application increases exudation of phenolic compounds, which form non-toxic complexes with Al and provides protection against Al toxicity in maize (Kidd et al. 2001).
3. According to Ma et al. (1997), Si forms biologically inactive and non-phytotoxic hydroxy aluminosilicates (HAS) with Al in the apoplast of root apex. This mechanism is supported by X-ray microanalysis by Hodson and Sangster (1993), which shows copresence of Al and Si in the epidermal cells of sorghum roots treated with Al and Si. Similar observations have been reported in wheat (Cocker et al. 1997). Wang et al. (2004) working with an aluminium-sensitive maize cultivar (Lixis) concluded that the formation of HAS due to Si treatment in the apoplast of root apex detoxified Al. They did not observe any Al-induced exudation of phenolic compounds or organic acids from root apices.

Radiation Damage

Silicon has been reported to protect rice from radiation damage. Plants treated with Si after exposure to γ-radiation recover faster than those without application of Si (Takahashi 1966).

Silicon and Water Stress

Under drought conditions, there is closure of stomata and decrease in rate of photosynthesis. Si is deposited under the cuticle forming a Si-cuticle double layer. This reduces transpiration from cuticle of rice leaves. Si can reduce transpiration rate by 30 % in rice, which has a thin cuticle (Ma et al. 2001; Ma 2004). Treatment with Si increases percentage of ripened grains in rice (7 % Si) and barley (1.5 % Si) under water-stressed conditions by maintaining favourable moisture content in the hull. In the hull, Si is deposited between the epidermal cell wall and the cuticle forming a Si-cuticle double layer similar to leaf blade. Since hull does not have stoma, water loss occurs through the cuticle. The Si-cuticle double layer prevents water loss to the extent of 20 % as compared to Si-untreated plants (Ma 2004).

19.3.2.2 Silicon and Biotic Stress

Presence of silica increases the resistance of plants to attack by pathogens. Increased concentration of Si in plants reduces epidemics of both leaf and panicle blast of rice (Datnoff et al. 1997; Datnoff and Rodrigues 2005) and decreases incidence of powdery mildew in cucumber, barley and wheat (Fauteux et al. 2005). Attack by insect pests such as stem borer, brown plant hopper and rice green leaf hopper is suppressed by Si (Savant et al. 1997). Higher concentration of Si in wheat leaves reduces damage caused by wild rabbits (Cotterill et al. 2007).

Application of Si effectively induces broad spectrum disease resistance. The prophylactic effect of Si is both passive and active defences. Two hypotheses have been proposed to explain the role of Si to protect from biotic stress (Ma 2004):

1. Si deposited on the surface of tissues acts as a physical barrier for infection and additionally may make the cells less susceptible to enzymatic degradation by fungal pathogens. The absorbed orthosilicic acid polymerises to form insoluble silica, which reinforces cell wall and prevents attempted penetration by fungi into the epidermal cells. Kauss et al. (2003) have reported that during induction of SAR in cucumber (systemic all acquired resistance, infection of one leaf of cucumber makes other leaves resistant to pathogens), expression of a gene encoding a proline-rich protein (PRP1) with a C-terminal repetitive sequence of unusually high lysine and arginine is enhanced. A synthetic peptide derived from the repetitive sequence has been found to polymerise orthosilicic acid into insoluble silica. This probably is the molecular mechanism, which reinforces cell walls of epidermal cells with silica to prevent fungal penetration in other leaves in SAR.
2. Si acts as a signal to induce production of phytoalexins (Cherif et al. 1994). Si application to cucumber after infection with *Pythium* spp. resulted in stimulation of chitinase activity and rapid activation of peroxidase and polyphenol oxidases. Phenolic compounds produced from hydrolysis of phenol glycosides from Si-treated plants were found to have strong fungistatic activity. Induction of similar defence mechanisms has also been observed in powdery mildew (*Blumeria graminis* f. sp. *tritici*) infected wheat (Bélanger et al. 2003) and sheath blight of rice (Rodrigues et al. 2003) in the presence of Si, including increased production of glycosylated phenolic compounds and antimicrobial products such as diterpenoid phytoalexins. There are however reports that phenolic compounds are produced in response to infection by *Blumeria graminis* in oats under Si-deficient conditions (Carver et al. 1998).

19.4 Mechanism of Silicon Uptake by Plants

19.4.1 Si Transporters

Si is present as an uncharged molecule H_4SiO_4 (orthosilicic acid) at a pH below 9.0. Plants take up Si primarily as orthosilicic acid. Transporters responsible for Si uptake have been identified in several plant species such as barley, maize, pumpkin, rice and wheat (Ma et al. 2011). Recently, Si transporter in horsetail (*Equisetum arvense*) EaNIP3s (Nod 26-like major intrinsic protein3) has been reported (Grégoire et al. 2012).

The Si uptake process involves two different types of transport, Si-permeable channel and efflux transporter (Yamaji et al. 2012).

19.4.1.1 LSi1 (Low Silicon 1)

*LS*i1 (low silicon 1) is a member of NIP2 (Nod 26-like major intrinsic protein2) subgroup of NIP subfamily of aquaporin-like proteins and functions as a Si-permeable channel (Yamaji et al. 2012). LSi1 is localised in the distal side of root exodermis and endodermis in rice (Ma et al. 2006) and is responsible for uptake of silicic acid from external solution into the root cortical cells (Ma and Yamaji 2006). LSi1s in barley maize and pumpkin are localised in the epidermis and cortex (Chiba et al. 2009; Mitani et al. 2009, 2011). The LSi1 gene mapped to chromosome 2 contains 5 exons and 4 introns. It encodes a 298 amino acid protein with six transmembrane domains and two Asn-Pro-Ala (NPA) motifs and has a high homology with aquaporins. Ala-132 is critical since any substitution results in altered conformation of the protein.

19.4.1.2 LSi2 (Low Silicon 2)

LSi2 functions as an efflux Si transporter and belongs to the anion transporter family without any similarity with LSi1. Similar to LSi1, LSi2 in rice is localised in the root exodermis and endodermis but has a polar localisation at the proximal side (Ma et al. 2007a; Yamaji et al. 2012). In barley and maize, LSi2 is localised only on the endodermis of roots and does not show polar localisation (Mitani et al. 2009). There is difference between rice and other plant species with respect to Si uptake mediated by LSi1 and Lsi2 (Ma et al. 2011). Knockout of either LSi1 or LSi2 caused significant reduction of Si uptake by rice plants (Ma et al. 2006; Mitani et al. 2009). A coordinated functioning of these two transporters in the exodermis and endodermis ensures effective flow of Si into the vascular bundle overcoming the barrier of casparian strips (Ma et al. 2006, 2007b). Similar to rice LSi2 of maize and barley are anion transporters with 11 transmembrane domains. Maize and barley LSi2 are membrane proteins and contain 477 and 474 amino acids, respectively (Mitani et al. 2009).

19.4.1.3 LSi6

LSi6 is a homologue of LSi1 and is involved in xylem unloading of Si in rice (Yamaji et al. 2008). LSi6 is localised in the adaxial side of the xylem parenchyma cells in the leaf sheath and leaf blade. It is highly expressed in the nodes below the panicles at the reproductive growth stage of rice and is probably involved in transfer of Si from the vascular bundle coming from the roots to the diffuse vascular bundles connected to the panicle (Yamaji and Ma 2009).

HvLSi6 of barley encodes a plasma membrane-localised transporter for silicic acid. The open reading frame (ORF) of barley HvLSi6 is 900 bp long and encodes a protein with 300 amino acids. At amino acid level, HvLSi6 is 88.2 % identical to rice LSi6. HvLSi6, similar to other Si transporters, has two Asn-Pro-Ala motifs and distinct aromatic/Arg selectivity filters Gly, Ser, Gly and Arg (Yamaji et al. 2012).

There appears to be differences in functions of HvLSi1 and HvLSI6, which are independent of Si supply in contrast to downregulation of rice OsLSi1 and OsLSi6 in the presence of Si (Yamaji et al. 2012). HvLSi6 is highly expressed in root tips than in mature root region. The protein has a polar localisation in the epidermis and cortex of root tips. HvLSi1 is primarily expressed in the mature root region. In barley, Si uptake capacity is higher by root tips than the mature root region (Chiba et al. 2009). In rice, OsLSi6 and OsLsi1 are also highly expressed in root tips and mature root region, but Si uptake capacity of root tips is lower than mature root region (Yamaji and Ma 2007). However, knockout of OsLSi1 results in significant reduction in Si uptake, while knockout of OsLSi6 has no such effect (Yamaji et al. 2008). Contribution of OsLSi1 for Si uptake is crucial and higher than LSi6 (Yamaji et al. 2012).

In maize, two genes ZmLSi1 and ZmLSi6, which are homologous to rice Si transporter gene OsLSi1, have been identified (Mitani et al. 2009). ZmLSi1 is primarily expressed in roots. ZmLSi6 is expressed more in leaf sheath and leaf blade. Contrary to OsLSi1, expression of maize ZmLSi1 and ZmLSi6 is independent of Si supply. ZmLSi1 is an influx transporter, which transports Si from external solution to the root cells, and ZmLSi6 unloads Si into the xylem (Mitani et al. 2009).

References

Barceló J, Guevara P, Poschenrieder C (1993) Silicon amelioration of aluminium toxicity in teosinte (*Zea mays* L. *ssp. Mexicana*). Plant Soil 154:249–255

Baylis AD, Gragopoulou C, Davidson KJ, Birchall JD (1994) Effect of silicon on the toxicity of aluminium to soybean. Commun Soil Sci Plant Anal 25:537–546

Bélanger RR, Benhamou N, Menzies JG (2003) Cytological evidence of an active role of silicon in wheat resistance to powdery mildew (*Blumeria graminis* f. sp. *tritici*). Phytopathology 93:402–412

Carver TLW, Robbins MP, Thomas BJ, Troth K, Raistrick N, Zeyen RJ (1998) Silicon deprivation enhances localized autofluorescent responses and phenylalanine ammonia-lyase activity in oat attacked by *Blumeria graminis*. Physiol Mol Plant Pathol 52:245–257

Cherif M, Asselin A, Belanger RR (1994) Defense responses induced by soluble silicon in cucumber roots infected by *Pythium* spp. Phytopathology 84:236–242

Chiba Y, Mitani N, Yamaji N, Ma JF (2009) HvLsi1 is a silicon influx transporter in barley. Plant J 57:810–818

Cocker KM, Hodson MJ, Evans DE, Sangster AG (1997) The interaction between silicon and aluminium in *Triticum aestivum* L. (cv. Celtic). Isr J Plant Sci 45:289–292

Corrales I, Poschenrieder C, Barceló J (1997) Influence of silicon pretreatment on aluminium toxicity in maize roots. Plant Soil 190:203–209

Cotterill JV, Watkins RW, Brennon CB, Cowan DP (2007) Boosting silica levels in wheat leaves reduces grazing by rabbits. Pest Manag Sci 63:247–253

Datnoff LE, Rodrigues FA (2005) The role of silicon in suppressing rice diseases. APSnet Features. http://dx.doi.org/10.1094/APSnetFeature-2005-0205

Datnoff LE, Deren CW, Snyder GH (1997) Silicon fertilisation for disease management of rice in Florida. Crop Prot 16:525–531

Deren CW, Datnoff LE, Snyder GN (2001) Variable silicon content of rice cultivars grown on everglades histosols. J Plant Nutr 15:2363–2368

Epstein E (1994) The anomaly of silicon in plant biology. Proc Natl Acad Sci 91:11–17

Epstein E (1999) Silicon. Annu Rev Plant Physiol Plant Mol Biol 50:641–664

Fauteux F, Rémus-Borel W, Menzies JG, Bélanger RR (2005) Silicon and plant disease resistance against pathogenic fungi. FEMS Microbiol Lett 249:1–6

Galvez L, Clark RB, Gourley LM, Maranville JW (1987) Silicon interactions with manganese and aluminum toxicity in sorghum. J Plant Nutr 10:1139–1147

Grégoire C, Rémus-Borel W, Vivancos J, Labbé C, Belzile F, Bélanger RR (2012) Discovery of a multigene family of aquaporin silicon transporters in the primitive plant *Equisetum arvense*. Plant J 72:320–330

Hammond KE, Evans DE, Hodson MJ (1995) Aluminium/silicon interactions in barley (*Hordeum vulgare* L.) seedlings. Plant Soil 173:89–95

Hodson MJ, Sangster AG (1993) The interaction between silicon and aluminium in *Sorghum bicolor* (L.) Moench: growth analysis and X-ray microanalysis. Ann Bot (Lond) 72:389–400

Hodson MJ, White PJ, Mead A, Broadley MR (2005) Phylogenetic variation in silicon composition of plants. Ann Bot (Lond) 96:1027–1046

Horst WJ (1995) The role of the apoplast in aluminium toxicity and resistance of higher plants: a review. Z Pflanzenernähr Bodenkd 158:419–428

Iwasaki K, Maier P, Fecht M, Horst WJ (2002a) Leaf apoplastic silicon enhances manganese tolerance of cowpea *(Vigna unguiculata)*. Plant Physiol 159:167–173

Iwasaki K, Maier P, Fecht M, Horst WJ (2002b) Effect of silicon supply on apoplastic manganese concentrations in leaves and their relation to manganese tolerance in cowpea *(Vigna unguiculata* (L.) Walp.). Plant Soil 238:281–288

Kauss H, Seehaus K, Franke R, Gilbert S, Dietrich RA, Kroger N (2003) Silica deposition by a strongly cationic proline-rich protein from systemically resistant cucumber plants. Plant J 33:87–95

Kidd PS, Llugany M, Poschenrieder C, Gunsé B, Barceló J (2001) The role of root exudates in aluminium resistance and silicon-induced amelioration of aluminium toxicity in three varieties of maize (*Zea mays* L.). J Exp Bot 52:1339–1352

Li P, Song A, Li Z, Fan F, Liang Y (2012) Silicon ameliorates manganese toxicity by regulating manganese transport and antioxidant reactions in rice (*Oryza sativa* L.). Plant Soil 354(1):407–419

Ma JF (2004) Role of silicon in enhancing the resistance of plants to biotic and abiotic stresses. Soil Sci Plant Nutr 50:11–18

Ma JF, Takahashi E (2002) Soil, fertilizer, and plant silicon research in Japan. Elsevier, Amsterdam

Ma JF, Yamaji N (2006) Silicon uptake and accumulation in higher plants. Trends Plant Sci 11:392–397

Ma JF, Sasaki M, Matsumoto H (1997) Al-induced inhibition of root elongation in corn, *Zea mays* L. is overcome by Si addition. Plant Soil 188:171–176

Ma JF, Miyake Y, Takahashi E (2001) Silicon as a beneficial clement for crop plants. In: (ed) Silicon in agriculture

Ma JF, Higashitani A, Sato K, Tateda K (2003) Genotypic variation in Si content of barley grain. Plant Soil 249:383–387

Ma JF, Tamai K, Yamaji N, Mitani N, Konishi S, Katsuhara M, Ishiguro M, Murata Y, Yano M (2006) A silicon transporter in rice. Nature 440:688–691

Ma JF, Yamaji N, Tamai K, Mitani N (2007a) Genotypic difference in silicon uptake and expression of silicon transporters genes in rice. Plant Physiol 145(3):919–924

Ma JF, Yamaji N, Mitani N, Tamai K, Konishi S, Fujiwara T, Katsuhara M, Yano M (2007b) An efflux transporter of silicon in rice. Nature 448:209–212

Ma JF, Yamaji N, Mitani-Ueno N (2011) Transport of silicon from roots to panicles in plants. Proc Jpn Acad Ser B Phys Biol Sci 87:377–385

Maksimović JD, Mojović M, Maksimović V, Römheld V, Nikolic M (2012) Silicon ameliorates manganese toxicity in cucumber by decreasing hydroxyl radical accumulation in the leaf apoplast. J Exp Bot 63(7): 2411–2420. doi:10.1093/jxb/err359

Mitani N, Ma JF (2005) Uptake system of silicon in different plant species. J Exp Bot 56:1255–1261

Mitani N, Chiba Y, Yamaji N, Ma JF (2009) Identification and characterization of maize and barley Lsi2-like silicon efflux transporters reveals a distinct silicon uptake system from that in rice. Plant Cell 21:2133–2142

Mitani N, Yamaji N, Ago Y, Iwasaki K, Ma JF (2011) Isolation and functional characterization of an influx silicon transporter in two pumpkin cultivars contrasting in silicon accumulation. Plant J 66:231–240

Rodrigues FÁ, Vale FXR, Korndörfer GH, Prabhu AS, Datnoff LE, Oliveira AMA, Zambolim L (2003) Influence of silicon on sheath blight of rice in Brazil. Crop Prot 22:23–29

Rogalla H, Römheld V (2002) Role of leaf apoplast in silicon-mediated manganese tolerance of *Cucumis sativus* L. Plant Cell Environ 25:549–555

Ryder M, Gérard F, Evans DE, Hodson MJ (2003) The use of root growth and modeling data to investigate amelioration of aluminium toxicity by silicon in *Picea abies* seedlings. J Inorg Biochem 97:52–58

Sangster AG, Hodson MJ, Tubb HJ (2001) Silicon deposition in higher plants. In: Datonoff LE, Snyder GH, Korndörfer GH (eds) Silicon in agriculture. Elsevier Science, New York, pp 85–114

Savant NK, Snyder GH, Datnoff LE (1997) Silicon management and sustainable rice production. In: Sparks DL (ed) Advances in agronomy, vol 58. Academic, San Diego, pp 151–199

Shimoyama S (1958) Effect of silicon on lodging and wind damage in rice. Report for the research funds granted by Ministry of Agriculture, Japan, p 82

Takahashi E (1966) Effect of silicon on resistance of rice to radiation. Jpn J Soil Sci Plant Nutr 37:183–188

Takahashi E, Ma JF, Miyake Y (1990) The possibility of silicon as an essential element for higher plants. Comments Agric Food Chem 2:99–122

Van Hoest PJ (2006) Rice straw, the role of silica and treatments to improve quality. Anim Feed Sci Technol 130:137–171

Wang Y, Stass A, Horst WJ (2004) Apoplastic binding of aluminium is involved in silicon-induced amelioration of aluminium toxicity in maize. Plant Physiol 136(3):3762–3770

Yamaji N, Ma JF (2007) Spatial distribution and temporal variation of the rice silicon transporter Lsi1. Plant Physiol 143:1306–1313

Yamaji N, Ma JF (2009) A transporter at the node responsible for intervascular transfer of silicon in rice. Plant Cell 21:2878–2883

Yamaji N, Mitatni N, Ma JF (2008) A transporter regulating silicon distribution in rice shoots. Plant Cell 20:1381–1389

Yamaji N, Chiba Y, Mitani-Ueno N, Ma JF (2012) Functional characterization of a silicon transporter gene implicated in silicon distribution in barley. Plant Physiol 160(3):1491–1497

20 Cobalt (Co), Selenium (Se), Vanadium (V), Cadmium (Cd), Lead (Pb) and Titanium (Ti)

20.1 Cobalt (Co)

Abstract Cobalt (Co) is not essential for plants. It is required by animals especially ruminants. Co is essential for the growth of nitrogen-fixing microorganisms. It has been reported that Co regulates the alkaloid accumulation in medicinal plants.

20.1.1 Occurrence of Cobalt (Co) and Soil Reactions

Cobalt concentration of the earth's crust is around 40 μg g^{-1}. Soils contain about 8 μg g^{-1} of Co with a range of 1–70 μg g^{-1}. Soils formed on granitic glacial materials are low in Co. Sandstones and shales are normally low in cobalt. Mg-rich ferro-magnesian minerals are relatively rich in Co (100–300 μg g^{-1}). While Co is not essential for plants, it is required by animals especially ruminants. Forages produced on soils containing < 5 μg g^{-1} of Co may cause Co deficiency in ruminants.

20.1.2 Co Content of Plants

Vitamin B_{12} (cyanocobalamin) of which Co is a constituent is neither absorbed nor produced by higher plants. It is synthesised by soil bacteria, intestinal microbes and algae. Co is essential for the growth of nitrogen-fixing microorganisms such as symbiotic *Rhizobia* of legumes, free-living nitrogen-fixing microorganisms and blue–green algae (*Cyanobacteria*). The Co content of plants is species dependent.

The Co content of leafy plants such as lettuce, cabbage and spinach is relatively high (0.6–3.5 μg g^{-1}) (Kloke 1980). The highest concentration of Co has been found in legumes, followed by perennial grasses and the lowest in cereals (2.2 mg kg^{-1}) (Roy et al. 1988). Rice contains 0.02–150 mg kg^{-1} of plant mass (Palit et al. 1994).

20.1.3 Co Toxicity in Plants

Excess Co produces visual symptoms similar to Fe and Mn deficiencies in plants.

20.1.4 Effects of Co on Alkaloid Accumulation

It has been reported that Co regulates the alkaloid accumulation in medicinal plants such as Downy thorn apple (*Datura innoxia* Mill) (Yadrov et al. 1978); *Atropa caucasica* (Koval'Skii et al. 1971), belladonna, *A. belladonna* L. (Petrishek et al. 1984); and horned poppy (*Glaucium flavum* Crantz.) (Lovkova et al. 1988). It also increases rutin (11.6 %) and cyanide (67 %) levels of different species of buck wheat (Grinkevich et al. 1971).

G.N. Mitra, *Regulation of Nutrient Uptake by Plants: A Biochemical and Molecular Approach*, DOI 10.1007/978-81-322-2334-4_20, © Springer India 2015

20.1.5 Effects of Co on Shelf and Vase Life of Flowers

Shelf and vase life of marigold, chrysanthemum (Chandra et al. 1981), rose (Venatarayappa et al. 1980) and maidenhair fern (Fujino and Reid 1983) is increased by Co. Application of Co preserves freshness of apple after its picking (Barbat et al. 1979).

20.2 Selenium (Se)

Abstract Se is not an essential element for plants but is required by animals and microorganisms. Occurrence of Se is widespread in surface waters due to a variety of natural and anthropogenic causes. Human death rates from cancer (Se supplement with pharmacological dose reduced incidence of cancer risk by 63 % for prostate cancer, 58 % for colon cancer and 46 % for lung cancer, development of mammary cancer) and heart disease have been found to be less in areas with higher Se in the environment. Crop plants such as, cabbage, mustard, onion and broccoli have relatively higher amounts of Se.

20.2.1 Occurrence of Selenium (Se) and Soil Reactions

Selenium (Se) is widely distributed on the earth's crust and generally occurs in association with sulphide minerals. It averages around 0.09 µg g^{-1} in rocks and is found primarily on sedimentary rocks.

Se content of most of the soils is between 0.1 and 0.2 µg g^{-1}. Calcareous soils formed from sedimentary shale in semiarid region with high pH have high Se and produce vegetation with toxic concentrations of Se for livestock.

Occurrence of Se is widespread in surface waters due to a variety of natural and anthropogenic causes. Natural causes are exposure of geological strata rich in Se and their weathering and leaching, absorption of Se from such strata by plants and making it available to animals and aquatic life resulting in introduction of Se into the ecosystem. Anthropogenic sources include the use of sewage sludge from cities and fly ash from coal-fired power plants in agriculture. Mining of phosphates and metal ores are other sources of Se contamination.

Selenium is present in soils as selenides (Se^{2-}), which are insoluble and occur along with sulphides. Elemental Se is present in some soils in traces. A large fraction of Se is present as selenites (SeO_3^{2-}) as stable complexes with Fe oxides in acid soils. Fe–selenite complexes are sparingly soluble and do not accumulate in toxic concentrations in plants. Selenates (SeO_4^{2-}) are highly soluble and are readily available to plants. They may accumulate in toxic concentrations in plants grown on soils with high pH. Se sorption is highest at lower pH values. Sorption of Se (+4, selenite) decreases above pH 6.0. Sorption of Se (+6, selenate) decreases within pH range of 2.5–10.

20.2.2 Se in Plants

Se is not an essential element for plants but is required by animals and microorganisms. Plants take up Se from soil primarily as selenates (SeO_4^{2-}) or selenites (SeO_3^{2-}) (Ellis and Salt 2003). Uptake of Se by plants is governed by its chemical form and concentration; soil factors such as its pH, salinity and $CaCO_3$ content; the identity and concentration of competing ions; and the ability of the plants to absorb and metabolise Se (Wu 2004; Germ et al. 2007). Increasing Ca^{2+} concentration increases Se sorption, but sulphate (SO_4^{2-}) suppresses Se uptake more than any increase by Ca^{2-}(Hyun et al. 2006). Sulphate has greater antagonising effect on Se uptake and accumulation by plants than chloride (Cl^{-}) (Germ et al. 2007).

Plant species differ in their Se uptake. Some plant species such as *Morinda reticulata* and *Neptunia amplexicaulis* can accumulate high concentrations of Se, when grown on high Se soils. They can accumulate 4,000 mg of Se kg^{-1} of dry matter.

Crop plants in many parts of the world such as Western Europe, Northern Africa and some parts of China are low in Se due to low availability of Se in soils (Hawkesford and Zhao 2007; Zhu et al. 2009). Se content of most of the plants is less than 25 mg kg^{-1} of dry matter (Terry et al. 2000; Ellis and Salt 2003; Tinggi 2003). Wheat crop can recover 20–35 % of Se from the applied Se fertiliser (Broadley et al. 2010; Stroud et al. 2010). Crop plants such as cabbage, mustard and onion absorb relatively higher amounts of Se along with S. Most of the cereal crops and fodders weakly absorb Se, even when grown on Se-rich soils (Nowak et al. 2004). Actively growing tissues contain the largest amount of Se (Kahakachchi et al. 2004; Sugihara et al. 2004).

20.2.3 Beneficial Effects of Se in Plants

A number of beneficial effects of Se application to plants have been reported. Se application in concentrations of 0.1–1.0 mg of Se kg^{-1} of soil inhibited lipid peroxidation in rye grass (*Lolium perenne*) (Hartikainen et al. 2000). Application of Se at a concentration of 1.5 mg L^{-1} increased yield of pumpkin (*Cucurbita pepo*) (Germ et al. 2005). Se has been reported to promote growth of plants subjected to UV-induced oxidative stress (Xue and Hartikainen 2000). Se can delay senescence and promote growth of ageing seedlings. Addition of Se in low doses strengthens antioxidative capacity by preventing reduction of tocopherol concentrations and by increasing superoxide dismutase (SOD) activity (Hartikainen and Xue 1999; Xue et al. 2001).

Se has a positive effect on accumulation of starch in potato (Turakainen et al. 2004). Se improves nutritive value of potato by increasing SeMet (S in methionine substituted by Se), which is very beneficial for human consumption (Smrkolj et al. 2006a; Turakainen et al. 2006). SeMet is primarily the Se-containing compound found in seeds of pea enriched with Se by foliar spray (Smrkolj et al. 2006a), in buckwheat and pumpkin seeds (Smrkolj et al. 2005, 2006a, b) and in seeds of wheat, barley and rye (Stadlober et al. 2001). Broccoli (*Brassica oleracea* var. *italica*), which has the ability to accumulate high levels of Se, has the majority of Se amino acids as Se-methyl-seleno-cystein (SeMeSeCys) (Lyi et al. 2005). Major components of seleno amino acids in soybean and kidney bean also are SeMeSeCys and a minor component SeMet.

20.2.3.1 Beneficial Effects of Se on Human Beings

Between 0.5 and 1.0 billon people around the world are estimated to have insufficient uptake of Se (Combs 2001). Selenium plays an important role in the prevention of atherosclerosis, certain specific cancers, arthritis and altered immunological function (Shamberger 1981; Glover et al. 1996). Human death rates from cancer have been found to be less in areas with higher Se in the environment. Se-methyl-seleno-cystein (SeMeSeCys) has been found to be twice as effective as selenomethionine (the primary component of the Se-yeast supplement) in preventing the development of mammary tumour (McKenzie et al. 2009). SeMeSeCys is easily converted to an anticancer agent, methyl selenol, which is less toxic and has low body accumulation (Medina et al. 2001; Finley et al. 2004). A clinical trial with 1,312 Americans showed that Se supplements in pharmacological amounts reduced incidence of cancer risk by 63 % for prostate cancer, 58 % for colon cancer and 46 % for lung cancer (Clark et al. 1996; Lyi et al. 2005). Se has also been reported to improve male fertility [the selenoprotein, phospholipid hydroperoxide glutathione peroxidase (PHGPx) accounts for almost the entire selenium content of mammalian testis] (Foresta et al. 2002) and immune function (Mckenzie et al. 2001) in reducing viral infection (Beck et al. 2003) and slowing ageing process (Soriano-Garcia 2004). The nonprotein amino acid SeMeSeCys is produced by some of the plants such as *Astragalus*,

Allium and Brassica (Clark et al. 1996). Human mortality from heart disease is also less in high Se areas (Shamberger 1981).

20.2.4 Mechanism of Se Uptake by Plants

Se is taken up by plants as selenate, selenite or organic Se compounds by sulphate transporters located in the plasma membrane of the roots. Se is assumed to be assimilated in the same pathway as S, since selenate molecule has the same size and charge as sulphate (Sors et al. 2005a, b; Barker et al. 2007). Uptake of selenite is probably related to phosphate transport pathway in the plasma membrane (Hopper and Parker 1999; Li et al. 2008). There is an enhancement of selenate uptake in S-deficient plants and of selenite in P-deficient plants (Li et al. 2008). A number of plants show antagonistic effects of S and Se uptake. It is reported that reduced sulphate pool due to S deficiency induces enhanced expression of sulphate transporters (Buchner et al. 2004; Shinmachi et al. 2010). Se and Mo, which are transported by the same transporters, are preferentially taken up at low S concentration. This does not happen if S concentration is high (Shinmachi et al. 2010).

20.3 Vanadium (V)

Abstract Vanadium is not an essential plant nutrient. The importance of V is due to the discovery in 1980 that it can act as an insulin-mimetic agent. The normative human requirement is estimated to be about 10 μg day^{-1}. Vanadium is a toxic element in its both cationic and anionic form for humans; the latter form has more serious side effects. The threshold toxic concentration for humans is about 10 mg day^{-1}. Wild thyme (*Thymus pulegioides*) contains the highest V content and can be used as a supplement in diabetes mellitus type II along with other plants known to have hypoglycaemic effects. The V content of this plant is at a much lower level than the human toxic limit.

20.3.1 Occurrence of Vanadium (V) and Soil Reactions

The concentration of vanadium in the outer earth's crust is reported to be 100 mg kg^{-1} (Anke 2004a). The average V content of soils worldwide has been estimated at 18–115 mgkg^{-1} (Anke 2004a). At acidic pH, vanadyl cation (VO^{3+}) predominates and is readily taken up by plants. In neutral and alkaline soils, anionic forms VO_3^- and HVO_4^{2-} dominate (Goodman and Cheshire 1975).

20.3.2 Vanadium in Plants

Vanadium is yet to be established as an essential nutrient for plants although V-dependent nitrogenase has been found in nitrogen-fixing bacteria (Eady 1995). Plants can easily take up V depending on the V status of soil. There appears to be no homeostatic mechanism involved in V uptake. Within plants, V is bio-transformed into V^{4+} (Morrell et al. 1986).

The highest V content on an average has been found in spices, 218 μg kg^{-1} of dry matter (DM). Fruits contain 23 and legumes 41 μg kg^{-1}of DM (Antal et al. 2009). Some of the medicinal plants analysed by Antal et al. (2009) contained on an average 502 μg kg^{-1}of DM.

20.3.3 Vanadium as an Insulin-Mimetic Agent

In the 1980s, V was identified as an insulin-mimetic agent (Anke 2004b). The normative human requirement is estimated to be about 10 μg day^{-1} (Antal et al. 2009). This small requirement is met from regular food intake in different regions of the world (Anke et al. 1998). Beer and wine account for 75 % and 40 % intake of V (Anke 2004a, b).

Wild thyme (*Thymus pulegioides*), which contains the highest V content, can be used as a supplement in diabetes mellitus type II along with other plants known to have hypoglycaemic effects. The V contents of these plants are at a much lower level than the human toxic limit of 10 mg day^{-1}.

20.3.4 Vanadium Toxicity

Vanadium is considered as a toxic element in its both cationic and anionic form; the latter form has more serious side effects. The threshold toxic concentration for humans is about 10 mg day^{-1} (Anke 2004b).

20.4 Cadmium, Lead and Titanium

Heavy metals have been studied more for their toxic effects on plants rather than any stimulating effects. Heavy metals are not toxic to plants per se. Only when their cellular concentrations exceed a certain threshold value they become toxic and they are commonly termed as 'heavy metals' (Appenroth 2010).

In Great Britain, Cd concentration of soils heavily polluted by use of sewage sludge is 150 mg kg^{-1}, as compared to 0.1–0.5 mg kg^{-1} in unpolluted soils (Jackson and Alloway 1991). However, Cd has been observed to have some stimulating effects on the growth of barley seedlings at a concentration of 5×10^{-8} M. Beneficial effects have also been observed for Pd and Ti at low concentrations on barley leaves (Kovacs et al. 2009; Nyitrai et al. 2007).

References

Cobalt

Barbat I, Tomsa M, Suciu T (1979) Influence of foliar nutrition with microelements on some physiological processes in apple tree. Bull Inst Agron Cluj-Napora Ser Agric 33:69–74

Chandra G, Reddy KS, Mohan Ram HY (1981) Extension of vase life of cut marigold and chrysanthemum flowers by the use of cobalt chloride. Indian J Exp Biol 19:150–154

Fujino DW, Reid MS (1983) Factors affecting the vase life of fronds of maidenhair fern (*Adiantum raddianum*). Sci Hortic 21:181–188

Grinkevich NL, Grivovaskaya LF, Shandova AN, Dinevich LS (1971) Concentration of cobalt in some medicinal plants and its effect on accumulation of flavonoids in buckwheat. Biol Nauk 14:88–91

Kloke A (1980) Effects of heavy metals on soil fertility, flora and fauna as well as its importance for the nutrition of fodder chain (in German). In: Proceedings of the utilisation of refuse and sewage sludge in Agriculture, Zurich, pp 58–87

Koval'Skii VV, Grinkevich NI, Gribovskaya IF, Dinevech LS, Shandova AN (1971) Cobalt in medicinal plants and its effects on accumulation of biologically active compounds. Rastit Resurcy 7:503–551

Lovkova MY, Buzuk GN, Sabirova NS, Kliment'eva NI, Grinkevich NI (1988) Pharmacognostic examination of *Glaucium flavum* Cr. Farmatsiya 37:31–34

Palit S, Sharma A, Talukder G (1994) Effects of cobalt on plants. Bot Rev 60:149–181

Petrishek IA, Lovkova MY, Grinkevich NI, Oriova LP, Poludennyi LV (1984) Effects of cobalt and copper on the accumulation of alkaloids in *Atropa belladonna*. Biol Bull Acad Sci USSR 10(6):509–516

Roy AK, Srivastava LL et al (1988) Removal of some micronutrients by forage crops in soils. J Indian Soc Soil Sci 36:133

Venatarayappa T, Tsujita MJ, Murr DP (1980) Influence of cobaltous ion on the post harvest behaviour of roses (*Rosa hybrida Samantha*). J Am Soc Hortic Sci 105:148–151

Yadrov BN, Donitruk SE, Baturin VG (1978) The effects of copper, manganese and cobalt on the productivity of a culture of isolated tissue of *Datura innoxia* Mill. Rastit Resursy 14:408–411

Selenium

Barker AE, Pilbeam DJ, Kopsell DA, Kopsell DE (2007) Selenium. In: Barker AE, Pilbeam DJ (eds) Handbook of plant nutrition. CRC Press, New York, pp 515–549

Beck MA, Levander OA, Handy J (2003) Selenium deficiency and viral infection. J Nutr 133:1463S–1467S

Broadley MR, Alcock A, Alford J, Cartwright P, Foot I, Fairweather-Tait SJ, Hart DJ, Hurst R, Knott P, McGrath SP et al (2010) Selenium bio-fortification of high-yielding winter wheat (*Triticum aestivum* L.) by liquid or granular Se fertilisation. Plant Soil 332:5–18

Buchner P, Prosser IM, Hawkesford MJ (2004) Phylogeny and expression of paralogous and orthologous sulphate transporter genes in diploid and hexaploid wheats. Genome 47:526–534

Clark LC, Combs GF, Turnbull BW, Slate EH, Chalker DK, Chow J, Davis LS, Glover RA, Graham GF, Gross EG et al (1996) Effects of selenium supplementation for cancer prevention in patients with carcinoma of the skin: a randomized controlled trial. JAMA 276:1957–1963

Combs GF (2001) (Selenium) in global food systems. Br J Nutr 85:517–547

Ellis DR, Salt DE (2003) Plants, selenium and human health. Curr Opin Plant Biol 6:273–279

Finley JW, Grusak MA, Keck AS, Gregoire BR (2004) Bioavailability of selenium from meat and broccoli as determined by retention and distribution of ^{75}Se. Biol Trace Elem Res 99:191–209

Foresta C, Flohe L, Garolla A, Roveri A, Ursini F, Maiorino M (2002) Male fertility is linked to the selenoprotein, phospholipid hydroperoxide glutathione peroxidise. Biol Reprod 67:967–971

Germ M, Kreft I, Osvald J (2005) Influence of UV-B exclusion and selenium treatment on photochemical efficiency of photosystem II, yield and respiratory potential in pumpkins (*Cucurbita pepo* L.). Plant Physiol Biochem 43:445–448

Germ M, Stibilj V, Kreft I (2007) Metabolic importance of selenium for plants. Eur J Plant Sci Biotechnol 1(1):91–97

Glover RA, Graham GF, Gross EG, Krongrad A, Lesher JL Jr, Park HK, Sanders BB Jr, Smith CL, Taylor JR (1996) Effects of selenium supplementation for cancer prevention in patients with carcinoma of the skin – a randomized controlled trial. J Am Med Assoc 276:1957–1963, ISSN 0098–7484

Hartikainen H, Xue T (1999) The promotive effect of selenium on plant growth as trigged by ultraviolet irradiation. J Environ Qual 28:1272–1275

Hartikainen H, Xue T, Piironen V (2000) Selenium as an antioxidant and prooxidant in ryegrass. Plant Soil 225:193–200

Hawkesford MJ, Zhao FJ (2007) Strategies for increasing the selenium content of wheat. J Cereal Sci 46:282–292

Hopper JL, Parker DR (1999) Plant availability of selenite and selenate as influenced by the competing ions phosphate and sulphate. Plant Soil 210:199–207

Hyun S, Burns PE, Murarka I, Lee LS (2006) Selenium(IV) and (VI) sorption by soils surrounding fly ash management facilities. Vadose Zone J 5:1110–1118

Kahakachchi C, Boakye HT, Uden PC, Tyson JF (2004) Chromatographic speciation of anionic and neutral selenium compounds in Se-accumulating *Brassica juncea* (Indian mustard) and in selenized yeast. J Chromatogr A 1054:303–312

Li HF, McGrath SP, Zhao FJ (2008) Selenium uptake, translocation and speciation in wheat supplied with selenate or selenite. New Phytol 178:92–102

Lyi SM, Heller LI, Rutzke M, Welch RM, Kochian LV, Li L (2005) Molecular and biochemical characterization of the selenocysteine Se-methyltransferase gene and Se-methylselenocysteine synthesis in broccoli. Plant Physiol 138:409–420

McKenzie RC, Rafferty TS, Beckett GJ, Arthur JR (2001) Effects of selenium on immunity and aging. In: Hatfield DL (ed) Selenium, its molecular biology and role in human health. Kluwer Academic Publishers, Boston, pp 257–272

McKenzie MJ, Hunter DA, Pathirana R, Watson LM, Joyce NI, Matich AJ, Rowan DD, Brummell DA (2009) Accumulation of an organic anticancer selenium compound in a transgenic *Solanaceous* species shows wider applicability of the selenocysteine methyltransferase transgene from selenium hyperaccumulators. Transgenic Res 18(3):407–424, ISSN 0962–8819

Medina D, Thompson H, Ganther H, Ip C (2001) Se-methylselenocysteine: a new compound for chemoprevention of breast cancer. Nutr Cancer 40:12–17

Nowak J, Kaklewski K, Ligocki M (2004) Influence of selenium on oxidoreductive enzymes activity in soil and in plants. Soil Biol Biochem 36:1553–1558

Shamberger RJ (1981) Selenium in the environment. Sci Total Environ 17:59–74

Shinmachi F, Buchner P, Stroud JL, Parmar S, Zhao F-J, McGrath SP, Hawkesford MJ (2010) Influence of sulphur deficiency on the expression of specific sulphate transporters and the distribution of sulphur, selenium, and molybdenum in wheat. Plant Physiol 153(1):327–336

Smrkolj P, Kreft I, Kapolna E, Stibilj V (2005) Selenium species determination in selenium enriched pumpkin (*Cucurbita pepo* L.) seeds by HPLC-UVHG-AFS. Anal Sci 21:1501–1504

Smrkolj P, Germ M, Kreft I, Stibilj V (2006a) Respiratory potential and Se compounds in pea (*Pisum sativum* L.) plants grown from Se-enriched seeds. J Exp Bot 57:3595–3600

Smrkolj P, Stibilj V, Kreft I, Germ M (2006b) Selenium species in buckwheat cultivated with foliar addition of Se(VI) and various levels of UV-B radiation. Food Chem 96:675–681

Soriano-Garcia M (2004) Organo-selenium compounds as potential therapeutic and chemo-preventive agents: a review. Curr Med Chem 11:1657–1669

Sors TG, Ellis DR, Na GN, Lahner B, Lee S, Leustek T, Pickering IJ, Salt DE (2005a) Analysis of sulphur and selenium assimilation in *Astragalus* plants with varying capacities to accumulate selenium. Plant J 42:785–797

Sors TG, Ellis DR, Salt DE (2005b) Selenium uptake, translocation, assimilation and metabolic fate in plants. Photosynth Res 86:373–389

Stadlober M, Sager M, Irgolic KJ (2001) Effects of selenate supplemented fertilisation on the selenium level of cereals-identification and quantification of selenium compounds by HPLC-ICP-MS. Food Chem 73:357–366

Stroud JL, Li HF, Lopez-Bellido FJ, Broadley MR, Foot I, Fairweather-Tait SJ, Hart DJ, Hurst R, Knott P, Mowat H et al (2010) Impact of sulphur fertilisation on crop response to selenium fertilisation. Plant Soil 332:19–30

Sugihara S, Kondo M, Chihara Y, Yuji M, Hattori H, Yoshida M (2004) Preparation of selenium enriched sprouts and identification of their selenium species by high-performance liquid chromatography-inductively coupled plasma mass spectrometry. Biosci Biotechnol Biochem 68:193–199

Terry N, Zayed AM, de Souza MP, Tarun AS (2000) Selenium in higher plants. Annu Rev Plant Physiol Plant Mol Biol 51:401–432

Tinggi U (2003) Essentiality and toxicity of selenium and its status in Australia: a review. Toxicol Lett 137:103–110

Turakainen M, Hartikainen H, Seppänen MM (2004) Effects of selenium treatments on potato (*Solanum tuberosum* L.) growth and concentrations of soluble sugars and starch. J Agric Food Chem 52:5378–5382

Turakainen M, Hartikainen H, Ekholm P, Seppänen MM (2006) Distribution of selenium in different biochemical fractions and raw darkening degree of potato (*Solanum tuberosum* L.) tubers supplemented with selenate. J Agric Food Chem 54:8617–8622

Wu L (2004) Review of 15 years of research on ecotoxicology and remediation of land contaminated by agricultural drainage sediment rich in selenium. Ecotoxicol Environ Saf 57:257–269

Xue TL, Hartikainen H (2000) Association of antioxidative enzymes with the synergistic effect of selenium and UV irradiation in enhancing plant growth. Agric Food Sci Finl 9:177–186

Xue T, Hartikainen H, Piironen V (2001) Antioxidative and growth-promoting effect of selenium in senescing lettuce. Plant Soil 237:55–61

Zhu YG, Pilon-Smits EAH, Zhao FJ, Williams PN, Meharg AA (2009) Selenium in higher plants: understanding mechanisms for bio-fortification and phytoremediation. Trends Plant Sci 14:436–442

V, Cd, Pb and Ti

Anke M (2004a) Vanadium. In: Merian E, Anke M, Ihnat M, Stoeppler M (eds) Elements and their compounds in the environment, 2nd edn. Wiley VCH Verlag, Weinheim, pp 1173–1191

Anke M (2004b) Vanadium – an element both essential and toxic to plants, animals and humans? Anales de la Real Academia Nacional de Farmacia Madrid 70(4):961–999

Anke M, Glei M, Groppel B, Rother C, Gonzales D (1998) Mengen-, Spuren- und Ultraspurenelemente in der Nahrungskette. Nova Acta Leopold 79:57–190

Antal DS, Dehelean CA, Canciu CM, Anke M (2009) Vanadium in medicinal plants: new data on the occurrence of an element both essential and toxic to plants and man. Analele Universitatii din Oradea, Fascicula Biologie, Tom. XVI / 2, pp 5–10

Appenroth K-J (2010) Definition of 'Heavy metals' and their role in biological system, *Soil heavy metals*, Soil biology, 19, Chapter-2.Springer-Verlag Berlin, Heidelberg, pp 19–60

Eady RR (1995) Vanadium nitrogenases of azotobacter. In: Sigel HA, Sigel A (eds) Metal ions in biological systems, vanadium and its role in life. Marcel Dekker, Inc., New York/Basl/Hong Kong, pp 363–405

Goodman BA, Cheshire MV (1975) The bonding of vanadium in complexes with humic acid, an electronic paramagnetic resonance study. Geochim Cosmochim Acta 39:1711–1715

Jackson AP, Alloway BJ (1991) The transfer of cadmium from sewage sludge amended soils into the edible component of food crops. Water Air Soil Pollut 57:873–881

Kovacs E, Nyitrai P, Czovek P, Ovari M, Keresztes A (2009) Investigation into the mechanism of stimulation by low-concentration stressors in barley seedlings. J Plant Physiol 166:72–79

Morrell B, Lepp N, Phipps DA (1986) Vanadium uptake by higher plants: some recent developments. Environ Geochem Health 8:14–19

Nyitrai P, Mayer M, Ovari M, Keresztes A (2007) Involvement of the phospho-inositide signalling pathway in the anti-senescence effect of low-concentration stressors on detached barley leaves. Plant Biol 9:420–426